Statistics and Neural Networks

Statistics and Neural Networks

Advances at the Interface

EDITED BY

J.W. Kay

and

D.M. Titterington

Department of Statistics
University of Glasgow

OXFORD

UNIVERSITY PRESS

OXFORD
UNIVERSITY PRESS

Great Clarendon Street, Oxford OX2 6DP

Oxford University Press is a department of the University of Oxford.
It furthers the University's objective of excellence in research, scholarship,
and education by publishing worldwide in

Oxford New York

Athens Auckland Bangkok Bogotá Buenos Aires Calcutta
Cape Town Chennai Dar es Salaam Delhi Florence Hong Kong Istanbul
Karachi Kuala Lumpur Madrid Melbourne Mexico City Mumbai
Nairobi Paris São Paulo Singapore Taipei Tokyo Toronto Warsaw

with associated companies in Berlin Ibadan

Oxford is a registered trade mark of Oxford University Press
in the UK and in certain other countries

Published in the United States
by Oxford University Press, Inc., New York

A catalogue record for this book is available from the British Library

Library of Congress Cataloging in Publication Data

Statistics and neural networks: advances at the interface/edited by
J.W. Kay and D.M. Titterington.
Includes bibliographical references and index.
1. Mathematical statistics.   2. Neural computers.   3. Neural
networks (Computer science)   I. Kay, J.W. (Jim W.)
II. Titterington, D.M.
QA276.S78343   1999   519.5'0285'632–dc21   99-41248

ISBN 0 19 852422 6

Typeset by
Newgen Imaging Systems (P) Ltd., Chennai, India

Printed in Great Britain
on acid-free paper by
Bookcraft (Bath) Ltd.,
Midsomer Norton, Avon

Preface

Background

Recent years have seen a burgeoning awareness of the interface between artificial neural network research and modern methods in statistics. This has been manifested in books, such as those by Bishop (1995) and, particularly, Ripley (1996); in published compilations, such as Cherkassky *et al.* (1994) and the present volume; in review articles, such as those by Cheng and Titterington (1994) and Ripley (1993, 1994); and in many particular instances of cross-fertilisation in the two literatures.

There is, for instance, increasing statistical activity in the neural-computation literature. This is not surprising if one takes a statistical view of the models and methods involved: multilayer perceptrons used for classification or prediction can be interpreted as classes of nonlinear classifiers or regression models; and various networks trained by unsupervised learning can be regarded as providing particular ways to carry out cluster analysis. Often, the models involved are highly parameterised, and the parameter-estimation problem ('learning') can become ill-posed and unstable. This leads to the use of techniques such as regularisation, familiar in many statistical contexts and employed in forms such as 'weight decay' in the analysis of multilayer perceptrons. A simple-minded way of interpreting regularisation is that, by imposing a little regularisation, otherwise called smoothing or shrinkage, one can greatly reduce variability (increase precision), at the expense of a small amount of bias. New ideas are being developed, at least partly within the neural-network literature, for dealing with the inadequate precision in other ways, such as bagging (Breiman, 1996), stacking (Wolpert, 1992) and averaging (Raviv and Intrator, 1996). These are all statistically based, are founded on modern ideas such as resampling, and are of interest to statisticians.

Model choice is also an issue, and concepts such as Akaike's AIC, Rissanen's MDL and Bayes factors have all found application. Another fertile area for interaction has been that of hierarchical/graphical models; this is hardly surprising in view of the underlying graphical structures, and the expanded interest is providing new insights into the theory and methodology, fresh computational ideas and new directions for the practical application of the research; see for instance Heckerman (1996) and Jordan and Jacobs (1994).

Much of the new work takes a Bayesian standpoint, including Buntine and Weigend's (1991) application of the Bayesian paradigm to multilayer perceptrons, Mackay's (1992) research on model choice based on Bayes factors, and Neal's (1992) implementation of Monte Carlo methods such as Gibbs sampling in the context of Bayesian networks and similar structures.

As is appropriate for this volume, the contributors have been drawn from both the statistical and neural-network communities. It is hoped that, as a result, the material indicates both the communalities and differences of the outlooks and approaches, but in particular emphasises that the former far outweigh the latter. Indeed, many of the differences are merely 'apparent' and amount to differences only in terminology and notation. We have not presumed to 'translate' all the contributions into a common terminology, but we have tried to unify the notation to some extent, and we give here a brief glossary of some of the more obvious differences in terminology, in the hope of easing the reader's path through the different chapters. There are of course some differences over and above terminology, resulting from the neurological genesis of artificial neural-network research and from the need, in that community, to try to solve real-life problems on a very large scale; not many mainstream statisticians are required to fit complex nonlinear models, involving thousands of parameters, to vast datasets, and yet this is the environment in which neural-network research has developed. Nevertheless, it is clear from many recent workshops and conferences, including the Edinburgh meeting that sowed the seed for this volume, that members of the two communities are coming together. Statisticians are willing to meet the challenge of large-scale problems, and they and neural-network researchers are communicating and collaborating to advance a number of frontier topics in what is indisputably statistical research, including latent structure analysis, graphical models, modern Bayesian methods and flexible (nonlinear) approaches to classification and regression.

Review of the contents

In Chapter 1, Hastie, Tibshirani and Buja use standard linear discriminant analysis as a springboard for the development of more versatile and nonparametric alternatives. They note a relationship between linear discriminant analysis and a sequence of linear regressions, and create more flexible versions by replacing the linear regressions by more general notions such as generalised additive models, splines and multivariate additive regression splines (MARS). This leads on to a regularisation formulation called penalised discriminant analysis. Finally, the underlying Gaussian genesis of linear discriminant analysis is generalised in a natural way to create mixture discriminant analysis. All the methods are applied to real datasets.

In Chapter 2, Kay uses information theory to define a class of objective functions for the contextual guidance of learning and processing in multilayer, multistream networks. The networks are composed of local processors which are connected in various architectures. The learning rules are derived in the case where the local processor has a single binary output and these results are extended to encompass mutual contextual guidance, contextually modulated competitive learning and the case of the multivariate binary processor. Two illustrations are provided and some approximations are discussed.

Lowe's contribution, in Chapter 3, reviews the methodology and scope for applications of radial basis networks. Although somewhat related to the multilayer perceptrons that have tended to dominate the impact of artificial neural networks on application areas, radial basis function networks have a distinct character, and bear a close similarity to kernel-based smoothing techniques in statistics.

In Chapter 4, Intrator examines in detail aspects of the need to compromise between bias and variance in modelling complex relationships using an inevitably finite amount of data. He discusses ways of controlling both aspects separately and describes strategies based on approaches such as ensemble averaging and noise injection, both of which can impose beneficial regularisation.

Bishop and Tipping's contribution on the theme of latent variable models shows first that prinicpal components have a direct probabilistic interpretation along the same lines as standard Gaussian-based factor analysis. This allows likelihood-based inference and motivates more flexible data-visualisation procedures based on mixtures of principal component analysers and hierarchical mixture models. The models can be applied interactively and the graphical output monitored in order to shape the way the models develop for a given complex dataset.

MacKay and Gibbs carry on the theme of latent structure modelling. Their density network model involves categorical observables and continuous latent variables, along the lines of latent trait analysis in the statistical literature. The two sets of variables are linked through a multicategory linear logistic model for the observables given the latent variables, and a prior distribution is proposed for the relevant parameters. Discussion of the Markov chain Monte Carlo aspects represents an important part of this contribution.

In Chapter 7, Dunmur and Titterington concentrate on the general topic of latent structure models, which represent a class of incomplete-data models that have attracted considerable attention in the statistical and neural-computing literatures. Not surprisingly, the maximum likelihood analysis that they develop relies on numerical techniques such as the EM algorithm, and computational complexity here motivates approximate methods such as the mean-field approach. They also review Bayesian approaches, which typically employ Markov chain Monte Carlo methods to overcome their particular computational hurdles.

The foundation of the contribution of Martin and Morris in Chapter 8 is a range of practical problems in the area of industrial processing, but these are used as vehicles for the exposition and application of various techniques in the area where multivariate statistics and neural-network research blend together, including principal components analysis, radial basis function networks, wavelets, bagging, stacking and partial least squares.

Glossary

In this glossary, we list some of the terms in the neural-computing jargon, along with their statistical counterparts. We perhaps betray our origins in this; authors from the neural-computing community would no doubt reverse the direction! In fact some of the terms arguably exist already in both literatures. For more detailed explanations see for instance the books by Bishop (1995) and Ripley (1996), and articles by Cheng and Titterington (1994), Ripley (1993, 1994, 1997) and Titterington (1999).

backpropagation network: see *feedforward network*.

bias: an added constant in a model, an *intercept*. Sometimes called a *threshold*.

clamped: of a variable, meaning fixed, usually in the context of a variable whose value is being conditioned on. For instance, input variables/covariates are typically treated as *clamped* when it comes to analysing the data.

error backpropagation: essentially a gradient-descent algorithm for finding the minimum of a surface that represents a criterion of fit of the proposed model to the data. In the most familiar version, the criterion of fit amounts to a sum-of-squares function and the optimisation problem is a nonlinear least squares exercise. The motivation for the nomenclature is that use of the chain rule to calculate derivatives with respect to the parameters associated with the different layers in the network leads to an iterative stage in which the 'layers of parameters' are updated consecutively and the changes made are propagated through the network in the opposite direction from that in which input data are processed.

feedforward network: a layered network representing a certain class of nonlinear regression or classification models, relating a set of input variables (covariates, independent variables, predictors) to one or more output variables (target variables, dependent variables, response variables). The network also includes one or more layers of hidden (latent) units that add flexibility to the model. Normally, the outputs from the hidden units are logistic (*sigmoid*) functions of linear combinations of the inputs to that node, and the final output(s), which provide the overall regression function(s), are linear combinations of the outputs from the final layer of hidden units. Also known as *multilayer perceptron* and *backpropagation network*, the latter because of the gradient-descent algorithm, called the *(error) backpropagation algorithm*, used to *train the network*, i.e. estimate the parameters.

generalisability: a measure of the ability of a model/network to perform its function on the members of the 'universe' of possible cases/patterns that might present themselves; often assessed empirically using a *test set* (q.v.).

inputs: values of independent variables/regressors/predictors/covariates.

multilayer perceptron: see *feedforward network*.

pattern: the data corresponding to one experimental unit, i.e. one *observation* or *case*, in statistical parlance.

(statistical) learning: using available data from the *training set* to estimate the *weights* (parameters) according to some statistical procedure, usually equivalent to least-squares estimation or maximum likelihood or Bayesian inference.

supervised learning: analysing a set of data (*training set*) in which each observation (*pattern*) contains values of both covariates (*inputs*) and responses (*targets*), with the aim of making inferences about a model relating the two types of variable, i.e. a regression/prediction model or, in the particular case where the target variable(s) is (are) categorical, a model for classification/discriminant analysis. Contrast *unsupervised learning*.

target(s): the dependent/response variable(s).

test set: a set of data, independent of the *training set*, used to assess the *generalisability* of an estimated model/network. Usually, assessment of performance using the training set is unreasonably optimistic, but use of an independent test set gives a good indication of how the model will perform when applied to members of the universe of possible observations/*patterns*.

training a network: see *statistical learning*.

training set: the data available for *training* a network, i.e. for estimating parameters in the associated model and also for selecting a suitable network/model.

unit: the node of a network and sometimes the (output) variable associated with it.

unsupervised learning: analysing a set of data (*training set*) in which some aspect (usually some latent quantity rather than a physically real variable) is missing. Often the missing quantity is envisaged as a categorical variable intended to summarise the similarities and dissimilarities among the observations in the training set, and the inferential process amounts to some form of *cluster analysis*. Contrast *supervised learning*.

weight decay: *training a network with weight decay* is operationally equivalent to estimating parameters subject to a penalty function. In the most standard version, a penalty function that is quadratic in the weights (in fact usually simply proportional to the sum of squares of the weights) is added on to the sum-of-squares function, resulting in a procedure equivalent to penalised least squares, or, if Gaussian errors are assumed within the regression model, to penalised maximum likelihood, or, if the penalty function is interpreted as minus the logarithm of a Gaussian 'prior' density on the weights, to the calculation of a posterior mode (MAP analysis). The motivation is the familiar one of trying to avoid overfitting the model to the data by dramatically reducing 'variance' at the expense of incurring a little bias. The trade-off is dictated by a smoothing parameter/hyperparameter in the penalty function, and the procedure has many parallels; see for instance Titterington (1985). Recent developments include a Bayesian treatment involving separate hyperparameters for all the weights, called *automatic relevance determination* (ARD) by Neal (1996).

weights: parameters associated with those variables that constitute inputs into a node. Usually the overall input into a node is a linear combination of variables/outputs associated with nodes feeding into the node in question, with an extra constant (*bias/intercept*) added on. The weights are the multiplying constants and are equivalent to *slopes/regression coefficients* in statistical parlance. Also called *connectivities* or *synaptic weights*, the latter motivated by the neurological associations of artificial neural networks. Sometimes the *weights* are assumed to include all parameters except for variance parameters but including *bias/intercept* parameters.

Acknowledgements

As editors we acknowledge most sincerely our gratitude to the contributors for their material. We are also grateful to them and all other participants for contributing to the success of the workshop in Edinburgh that led eventually to this volume. We also acknowledge the International Centre for Mathematical Sciences for providing the venue and administrative support and the UK Engineering and Physical Science Research Council for financial backing.

References

Bishop, C.M. (1995). *Neural Networks for Pattern Recognition*. Clarendon Press, Oxford.

Breiman, L. (1996). Bagging predictors. *Machine Learning* **24**, 123–140.

Buntine, W.L. and Weigend, A.S. (1991). Bayesian backpropagation. *Complex Systems* **5**, 603–643.

Cheng, B. and Titterington, D.M. (1994). Neural networks: a review from a statistical perspective (with discussion). *Statist. Science* **9**, 2–54.

Cherkassky, V., Friedman, J.H. and Wechsler, H. (Eds.) (1994). *From Statistics to Neural Networks. Theory and Pattern Recognition Applications*. Springer-Verlag, Berlin.

Heckerman, D. (1995). A tutorial on learning Bayesian networks. Tech. Report MSR-TR-95-06, Microsoft, Redmond, WA.

Jordan, M.I. and Jacobs, R.A. (1994). Hierarchical mixtures of experts and the EM algorithm. *Neural Computation* **6**, 181–214.

MacKay, D.J.C. (1992). Bayesian interpolation. *Neural Computation* **4**, 415–447.

Neal, R.M. (1992). Connectionist learning of belief networks. *Artificial Intell.* **56**, 71–113.

Neal, R.M. (1996). *Bayesian Learning for Neural Networks. Lecture Notes in Statistics 118*. Springer, New York.

Raviv, Y. and Intrator, N. (1996). Bootstrapping with noise: an effective regularisation technique. *Connect. Sci.* **8**, 356–372.

Ripley, B.D. (1993). Statistical aspects of neural networks. In *Networks and Chaos—Statistical and Probabilistic Aspects*, Eds. O.E. Barndorff-Nielsen, J.L. Jensen and W.S. Kendall, pp. 40–123. Chapman and Hall, London.

Ripley, B.D. (1994). Neural networks and related methods for classification (with discussion). *J. R. Statist. Soc.* B **56**, 409–456.

Ripley, B.D. (1996). *Pattern Recognition and Neural Networks.* Cambridge Univ. Press.

Ripley, B.D. (1997). Classification (update). In *Encyclopedia of Statistical Science Update*, Volume 1, Eds. S. Kotz, C.B. Read and D. Banks, pp. 110–116. Wiley, New York.

Titterington, D.M. (1985). Common structure of smoothing techniques in Statistics. *Int. Statist. Rev.* **53**, 141–170.

Titterington, D.M. (1999). Neural networks. In *Encyclopedia of Statistical Science Update*, Volume 3, Eds. S. Kotz, C.B. Read and D. Banks, pp. 528–535. Wiley, New York.

Wolpert, D.H. (1992). Stacked generalisation. *Neural Networks* **5**, 241–259.

Contents

Contents

3　Radial Basis Function Networks and Statistics　65

4　Robust Prediction in Many-parameter Models　97

7 Analysis of Latent Structure Models with Multidimensional Latent Variables   165

A.P. Dunmur

D.M. Titterington

8 Artificial Neural Networks and Multivariate Statistics   195

E.B. Martin

A.J. Morris

Contributors

Christopher Bishop	Microsoft Research
Andreas Buja	AT&T Laboratories, Research
Alan Dunmur	University of Glasgow
Mark Gibbs	University of Cambridge
Trevor Hastie	Stanford University
Nathan Intrator	University of Tel Aviv
Jim Kay	University of Glasgow
David Lowe	Aston University
David MacKay	University of Cambridge
Elaine Martin	University of Newcastle
Alan Morris	University of Newcastle
Robert Tibshirani	University of Toronto
Michael Tipping	Microsoft Research
Michael Titterington	University of Glasgow

1

Flexible Discriminant and Mixture Models

Trevor Hastie
Stanford University
http://stat.stanford.edu/~trevor

Robert Tibshirani
University of Toronto
http://utstat.toronto.edu/~tibs

Andreas Buja
AT&T Laboratories, Research
http://www.research.att.com/~andreas

Abstract

Fisher's linear discriminant analysis is a useful and graphical tool for multi-group classification. With a large number of predictors, one can find a reduced number of discriminant coordinate functions that are 'optimal' for separating the groups. With two such functions one can produce a classification map that partitions the reduced space into regions that are identified with group membership, and the decision boundaries are linear.

Often this formulation is inadequate:

- the linear decision boundaries can be too restrictive to capture the class separation;
- if there are too many (correlated) predictors, for example, grey-scale pixel values from a digitised image, LDA tends to overfit the training data, and the discriminant functions are too noisy to be interpretable.

These deficiencies occupy opposite sides of the spectrum: in one case LDA is too restrictive, in the other it is too flexible.

In this chapter we represent LDA as an optimal scaling problem in multiple regression, which we generalise to forms of nonparametric regression suitable for the two scenarios outlined above. A natural extension of LDA is to model each class by a mixture of Gaussians. Our approach adapts seamlessly, and for example we can fit a mixture model, confined to an optimal lower dimensional subspace, by regression methods.

1   Introduction

In the generic classification or discrimination problem, the outcome of interest G falls into J unordered classes, which for convenience we denote by the set $\{1, 2, \ldots, J\}$.

We wish to build a rule for predicting the class membership of an item based on p measurements of predictors or features $X \in R^p$. Our training sample consists of the class membership and predictors for N items. This is an important practical problem with applications in many fields. Traditional statistical methods for this problem include linear discriminant analysis (LDA), multiple logistic regression, nearest neighbour methods and classification trees. Neural network classifiers have become a powerful alternative, with the ability to incorporate a very large number of features in an adaptive nonlinear model. Ripley (1994) gives an informative review from a statistician's viewpoint.

This chapter is about linear discriminant analysis, and a variety of ways of enhancing it as a tool for classification and data analysis. Some of the virtues of LDA are as follows:

- It is a simple prototype classifier. A new observation is classified to the class with closest centroid. A slight twist is that distance is measured in the Mahalanobis metric, using a pooled covariance estimate.
- LDA is the estimated Bayes classifier if the observations are multivariate Gaussian in each class, with a common covariance matrix. Since this assumption is unlikely to be true, this might not seem to be much of a virtue.
- The decision boundaries created by LDA are linear, leading to decision rules that are simple to describe and implement.
- LDA provides natural low-dimensional views of the data. For example, Figure 7 below is an informative two-dimensional view of data in 256 dimensions with 10 classes.
- Often LDA produces the best classification results, because of its simplicity and low variance. LDA was among the top three classifiers for 11 of the 22 datasets studied in the STATLOG project (Michie *et al.*, 1994).

Unfortunately the simplicity of LDA causes it to fail in a number of situations as well:

- Often linear decision boundaries do not adequately separate the classes. When N is large, it is possible to estimate more complex decision boundaries. Quadratic discriminant analysis (QDA) is often useful here, and allows for quadratic decision boundaries. More generally we would like to be able to model irregular decision boundaries.
- The aforementioned shortcoming of LDA can often be paraphrased by saying that a single prototype per class is insufficient. LDA uses a single prototype (class centroid) plus a common covariance matrix to describe the spread of the data in each class. In many situations, several prototypes are more appropriate.
- At the other end of the spectrum, we may have way too many (correlated) predictors, for example in the case of digitised analogue signals and images. In this case LDA uses too many parameters, which are estimated with high variance, and its performance suffers. In cases such as this we need to restrict or regularise LDA even further.

In this chapter we describe a class of techniques that attend to all these issues by generalising the LDA model. This is achieved largely by three different ideas.

- The first idea is to recast the LDA problem as a linear regression problem. Many techniques exist for generalising linear regression to more flexible, nonparametric forms of regression. This in turn leads to more flexible forms of discriminant analysis, which we call FDA. In most cases of interest, the regression procedures can be seen to identify an enlarged set of predictors via basis expansions. FDA amounts to LDA in this enlarged space.

- In the case of too many predictors, such as the pixels of a digitised image, we do not want to expand the set: it is already too large. The second idea is to fit an LDA model, but penalise its coefficients to be smooth or otherwise coherent in the spatial domain, i.e. as an image. We call this procedure *penalised discriminant analysis* or PDA. With FDA itself, the expanded basis set is often so large that regularisation is also required. Both of these can be achieved via a suitably regularised regression in the context of the FDA model.

- The third idea is to model each class by a mixture of two or more Gaussians with different centroids, but with every component Gaussian, both within and between classes, sharing the same covariance matrix. This allows for more complex decision boundaries, and allows for subspace reduction as in LDA. We call this extension *mixture discriminant analysis* or MDA.

The ideas behind FDA were originally proposed in Breiman and Ihaka (1984), and are developed further in Hastie *et al.* (1994) and Ripley (1996), and along with PDA in Hastie *et al.* (1995). The MDA model is developed in Hastie and Tibshirani (1996). In this chapter we describe the essential aspects of each of these techniques, and illustrate each with examples. We also describe in some detail software we provide in S-Plus for fitting all these models. We also discuss some further extensions and modifications not contained in the previously published work.

2   Linear discriminant analysis

Here we review some relevant details of LDA. We assume that the conditional density of the predictors in each class, denoted by $P(X|G)$, is multivariate Gaussian with each class having its own mean vector, but sharing a common covariance matrix. The density in class j is

$$\phi(X; \mu_j, \Sigma) = \frac{1}{(2\pi)^{\frac{p}{2}}|\Sigma|^{\frac{1}{2}}} e^{-\frac{1}{2}(X-\mu_j)^T \Sigma^{-1}(X-\mu_j)}. \tag{2.1}$$

The class prior probabilities are $P(G = j) = \Pi_j$. In this idealised setting, where everything is known, we can also obtain the ideal or *Bayes optimal* classifier. We will use Bayes' formula to flip the densities into class posterior probabilities $P(G|X)$. Knowing $P(G|X)$ exactly is the best one can do in classification. If the new observations to be classified arise from this same joint distribution, the rule

$$C(x) = j \quad \text{if} \quad P(G = j|x) = \max_{\ell} P(G = \ell|x) \tag{2.2}$$

achieves the minimum misclassification rate. In this case we have

$$P(G = j|X = x) = \frac{\phi(x; \mu_j, \Sigma)\Pi_j}{\sum_{\ell} \phi(x; \mu_\ell, \Sigma)\Pi_\ell}$$

$$\propto \exp\left(x^T \Sigma^{-1}\mu_j - \frac{1}{2}\mu_j^T \Sigma^{-1}\mu_j + \log \Pi_j\right)$$

$$= \exp(x^T \beta_j + \alpha_j). \tag{2.3}$$

The $\propto$ in the second line denotes proportionality; since we are interested in which is largest, we are concerned only with the numerators since the denominators do not depend on the class label. Note also that the quadratic terms cancel.

The decision boundary between class i and class j is defined as the set of points having equal posterior probability: $\{x \in R^p : P(G = i|x) = P(G = j|x)\}$. From (2.3), we see that, in the case of LDA, this is linear.

The *discriminant function* for class j is denoted by

$$\delta_j(x) = -x^T \Sigma^{-1}\mu_j + \frac{1}{2}\mu_j^T \Sigma^{-1}\mu_j - \log \Pi_j \tag{2.4}$$

$$= -(x^T \beta_j + \alpha_j) \quad \text{(bias-corrected),} \tag{2.5}$$

and the equivalent rule is to classify to the class for which $\delta_j(x)$ is smallest.

In practice we have to estimate the parameters $\{\beta_j\}$ and $\{\alpha_j\}$ using the training data. LDA uses the maximum-likelihood estimates (MLEs), which amounts to plugging the MLEs $\hat{\mu}_j$, $\hat{\pi}_j$ and $\hat{\Sigma}$ into the formula (2.4). Here

$$\hat{\Sigma} = \frac{1}{N - p} \sum_{j=1}^{J} \sum_{i:g_i=j} (x_i - \hat{\mu}_j)(x_i - \hat{\mu}_j)^T. \tag{2.6}$$

Some remarks:

- Even if the Gaussian assumptions are correct, the *estimated* Bayes classifier need not be optimal, since it will have bias and variance.
- When the dimension p of the space is large, it might seem that the number of parameters in LDA can become a problem, since Σ has $O(p^2)$ parameters. From (2.5), however, we see that only $J(p+1)$ parameters, nonlinear transformations of the originals, are required. In fact, since we *compare* discriminant functions, we can look at differences, and a more precise number of parameters is $(J - 1)(p + 1)$.
- Examination of (2.3) shows that the quadratic term $x^T \Sigma x$ plays no role, since it is common to all classes. If each of the classes has a different covariance matrix, this no longer holds. In this case the discriminant functions have quadratic components as well, and the decision boundaries are quadratic surfaces. This is known as *quadratic discriminant analysis*.

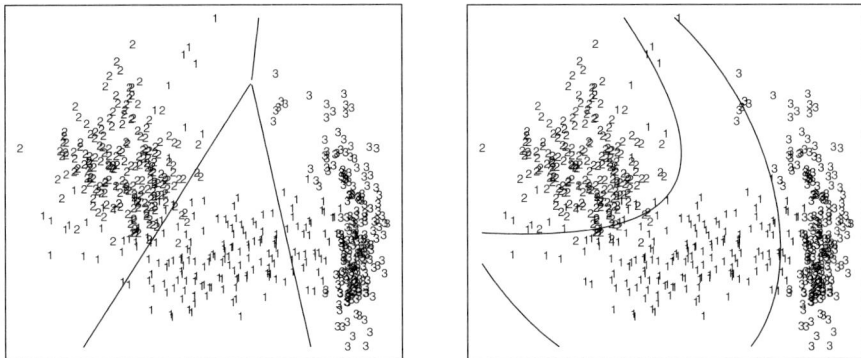

Fig. 1: The data in these plots are generated from a mixture of Gaussians in each class. The left plot shows the linear decision boundaries produced by the LDA model. The right plot shows the quadratic decision boundaries produced by QDA.

Figure 1 shows the linear and quadratic decision boundaries of LDA and QDA for a simulated example. In this case, the true density is a mixture of Gaussians in each class, so neither LDA nor QDA is ideal.

2.1   Reduced rank LDA

LDA operates by comparing Mahalanobis distances from the target point to the estimated class centroids, with an adjustment for unequal class priors. All the relevant distance information is contained in the at most $J - 1$ dimensional subspace of R^p spanned by the J group centroids. A reduced form of LDA due to Fisher and Rao adds a graphical component to the procedure. One finds the $K < J - 1$ dimensional subspace of R^p in which the group centroids are most separated, once again using the Mahalanobis metric confined to this subspace, and then classifies new data to the closest centroid in the reduced space. For small K the data can be plotted in the reduced space, giving a graphical representation of the group separation. Figure 7 below shows such a projection for digitised handwritten images, where $p = 256$ and $J = 10$.

Without going into the details, we note that the reduced subspace is obtained from a generalised principal component analysis of the J class centroids, using as metric the within-class covariance matrix:

$$\max u^T \Sigma_{\mathrm{Bet}} u \text{ subject to } u^T \Sigma_W u = 1. \tag{2.7}$$

Here Σ_{Bet} denotes the covariance matrix of the class centroids, and $\Sigma_W = \Sigma$, the common or *within class* covariance matrix; see Hastie *et al.* (1994, 1995) for details. By successively optimising (2.7) we get a sequence of u_k, orthogonal in Σ, known as the *discriminant* or *canonical coefficients*. The coordinates or *canonical variates* in the reduced space are given by $z_k = u_k^T x$, $k = 1, \ldots, K$. In fact the procedure is a version of generalised canonical correlation analysis (Hastie *et al.*, 1995). Hastie and Tibshirani (1996) show that this reduced rank formulation of LDA is equivalent to the Gaussian/Bayes procedure above, but where the class densities are estimated by maximum likelihood subject to rank constraints on the centroids: rank $\{\mu_j, \ j = 1, \ldots, J\} = K$.

3   Flexible discriminant analysis

In this section we describe a method for performing LDA using linear regression on derived responses. This in turn leads to nonparametric and flexible alternatives to LDA.

Suppose $\theta : \{1, \ldots, J\} \mapsto R^1$ is a function that assigns scores to the classes, such that the transformed class labels are optimally predicted by linear regression on X. This produces a one-dimensional separation between the classes. More generally, we can find K sets of independent scorings for the class labels, $\theta_1, \theta_2, \ldots, \theta_K$, and K corresponding linear maps $\eta_k(X) = X^T \beta_k$, $k = 1, \ldots, K$, chosen to be optimal for multiple regression in R^p. If our training sample has the form (g_i, x_i), $i = 1, 2, \ldots, N$, then the scores $\theta_k(g)$ and the maps β_k are chosen to minimise the average squared residual,

$$\text{ASR} = \frac{1}{N} \sum_{k=1}^{K} \left[\sum_{i=1}^{N} \left(\theta_k(g_i) - x_i^T \beta_k \right)^2 \right]. \tag{3.1}$$

The set of scores is assumed to be mutually orthogonal and normalised with respect to an appropriate inner product to prevent trivial zero solutions. It can be shown that the sequence of canonical vectors $\{u_k\}$ is identical to the sequence $\{\beta_k\}$ up to a constant (Mardia *et al.*, 1979; Hastie *et al.*, 1994).

Moreover, the Mahalanobis distance of a test point x to the jth class centroid $\hat{\mu}_k$, confined to the subspace defined by the first K canonical vectors, is given by

$$\delta_K(x, \hat{\mu}_j) = \sum_{k=1}^{K} w_k (\eta_k(x) - \bar{\eta}_k^j)^2, \tag{3.2}$$

where $\bar{\eta}_k^j$ is the mean of $\eta_k(x_i)$ for $i : g_i = j$. Here w_k are coordinate weights that are defined in terms of the mean squared residual r_k^2 of the kth optimally scored fit:

$$w_k = \frac{1}{r_k^2(1 - r_k^2)}. \tag{3.3}$$

To summarise: LDA can be performed by a sequence of linear regressions, followed by classification to the closest class centroid in the space of fits. The analogy applies both to the reduced rank version and to the full rank case when $K = J - 1$.

The real power of this result is in the generalisations that it invites. We can replace the linear regression fits $\eta_k(x) = x^T \beta_k$ by far more flexible, nonparametric fits, and by analogy achieve a more flexible classifier than LDA. We have in mind generalised additive fits, spline functions, MARS models, and the like. In this more general form the regression problems are defined via the criterion

$$\text{ASR}(\{\theta_k, \eta_k\}_{k=1}^K) = \frac{1}{N} \sum_{k=1}^{K} \left[\sum_{i=1}^{N} (\theta_k(g_i) - \eta_k(x_i))^2 + \lambda L(\eta_k) \right], \qquad (3.4)$$

where L is a regulariser appropriate for some forms of nonparametric regression, such as smoothing splines, additive splines and lower-order ANOVA spline models.

Before we describe the computations involved in this generalisation, let us consider a very simple example. Suppose we use degree-two polynomial regression for each η_k. The decision boundaries implied by (3.2) will be quadratic surfaces, since each of the fitted functions is quadratic, and as in LDA their squares cancel out when comparing distances. We could have achieved *identical* quadratic boundaries in a more conventional way, by augmenting our original predictors with their squares and cross-products. In the enlarged space one performs an LDA, and the linear boundaries in the enlarged space map down to quadratic boundaries in the original space. A classic example is a pair of multivariate Gaussians centred at the origin, one having covariance matrix I, and the other cI for $c > 1$. The Bayes decision boundary is the sphere $\|x\|^2 = (pc \log c)/(c - 1)$, which is a linear boundary in the enlarged space.

Many nonparametric regression procedures operate by generating a basis expansion of derived variables, and then performing a linear regression in the enlarged space. Friedman's MARS procedure (Friedman, 1991) is exactly of this form. Smoothing splines and additive spline models generate an extremely large basis set ($N \times p$ basis functions for additive splines), but then perform a penalised regression fit in the enlarged space. FDA in this case can be shown to perform a *penalised linear discriminant analysis* in the enlarged space. We elaborate in Section 4. Hastie *et al.* (1994) illustrate FDA on a tough speech recognition problem, with $J = 11$ classes and $p = 10$ predictors. The classes correspond to 11 vowel sounds, each contained in 11 different words. Here are the words, preceded by the symbols that represent them:

i	heed	O	hod	I	hid	C:	hoard	E	head	U	hood
A	had	u:	who'd	a:	hard	3:	heard	Y	hud		

Each of eight speakers spoke each word six times in the training set, and likewise seven speakers in the test set. The 10 predictors are derived from the digitised speech in a rather complicated way, but standard in the speech recognition world. There are thus 528 training observations, and 462 test observations. Figure 2 shows two-dimensional projections produced by LDA and FDA. The FDA model used adaptive additive-spline regression functions to model the $\eta_k(x)$, and the two coordinates plotted in the right plot correspond to $k = 1, 2$. The routine used in S-Plus is called `bruto`, hence the heading on the plot and in Table 1. We see that flexible modelling has helped to separate the classes in this case.

Table 1 shows training and test error rates for a number of classification techniques. FDA/MARS refers to Friedman's multivariate adaptive regression splines; degree $= 2$

Linear Discriminant Analysis   Flexible Discriminant Analysis -- Bruto

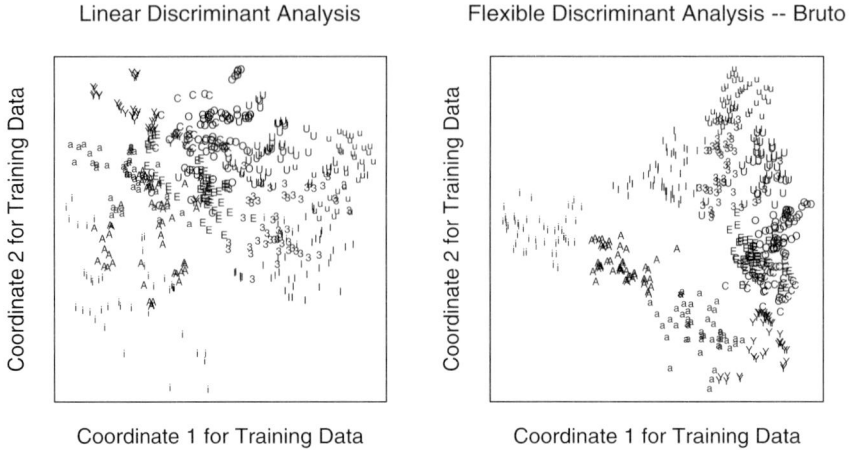

Coordinate 1 for Training Data   Coordinate 1 for Training Data

Fig. 2:   The left plot shows the first two LDA canonical variates for the vowel training data. The right plot shows the corresponding projection when FDA/BRUTO is used to fit the model. Notice the improved separation. The letters label the vowel sounds.

Table 1 Vowel recognition data performance results. The results for neural networks are the best among a much larger set, taken from a neural network archive. The notation FDA/BRUTO refers to the regression method used with FDA.

	Technique	Error rates	
		Training	Test
(1)	LDA	0.32	0.56
	Softmax	0.48	0.67
(2)	QDA	0.01	0.53
(3)	CART	0.05	0.56
(4)	CART (linear combination splits)	0.05	0.54
(5)	Single-layer perceptron		0.67
(6)	Multi-layer perceptron (88 hidden units)		0.49
(7)	Gaussian node network (528 hidden units)		0.45
(8)	Nearest neighbour		0.44
(9)	FDA/BRUTO	0.06	0.44
	Softmax	0.11	0.50
(10)	FDA/MARS (degree = 1)	0.09	0.45
	Best reduced dimension (= 2)	0.18	0.42
	Softmax	0.14	0.48
(11)	FDA/MARS (degree = 2)	0.02	0.42
	Best reduced dimension (= 6)	0.13	0.39
	Softmax	0.10	0.50

means second-degree tensor products are permitted. Notice that, for FDA/MARS, the best classification results are obtained in a reduced rank subspace.

3.1 Computing the FDA model

The computations for the FDA coordinates can be simplified in many important cases, in particular when the nonparametric regression procedure can be represented as a linear operator. We will denote this operator by S_λ; i.e. $\hat{y} = S_\lambda y$, where y is the vector of responses and $\hat{y}$ the vector of fits. Additive splines have this property, if the smoothing parameters are fixed, as does MARS once the basis functions are selected. The subscript λ denotes the entire set of smoothing parameters. We create an $N \times J$ *indicator response matrix* Y from the responses g_i, such that $Y_{ij} = 1$ if $g_i = j$, otherwise $Y_{ij} = 0$. For a five-class problem Y might look like

$$
\begin{array}{c}
\\
g_1 = 2 \\
g_2 = 1 \\
g_3 = 1 \\
g_4 = 5 \\
g_5 = 4 \\
\vdots \\
g_N = 3
\end{array}
\begin{array}{ccccc}
C_1 & C_2 & C_3 & C_4 & C_5 \\
\left(\begin{array}{ccccc}
0 & 1 & 0 & 0 & 0 \\
1 & 0 & 0 & 0 & 0 \\
1 & 0 & 0 & 0 & 0 \\
0 & 0 & 0 & 0 & 1 \\
0 & 0 & 0 & 1 & 0 \\
& & \vdots & & \\
0 & 0 & 1 & 0 & 0
\end{array} \right)
\end{array}
$$

(1) *Multivariate nonparametric regression.* Fit a multiresponse, adaptive nonparametric regression of Y on X, giving fitted values $\hat{Y}$. Let S_λ be the linear operator that fits the final chosen model, and $\eta(x)$ be the vector of fitted regression functions.

(2) *Optimal scores.* Compute the eigendecomposition of $Y^T \hat{Y} = Y^T S_\lambda Y$, where the eigenvectors Θ are normalised: $\Theta^T D_\pi \Theta = I$. Here $D_\pi = Y^T Y / N$ is a diagonal matrix of the class priors, often estimated by $Y^T Y / N$.

(3) *Update* the final model from step (1) using the optimal scores: $\eta(x) \leftarrow \Theta^T \eta(x)$.

Again S_λ can correspond to any regression method. When $S_\lambda = H_X$, the linear regression projection operator, then FDA is LDA. The software we describe in Section 6 makes good use of this modularity; the `fda` function has a `method` = argument which allows one to supply *any* regression function, as long as it follows some natural conventions. The regression functions we provide allow for polynomial regression, adaptive additive models and MARS. They all efficiently handle multiple responses, so step (1) is a single call to a regression routine. The eigendecomposition in step (2) simultaneously computes all the optimal scoring functions.

3.2 Indicator matrix regression versus FDA

For jointly distributed random variables Y and X, $E(Y|X)$ is known as the regression function. Classical parametric regression procedures, such as linear regression, assume this function has some parametric form and then estimate the parameters. Many nonparametric regression procedures are motivated as ways to estimate this function directly and with minimal assumptions.

In our context, if $Y = Y(G)$ is a coding of the random class variable G, then $E(Y|X) = P(G|X)$, the vector of posterior class probabilities. This suggests that one can apply regression methods directly to the indicator response matrix, and classify to the class corresponding to the largest fitted values, a procedure known in the machine-learning community as *Softmax*. In the context of the FDA model, this raises the question: why go beyond step 1: $\hat{Y} = S_\lambda Y$?

Figure 3 shows how disastrous this can be for linear regression, in that one class is completely masked. Figure 4 shows what goes wrong; without loss of generality we have projected the data on to the diagonal line passing through the centroids of the three classes.

The classes are perfectly separated, yet when we perform the indicator variable regressions, using linear regression, we see that the middle class never dominates. Of course, this problem can be easily solved by using quadratic regressions rather than linear, and, since we anticipate adaptive regression procedures, why the concern?

- Suppose there are 10 predictors and the three classes line up along a particular direction α in predictor space. In order to solve the problem via quadratic polynomials, we would need to fit a general quadratic surface with all the bilinear terms included. Of course, if we knew about projection pursuit regression, we could be a bit smarter than that.

- If four classes line up, then quadratic curves do not drop down sufficiently fast, and cubic curves are more appropriate. In general, if M classes line up, order $M - 1$ polynomials tend to be needed to completely untangle them.

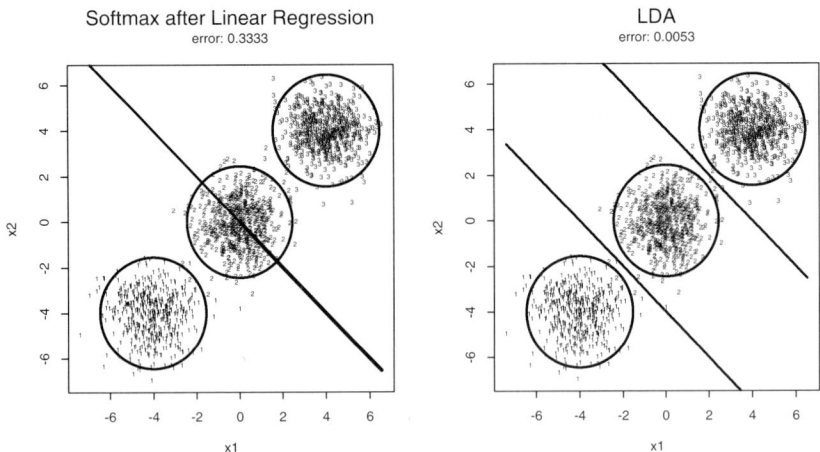

Fig. 3: Masking effects of Softmax. The data consist of 500 samples each from three spherical bivariate Gaussian distributions, whose centroids line up along a line. The centre class is completely masked by the outside two when Softmax is used, while LDA has no such problem.

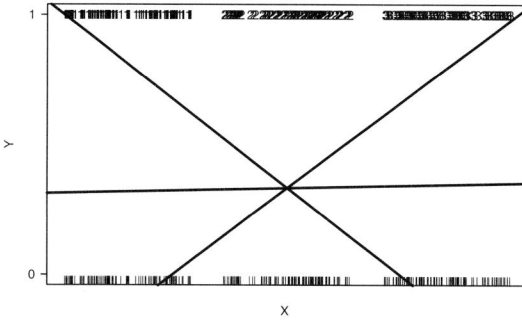

Fig. 4: The three classes are perfectly separated by the single predictor X (the rug plot shows the distribution of the data). The three lines represent the linear regression fits of each of the three columns of the indicator response matrix Y on X. The 1s, 2s and 3s at the top of the plot indicate the three response indicators Y_i, $i = 1, \ldots, 3$, each to be matched with the zeros for each of the other two classes. The middle class is completely masked, in that its regression line (fitted probabilities) never dominates.

When the number of classes is large relative to the number of predictors, masking or partial masking of this kind is relatively frequent. Procedures like MARS will struggle in general to achieve the untangling, because they have difficulty in creating the type of general interaction terms required here.

The post-processing of FDA overcomes this masking without the need for unnecessary transformations. See Hastie *et al.* (1994) for more details.

4 Penalised discriminant analysis

Although FDA is motivated by generalising optimal scoring, it can also be viewed directly as a form of regularised discriminant analysis. Suppose the regression procedure used in FDA amounts to a linear regression on to a basis expansion $h(X)$, with a quadratic penalty on the coefficients:

$$\text{ASR}(\{\theta_k, \beta_k\}_{k=1}^K) = \frac{1}{N} \sum_{i=1}^N (\theta_k(g_i) - h^T(x_i)\beta_k)^2 + \beta_k^T \Omega \beta_k. \qquad (4.1)$$

The choice of Ω depends on the problem. If $\eta_k(X) = h(X)\beta_k$ is an expansion on spline basis functions, Ω might constrain η_k to be smooth over the domain of X. In the case of additive splines, there are N spline basis functions for each coordinate, resulting in a total of Np basis functions in $h(X)$; Ω in this case is $Np \times Np$ and block diagonal.

The steps in FDA can then be viewed as a generalised form of LDA, which we call *penalised discriminant analysis* or PDA:

- Enlarge the set of predictors X via a basis expansion $h(X)$.
- Use (penalised) LDA in the enlarged space, where the penalised Mahalanobis

distance is given by

$$D(x, \mu) = (h(x) - h(\mu))^T (\Sigma_W + \Omega)^{-1} (h(x) - h(\mu)).$$

Σ_W is the within-class covariance matrix of the derived variables $h(x_i)$.

- Decompose the classification subspace using a penalised metric:

$$\max u^T \Sigma_{\text{Bet}} u \quad \text{subject to} \quad u^T (\Sigma_W + \Omega) u = 1.$$

Loosely speaking, the penalised Mahalanobis distance tends to give less weight to 'rough' coordinates, and more weight to smooth ones; since the penalty is not diagonal, the same applies to linear combinations that are rough or smooth.

For some classes of problems, the first step, involving the basis expansion, is not needed; we already have far too many (correlated) predictors. A leading example is when the objects to be classified are digitised analogue signals:

- the log-periodogram of a fragment of spoken speech, sampled at a set of 256 frequencies;
- the grey-scale pixel-values in a digitised image of a handwritten digit.

It is also intuitively clear in these cases why regularisation is needed. Take the digitised image as an example. Neighbouring pixel values will tend to be correlated, being often almost the same. This implies that the pair of corresponding LDA coefficients for these pixels can be wildly different and opposite in sign, and thus cancel when applied to similar pixel values. Positively correlated predictors lead to noisy, negatively correlated coefficient estimates, and this noise results in unwanted sampling variance. A reasonable strategy is to regularise the *coefficients* to be smooth over the spatial domain, as with images. This is what PDA does. The computations proceed just as for FDA, except that an appropriate penalised regression method is used. Here $h^T(X)\beta_k = X\beta_k$, and Ω is chosen so that $\beta_k^T \Omega \beta_k$ penalises roughness in β_k when viewed as an image. Figure 5 shows some examples of handwritten digits. Figure 6 shows the discriminant variates using LDA and PDA. Those produced by LDA appear as *salt and pepper* images, while those produced by PDA are smooth images. The first smooth image can be seen as the coefficients of a linear contrast functional for separating images with a dark central vertical strip (ones, possibly sevens) from images that are hollow in the middle (zeros, some fours). Figure 7 supports this interpretation, and with more difficulty allows an interpretation of the second coordinate. This and other examples are discussed in more detail in Hastie *et al.* (1995), who also show that the regularisation improves the classification performance of LDA on independent test data, by a factor of around 25%, in the cases they tried.

5   Mixture discriminant analysis

LDA can be viewed as a *prototype* classifier; each class is represented by its centroid, and we classify to the closest using an appropriate metric. In many situations a single

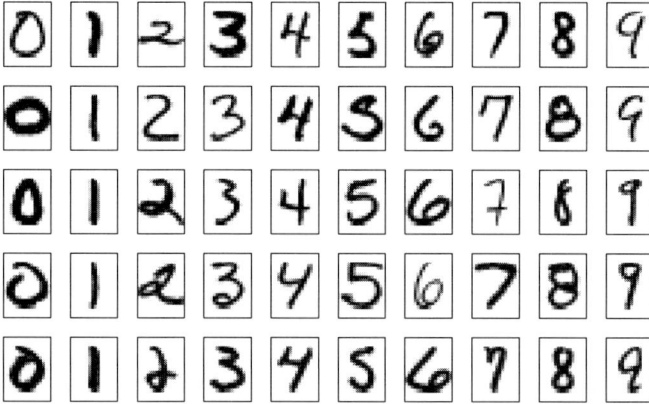

Fig. 5: Some examples of handwritten digits.

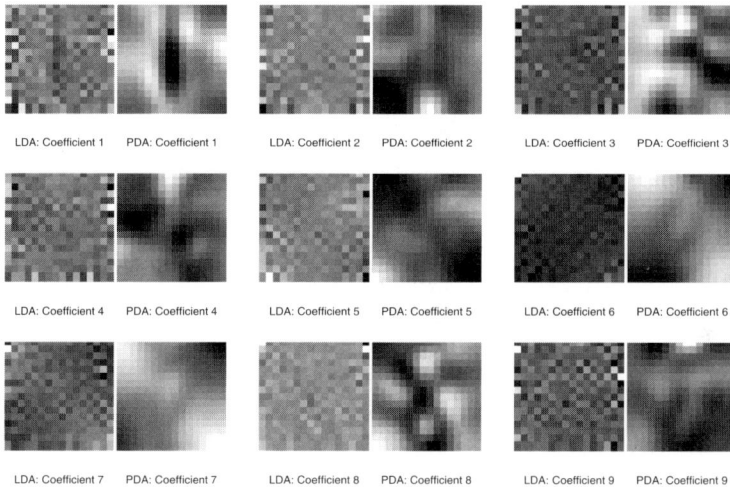

Fig. 6: The images appear in pairs, and represent the nine discriminant coefficient functions for the digit recognition problem. The left member of each pair is the LDA coefficient, while the right member is the PDA coefficient, regularised to enforce spatial smoothness.

prototype is not sufficient to represent inhomogeneous classes, and mixture models are more appropriate. In this section we review Gaussian mixture models and show how they can be generalised via the FDA and PDA methods discussed earlier. A Gaussian mixture model for the jth class has density

$$P(X|G = j) = \sum_{r=1}^{R_j} \pi_{jr} \phi(X; \mu_{jr}, \Sigma), \qquad (5.1)$$

Canonical Variate Plot --- Digit Test Data

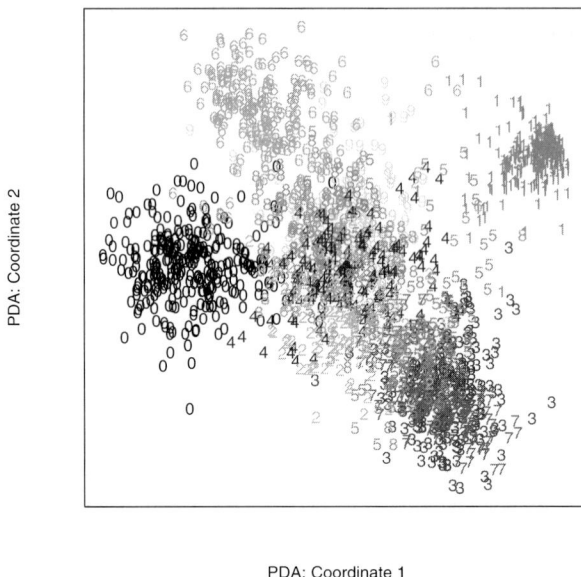

Fig. 7: The first two penalised canonical variates, evaluated for the test data. The circles indicate the class centroids. The first coordinate contrasts mainly 0s and 1s, while the second contrasts 6s and 7/9s.

where the *mixing proportions* $\{\pi_{jr} : r = 1, 2, \ldots, R_j\}$ sum to one for each j. This has R_j prototypes for the jth class, and, in our specification, the same covariance matrix Σ is used as the metric throughout. Given such a model for each class, the class posterior probabilities are given by

$$P(G = j | X = x) = \frac{\sum_{r=1}^{R_j} \pi_{jr} \phi(X; \mu_{jr}, \Sigma) \Pi_j}{\sum_{\ell=1}^{J} \sum_{r=1}^{R_\ell} \pi_{\ell r} \phi(X; \mu_{\ell r}, \Sigma) \Pi_\ell}. \tag{5.2}$$

As in LDA, we estimate the parameters by maximum likelihood; again the joint log-likelihood based on $P(G, X)$ is used:

$$\sum_{j=1}^{J} \sum_{i:g_i=j} \log \left(\sum_{r=1}^{R_j} \pi_{jr} \phi(x_i; \mu_{jr}, \Sigma) \Pi_j \right). \tag{5.3}$$

The sum within the log makes this a rather messy optimisation problem if tackled directly. The classical and natural method for computing the MLEs for mixture distributions is the EM algorithm (Dempster *et al.*, 1977) which is known to possess good convergence properties. EM alternates between the two steps:

E-step: Given the current parameters, compute the *responsibility* of subclass c_{jr} within class j for each of the class-j observations ($i : g_i = j$):

$$W(c_{jr}|x_i, g_i) = \frac{\pi_{jr}\phi(x_i; \mu_{jr}, \Sigma)}{\sum_{k=1}^{R_j} \pi_{jk}\phi(x_i; \mu_{jk}, \Sigma)}. \tag{5.4}$$

M-step: Compute the weighted MLEs for the parameters of each of the component Gaussians within each of the classes, using the weights from the E-step.

In the E-step, the algorithm apportions the unit weight of an observation in class j to the various subclasses assigned to that class. If it is close to the centroid of a particular subclass, and far from the others, it will receive a mass close to one for that subclass. On the other hand, observations halfway between two subclasses will get approximately equal weight for each.

In the M-step, an observation in class j is used R_j times, to estimate the parameters in each of the R_j component densities, with a different weight for each.

The algorithm requires initialisation, which can have an impact, since mixture likelihoods are generally multimodal. Our software allows several strategies; here we describe the default. The user supplies the number R_j of subclasses per class. Within class j a *k-means clustering model* (Hartigan and Wong, 1979), with multiple, random starts, is fitted to the data. This partitions the observations into R_j disjoint groups, from which an initial weight matrix, consisting of zeros and ones, is created.

Our assumption of equal component covariance matrix Σ throughout buys an additional simplicity; we can incorporate the rank restrictions in the mixture formulation. We thus maximise the log-likelihood (5.3) subject to rank constraints on *all* the $\sum_j R_j$ centroids: rank$\{\mu_{jk}\} = K$. Again the EM algorithm is available, and the M-step turns out to be a weighted version of LDA, with $R = \sum_{j=1}^{J} R_j$ 'classes'. Furthermore, we can use optimal scoring as before to solve the weighted LDA problem, which allows us to use a weighted version of FDA or PDA at this stage. One would expect, in addition to an increase in the number of 'classes', a similar increase in the number of 'observations' in the jth class by a factor of R_j. It turns out that, if linear operators are used for the optimal scoring regression, we use a *blurred* response matrix Z rather than the indicator matrix Y. For example, suppose there are $J = 3$ classes, and $R_j = 3$ subclasses per class. Then Z might be

$$
\begin{array}{c}
\\
g_1 = 2 \\
g_2 = 1 \\
g_3 = 1 \\
g_4 = 3 \\
g_5 = 2 \\
\vdots \\
g_N = 3
\end{array}
\begin{array}{c}
\begin{array}{ccccccccc}
c_{11} & c_{12} & c_{13} & c_{21} & c_{22} & c_{23} & c_{31} & c_{32} & c_{33}
\end{array} \\
\left(
\begin{array}{ccccccccc}
0 & 0 & 0 & 0.3 & 0.5 & 0.2 & 0 & 0 & 0 \\
0.9 & 0.1 & 0.0 & 0 & 0 & 0 & 0 & 0 & 0 \\
0.1 & 0.8 & 0.1 & 0 & 0 & 0 & 0 & 0 & 0 \\
0 & 0 & 0 & 0 & 0 & 0 & 0.5 & 0.4 & 0.1 \\
0 & 0 & 0 & 0.7 & 0.1 & 0.2 & 0 & 0 & 0 \\
& & & \vdots & & & & & \\
0 & 0 & 0 & 0 & 0 & 0 & 0.1 & 0.1 & 0.8
\end{array}
\right),
\end{array}
$$

where the entries in a class-j row correspond to $W(c_{jr}|x, g_i)$.

The remaining steps are the same:

$$\left.\begin{array}{l} \hat{Z} = SZ \\ Z^T \hat{Z} = \Theta D \Theta^T \\ \text{Update } \pi s \text{ and } \Pi s \end{array}\right\} \Leftrightarrow \text{M-step of MDA.}$$

These simple modifications to the mixture model add considerable flexibility:

- The dimension reduction step in LDA, FDA or PDA is limited by the number of classes; in particular, for $J = 2$ classes no reduction is possible. MDA substitutes subclasses for classes, and then allows us to look at low-dimensional views of the subspace spanned by these subclass centroids. This subspace will tend to be important for discrimination.

- By using FDA or PDA in the M-step, we can adapt even more to particular situations. For example, we can fit MDA models to digitised analogue signals and images, with smoothness constraints built in.

Figure 8 shows the two-dimensional discriminant plot for a language recognition task. The data are Japanese characters which are grouped as Hiragana or not. The predictors are eight features extracted from the images of each character. For these data the test error using 10 mixture centres per class is 10.7%, and with five per class

Discriminant Plot for true classes

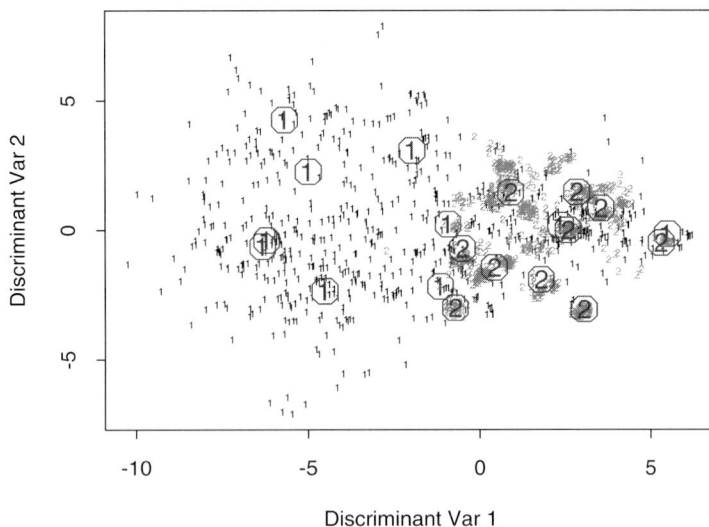

Fig. 8:　Optimal two-dimensional subspace for representing the mixture model fit to the two-class Japanese character recognition task. The numbered disks represent the subclass centres. The Hiragana class tend to occur in many clusters, and mixture models seem suitable for representing them.

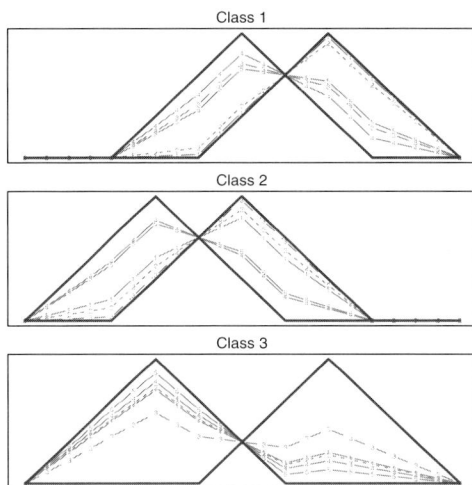

Fig. 9: Some examples of the waveforms generated from model (5.5) before the Gaussian noise is added.

is 16%. When combined with adaptive additive modelling in the regression step, the test error drops to 8.2% with five centres per class.

We now illustrate some of these ideas on a popular simulated example, taken from Breiman *et al.* (1984, pp. 49–55), and used in Hastie and Tibshirani (1994) and elsewhere. It is a three-class problem with 21 variables, and is considered to be a difficult pattern recognition problem. The predictors are defined by

$$x_i = uh_1(i) + (1-u)h_2(i) + \epsilon_i \qquad \text{Class 1}$$
$$x_i = uh_1(i) + (1-u)h_3(i) + \epsilon_i \qquad \text{Class 2} \qquad (5.5)$$
$$x_i = uh_2(i) + (1-u)h_3(i) + \epsilon_i \qquad \text{Class 3,}$$

where $i = 1, 2, \ldots, 21$, u is uniform on $(0, 1)$, ϵ_i are standard normal variates, and the h_i are the shifted triangular waveforms: $h_1(i) = \max(6 - |i - 11|, 0)$, $h_2(i) = h_1(i - 4)$ and $h_3(i) = h_1(i + 4)$; see Figure 9.

Each training sample has 300 observations, and equal priors were used, so there are roughly 100 observations in each class. We used test samples of size 500. The two MDA models are described in the caption of Table 2, where methods are compared.

Figure 10 shows the leading canonical variates for the penalised MDA model, evaluated at the test data. As we might have guessed, the classes appear to lie on the edges of a triangle. This is because the $h_j(i)$ are represented by three points in 21-space, thereby forming vertices of a triangle, and each class is represented as a convex combination of a pair of vertices, and hence lie on an edge. Also it is clear visually that all the information lies in the first two dimensions; the percentage of variance explained by the first two coordinates is 99.8%, and we would lose nothing

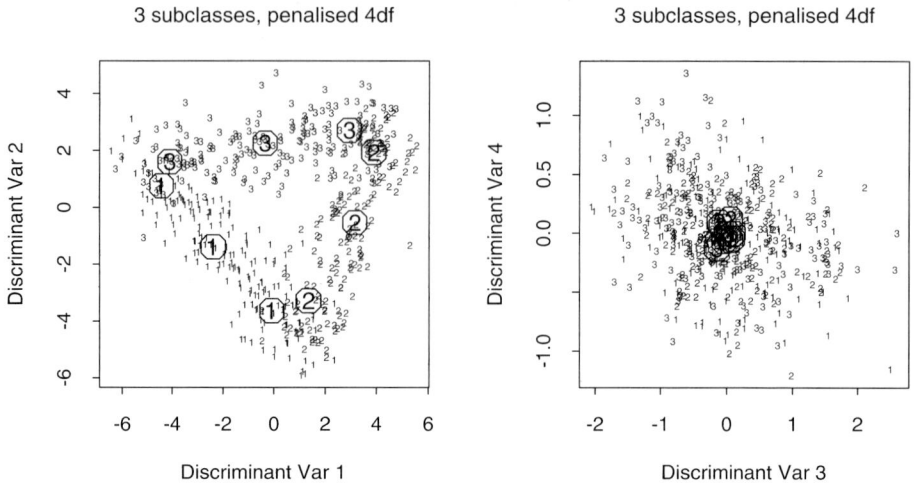

Fig. 10:   Some two-dimensional views of the MDA model fitted to a sample of the wave-form model. The points are independent test data, projected on to the leading two canonical coordinates (left panel), and the third and fourth (right panel). The subclass centres are indicated.

by truncating the solution there. The Bayes risk for this problem is about 0.14 (Breiman *et al.*, 1984); MDA comes close to the optimal rate, which is not surprising since the structure of the MDA model is similar to the generating model.

6 Software for fitting FDA, PDA and MDA models in S-Plus

Software for fitting FDA, PDA and MDA models is available in the S/S-Plus language (Becker *et al.*, 1988; Chambers and Hastie, 1991). The function fda() fits FDA and PDA models. A method argument allows the user to specify the multiresponse regression method to be used; the default is linear regression and thus LDA. Other regression methods provided for FDA are polynomial regression, ridge regression, BRUTO and MARS. The following examples illustrate the usage of fda, and the associated functions for producing the plots and making predictions. The data are supplied as a list or data-frame, in this case including components y, the class variable, and x.

```
fit.lda   <- fda(y ~ x, data = vowel.train)
fit.coef  <- coef(fit.lda)
fit.coord <- predict(fit.lda, type = 'variates')
```

fits an LDA model, and computes the canonical coefficients and coordinates. Simply typing fit.lda will print out a short summary of the model.

```
fit.disc <- coef(fit.lda, type = 'discriminant')
```

Table 2 Results for waveform data. The values are averages over 10 simulations, with the standard error of the average in parentheses. The five entries above the line are taken from Hastie *et al.* (1994). The first model below the line is MDA with three subclasses per class. The next line is the same, except that the discriminant coefficients are penalised via a roughness penalty to effectively 4df. The third is the corresponding penalised LDA or PDA model.

Technique	Error rates	
	Training	Test
LDA	0.121(0.006)	0.191(0.006)
QDA	0.039(0.004)	0.205(0.006)
CART	0.072(0.003)	0.289(0.004)
FDA/MARS (degree = 1)	0.100(0.006)	0.191(0.006)
FDA/MARS (degree = 2)	0.068(0.004)	0.215(0.002)
MDA (3 subclasses)	0.087(0.005)	0.169(0.006)
MDA (3 subclasses, penalised 4df)	0.137(0.006)	0.157(0.005)
PDA (penalised 4df)	0.150(0.005)	0.171(0.005)

computes instead the discriminant function coefficients.

```
plot(fit.lda, fit.train)
```

produces a plot of the first two canonical variates for the training data, while

```
plot(fit.lda, vowel.test, coords = c(3,4))
```

plots coordinates 3 and 4 for the test data; note that these functions will look for components x and y in vowel.test.

```
fit.bruto <- fda(y ~ x, data = vowel.train, method = bruto)
```

or simply

```
fit.brut <- update(fit.lda, method = bruto)
```

fits an FDA/BRUTO model, and assumes the function bruto() is available. Likewise method = 'mars' will fit FDA/MARS models, and method = 'polyreg' will fit polynomial LDA models. Each of these functions has additional, optional arguments, which can also be supplied by name in the call to fda(). The update() function is very useful for modifying a model by adding, changing or deleting one of the arguments that created its first argument.

```
confusion(fit.brut, vowel.test)
```

computes the $J \times J$ confusion matrix which results from applying the classification rule in fit.brut to the test data. One can produce the predicted classes for new data:

```
fit.predict <- predict(fit.brut, vowel.test)
```

where the default `type` = `'class'` is implicit. To compute the confusion matrix in a more direct fashion, use

```
confusion(fit.predict, vowel.test$g)
```

There are also `plot()` methods for MDA objects, and these label the subclasses as in Figure 10. For PDA models a generalised form of ridge regression is provided, called `gen.ridge()`. This is simply used in the `method` = argument to `fda()`. Users supply a penalty matrix and target `df`, and the procedure derives the appropriate penalty constant. The command

```
fit.pda <- fda(g ~ x, data = zip.train, method = gen.ridge,
                omega = Omega.zip, df = 80)
```

fits the PDA model for the zip-code data. The user has to supply an appropriate positive semi-definite penalty matrix `omega` in factored form. The function `laplacian()` produced the penalty matrix `Omega` for this image example, and it is provided with the software. The user also supplies the `df`, which generates a ridge penalty to achieve that many equivalent degrees of freedom.

The `mda()` function has additional arguments for controlling the number of subclasses and the initialisation.

```
fit.wave <- mda( g ~ x, data = wave.train, subclasses = 3,
                method = gen.ridge, omega = Omega.wf,
                df = 4, trace = T)
    plot(fit.wave, wave.test)
```

Here we fit an MDA model with three subclasses per class. By default the `kmeans()` clustering algorithm is used to initialise the weights, and also by default five random starts to `kmeans` are used. The `trace` = T argument traces the iterations, and prints out the *conditional* log-likelihood at each iteration; in general the full log-likelihood is not easily available for MDA models. Another choice for initialisation is `start` = `'lvq'`, which uses *learning vector quantisation*, a variant of k-means focussed more on classification.

Additional features of the software can be found in the online documentation. The `mda` software is publicly available from the statistics archive at Carnegie–Mellon University with URL: `http://lib.stat.cmu.edu/S/mda`. The software and technical report are also available from the first author's home page:
`http://stat.stanford.edu/~trevor`

7   Further extensions

Our mixture formulation uses a separate mixture of Gaussians for each class. Here we propose a variation that attempts to allocate mixture centres where they are needed for representing the joint distribution $P(G, X)$. We refer to our first model as MDA1 and this new formulation as MDA2. We consider the mixture model

$$P(G, X) = \sum_{r=1}^{R} \pi_r P_r(G, X), \tag{7.1}$$

a mixture of *joint densities*. Furthermore we assume

$$P_r(G, X) = P_r(G)\phi(X; \mu_r, \Sigma). \tag{7.2}$$

This model consists of regions centred at μ_r, and for each there is a class profile $P_r(G)$. The posterior class distribution is given by

$$P(G = j|X = x) = \frac{\sum_{r=1}^{R} \pi_r P_r(G = j)\phi(x; \mu_r, \Sigma)}{\sum_{r=1}^{R} \pi_r \phi(x; \mu_r, \Sigma)}, \tag{7.3}$$

where the denominator is the marginal distribution $P(X)$. MDA2 can also be viewed as a version of MDA1, since

$$P(X|G = j) = \frac{\sum_{r=1}^{R} \pi_r P_r(G = j)\phi(x; \mu_r, \Sigma)}{\sum_{r=1}^{R} \pi_r P_r(G = j)}, \tag{7.4}$$

where $\pi_{rj} = \pi_r P_r(G = j)/\sum_{r=1}^{R} \pi_r P_r(G = j)$ corresponds to the mixing proportions for the j class, and all classes use *the same Gaussian distributions*. In fact, it turns out that MDA2 contains MDA1; see the comments on initialisation below.

Once again there is a natural EM algorithm for fitting this model:

E-step: Each mixture component is assigned a responsibility for the ith observation:

$$W(c_r|x_i, g_i) = \frac{\pi_r P_r(g_i, x_i)}{\sum_{k=1}^{R} \pi_k P_k(g_i, x_i)}. \tag{7.5}$$

Note that, in addition to the distance of x_i from μ_r, here the class label of the observation contributes to the weight determination; the log of the numerator is

$$-\frac{1}{2}\|x_i - \mu_r\|_\Sigma^2 + \log P_r(g_i), \tag{7.6}$$

after cancellation of common factors. This encourages centroids to favour particular classes.

M-step: These compute appropriately weighted MLEs for the parameters of each of the component Gaussians. $\hat{P}_r(j) = \sum_{g_i=j} W(c_r|x_i, g_i)/\sum_{i=1}^{N} W(c_r|x_i, g_i)$, and the π_r are estimated similarly.

Again we can fruitfully resort to optimal scoring to perform the M-step. This allows us to fit the models with rank constraints on the space spanned by the centroids, and also permits more flexible versions of the procedure by using nonparametric regression. In this case the blurred response matrix Z is slightly different, since each observation

$$Z = \begin{bmatrix} \ \end{bmatrix} \qquad Z = \begin{bmatrix} \ \end{bmatrix}$$

Separate Centres per Class   Common Centres

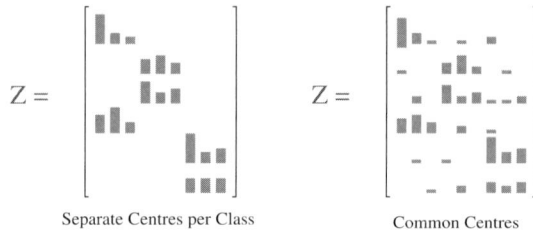

Fig. 11:   Depiction of the blurred response matrices Z for the two types of mixture model. On the left, each class has its own mixture components, while on the right, classes compete for and share mixture components.

has potentially some weight for each mixture component; Figure 11 illustrates the difference. We fitted this model to the Japanese character data. The results were similar to those obtained for the original mixture formulation and are not presented here.

We initialise this algorithm by using an unsupervised k-means clustering algorithm on all the data. This provides an initial weight matrix W, and the algorithm starts at the M-step.

It is not hard to see that, if this weight matrix is constructed as in MDA1, which involves separate k-means cluster models in each class, then MDA2 is identical to MDA1. In this case the initial versions of $P_k(g)$ will each be devoted to a single class (all zeros, with a one in the appropriate position.) Furthermore, once in this state, it remains like this through the iterations.

Another variation that we are currently exploring is to use *discriminative learning* to fit the model. This amounts to using a conditional likelihood based on $P(G|X)$ to fit the model. It turns out that the resulting algorithm is very similar to the one presented here. Instead of a multiple regression with Z as the response in the M-step, one uses a polychotomous logistic regression, with observations the rows of the weight matrix. Details of this will appear elsewhere.

8   References

Becker, R., Chambers, J. and Wilks, A. (1988). *The New S Language*. Wadsworth International Group, Pacific Grove, California.

Breiman, L. and Ihaka, R. (1984). Nonlinear discriminant analysis via scaling and ACE. Technical Report. Univ. of California, Berkeley.

Breiman, L., Friedman, J.F., Olshen, R. and Stone, C.J. (1984). *Classification and Regression Trees*. Wadsworth, Belmont, California.

Chambers, J. and Hastie, T. (1991). *Statistical Models in S*. Wadsworth/Brooks Cole, Pacific Grove, California.

Dempster, A.P., Laird, N.M. and Rubin, D.B. (1977). Maximum likelihood estimation from incomplete data via the EM algorithm (with discussion). *J. R. Statist. Soc.* B **39**, 1–38.

Friedman, J.H. (1991). Multivariate adaptive regression splines (with discussion). *Annals of Statistics* **19**, 1–141.

Hartigan, J.A. and Wong, M.A. (1979). A k-means clustering algorithm. *Applied Statistics* **28**, 100–108.

Hastie, T., Buja, A. and Tibshirani, R. (1995). Penalized discriminant analysis. *Annals of Statistics* **23**, 73–102.

Hastie, T. and Tibshirani, R. (1994). Discriminant adaptive nearest neighbour classification. Technical Report. Department of Statistics, Stanford University.

Hastie, T. and Tibshirani, R. (1996). Discriminant analysis by gaussian mixtures. *J. R. Statist. Soc.* B **58**, 155–176.

Hastie, T., Tibshirani, R. and Buja, A. (1994). Flexible discriminant analysis by optimal scoring. *J. Am. Statist. Assoc.* **89**, 1255–1270.

Mardia, K.V., Kent, J. and Bibby, J.M. (1979). *Multivariate Analysis*. Academic Press, London.

Michie, D., Spiegelhalter, D.J. and Taylor C.C. (Eds.) (1994). *Machine Learning, Neural and Statistical Classification*. Ellis Horwood, New York.

Ripley, B.D. (1996). *Pattern Recognition and Neural Networks*. Cambridge University Press.

Ripley, B.D. (1994). Neural networks and related methods for classification (with discussion). *J. R. Statist. Soc.* B **56**, 409–456.

2
Neural Networks for Unsupervised Learning Based on Information Theory

Jim Kay

Department of Statistics, University of Glasgow
Glasgow G12 8QQ, Scotland, UK

Abstract

We consider the statistical and information-theoretic basis of a class of artificial neural networks for unsupervised learning and processing. Taking a single processing unit as the basic component of the networks, we define information-theoretic objective functions and a class of activation functions, and derive the gradient-ascent learning rules. Several versions of the processor are considered, namely, the case in which the processor has a single binary output, the case in which the outputs are multinomial-winner-take-all, and that in which they are multivariate binary. In this latter case, local versions of the objective function are developed and these lead to local learning rules. Finally, the issue of computational complexity is briefly discussed, an alternative approach to the modelling is developed, and approximations are mentioned.

1 Introduction

Information theory has been used in various ways in psychology, statistics and neural computation stemming from the seminal work of Shannon (Shannon and Weaver, 1949). In psychology, it has been used to measure the uncertainty and redundancy in sequences of symbols, e.g. words and sentences, and also to measure the transmission of information in experiments in perception; see, for example, Attneave (1959). In statistics it has been used: as a measure of the amount of information provided by an experiment (Lindley, 1956); in the analysis of categorical data (Gokhale and Kullback, 1978); in feature selection in discrimination (Aitchison and Kay, 1975); and in other statistical problems (Kullback, 1959). Ideas from information theory have also been used in research in neural computation. Some examples are: the development of learning rules in synaptic plasticity (Intrator and Cooper, 1995); the study of measures of functional complexity in the nervous system (Tononi *et al.*, 1994); the unbiased measurement of transmitted information in monkey striate cortex (Optican *et al.*, 1991); the assessment of bias in measures of information (Treves and Panzeri, 1995); the use of information-theoretic objective functions in active-data selection (Mackay, 1992); the use of an information-maximisation approach to blind separation and blind deconvolution in signal processing (Bell and Sejnowski, 1995); the exploration of neural population coding for movement (Sanger, 1997); and the development of optimisation principles for the neural code (DeWeese, 1996). A good discussion of the role of information theory in neural coding and the measurement of information transmission in neural systems is provided by Rieke *et al.* (1997, chapter 3); in particular, they discuss theoretical upper limits for the quantity of information which can be transmitted, and show in experiments with real organisms that these limits can be close to realisation. See also Zador (1998). There has also been much use of information theory in the development of artifical neural networks for various purposes; see Atick (1992), Redlich (1993), Taylor and Plumbley (1993) and Becker (1992, 1996). It is in this area of work that the present article belongs and we now provide a brief description of the work of Becker and Hinton and that of Linsker which provided partial motivation. Linsker (1988) developed networks in which the goal was to maximise the transmission of information and he used the mutual information between the input and output distributions as an objective function, and this approach has been termed *infomax*. This approach may be viewed as an unsupervised attempt to discover features within the input field which exhibit the most variation and it is rather akin to principal component analysis. The work reported by Becker and Hinton (1992, 1995), however, was concerned with the information shared between the outputs of units which received input from different receptive fields, the aim being to maximise the *spatial coherence* and to use the units to 'supervise' each other. The objective function used was the mutual information between the output distributions. Information-theoretic objective functions were also used subsequently in the categorisation of objects using temporal coherence (Becker, 1993) and in the recognition of moving objects (Becker, 1995).

The present chapter describes the statistical basis of an approach to the contextual guidance of processing and learning within multistream, multilayered artificial neural networks. For further details, see Kay (1994), Kay and Phillips (1994, 1997), Phillips *et al.* (1995) and Kay *et al.* (1998), and for the neurobiological and neuropsychological motivation see Phillips and Singer (1997). The aim is to fuse, within a single objective function, the goals of unsupervised feature discovery and the learning of predictive relationships between different data sets; it may be viewed as a hybrid of the approaches of Linsker and Becker and Hinton. The basic component of the development is that of the local processor and it is envisaged that many such components can be connected together within a multilayer, multistream architecture, within which the computation is performed locally. In Section 2 we present a brief description of the aspects of information theory which are used in the chapter. We turn our attention in Section 3 to the case of a single local processor which has a binary output. We define the basic objective function used in our approach, define the activation function through which the different inputs are passed to produce an output and then derive the learning rules. In Section 4, we consider three extensions of this basic single processor to more complicated architectures, namely, (i) the case in which there are two streams which offer contextual guidance to each other, (ii) the case in which the local processor contains a multinomial winner-take-all output, leading to the contextual guidance of competitive learning, and (iii) the case in which the local processor has a multivariate binary output. We define the multivariate version of our objective function in Section 5 and derive the learning rules; it is shown that these are nonlocal in nature and so in Section 6 we define local versions of the objective functions and also define local conditional versions of *infomax* and *coherent infomax*. Two illustrations are presented in Section 7. Finally, in Section 8, the computational complexity of the approach is discussed, the basic approach is remodelled leading to improvements, and also some approximations are mentioned.

2   Some information theory

In this secton we describe the basic information-theoretic concepts of *entropy* and *mutual information*. We employ the usual distinction between random variables and their realised values by using capital letters to denote the former while the corresponding lower-case letters denote the latter. We use a generic 'p' to denote a probability density function, with the argument of the function signifying which random variable is being described; so $p(\mathbf{y})$ denotes the probability density function associated with the random vector $\mathbf{Y}$. In the discrete case $p(\mathbf{y})$ will denote the probability that the random vector $\mathbf{Y}$ takes the value $\mathbf{y}$ in a particular realisation. We denote the conditional probability density function of $\mathbf{Y}$, given that $\mathbf{X} = \mathbf{x}$, by $p(\mathbf{y}|\mathbf{x})$.

For an excellent discussion of basic information-theoretic concepts, see Hamming (1980, chapter 6). The mutual information shared between two random vectors $\mathbf{X}$ and $\mathbf{Y}$ is defined by

$$
\begin{aligned}
I(\mathbf{X}; \mathbf{Y}) &= H(\mathbf{X}) - H(\mathbf{X}|\mathbf{Y}) \\
&= H(\mathbf{Y}) - H(\mathbf{Y}|\mathbf{X}) \\
&= H(\mathbf{X}) + H(\mathbf{Y}) - H(\mathbf{X}, \mathbf{Y}).
\end{aligned}
\tag{2.1}
$$

Here $H(\mathbf{Y})$ is the Shannon entropy associated with the distribution of $\mathbf{Y}$, while $H(\mathbf{Y}|\mathbf{X})$ denotes the Shannon entropy associated with the conditional distribution of $\mathbf{Y}$ given $\mathbf{X}$. The mutual information is always nonnegative and is zero when the random vectors are statistically independent; hence it may be used as a general measure of correlation. In general we may write

$$
H(\mathbf{X}, \mathbf{Y}) = H(\mathbf{X}|\mathbf{Y}) + I(\mathbf{X}; \mathbf{Y}) + H(\mathbf{Y}|\mathbf{X}),
\tag{2.2}
$$

which shows that the entropy in the joint distribution of $\mathbf{X}$ and $\mathbf{Y}$ may be decomposed into the sum of three terms denoting, respectively, the information contained in the distribution of $\mathbf{X}$ that is not shared with $\mathbf{Y}$, the information shared by $\mathbf{X}$ and $\mathbf{Y}$, and the information in the distribution of $\mathbf{Y}$ that is not shared with $\mathbf{X}$. Using the relationships in equation (2.1) it is possible to decompose the amount of information in the distribution of $\mathbf{X}$ as follows:

$$
H(\mathbf{X}) = I(\mathbf{X}; \mathbf{Y}) + H(\mathbf{X}|\mathbf{Y}).
\tag{2.3}
$$

This equation may be interpreted to mean that the information in the distribution of **X** can be computed from the sum of two terms: the information shared between **X** and **Y** and the information in the distribution of **Y** that is not shared with **X**. Equation (2.3) is analogous to the 'data = signal + noise' metaphor which is often used in data analysis. This equation may be represented in an information diagram, as illustrated in Figure 1.

The idea of mutual information may be extended to more than two random vectors (McGill, 1954) and for our purposes here we consider the *three-way mutual information* that is shared among three random vectors, **X**, **Y** and **Z**, defined by

$$\begin{aligned} I(\mathbf{X}; \mathbf{Y}; \mathbf{Z}) &= I(\mathbf{X}; \mathbf{Y}) - I(\mathbf{X}; \mathbf{Y}|\mathbf{Z}) \\ &= I(\mathbf{X}; \mathbf{Z}) - I(\mathbf{X}; \mathbf{Z}|\mathbf{Y}) \\ &= I(\mathbf{Y}; \mathbf{Z}) - I(\mathbf{Y}; \mathbf{Z}|\mathbf{X}). \end{aligned} \quad (2.4)$$

In equation (2.4), $I(\mathbf{X}; \mathbf{Y}|\mathbf{Z})$ is the conditional mutual information shared between **X** and **Y**, having observed **Z**. We interpret this as being that information which is shared between **X** and **Y** but not shared with **Z**. Applying equations (2.1) and (2.4) we may obtain three equivalent expressions for the three-way mutual information as follows:

$$\begin{aligned} I(\mathbf{X}; \mathbf{Y}; \mathbf{Z}) &= H(\mathbf{X}) - H(\mathbf{X}|\mathbf{Y}) - H(\mathbf{X}|\mathbf{Z}) + H(\mathbf{X}|\mathbf{Y}, \mathbf{Z}) \\ &= H(\mathbf{Y}) - H(\mathbf{Y}|\mathbf{X}) - H(\mathbf{Y}|\mathbf{Z}) + H(\mathbf{Y}|\mathbf{X}, \mathbf{Z}) \\ &= H(\mathbf{Z}) - H(\mathbf{Z}|\mathbf{X}) - H(\mathbf{Z}|\mathbf{Y}) + H(\mathbf{Z}|\mathbf{X}, \mathbf{Y}). \end{aligned} \quad (2.5)$$

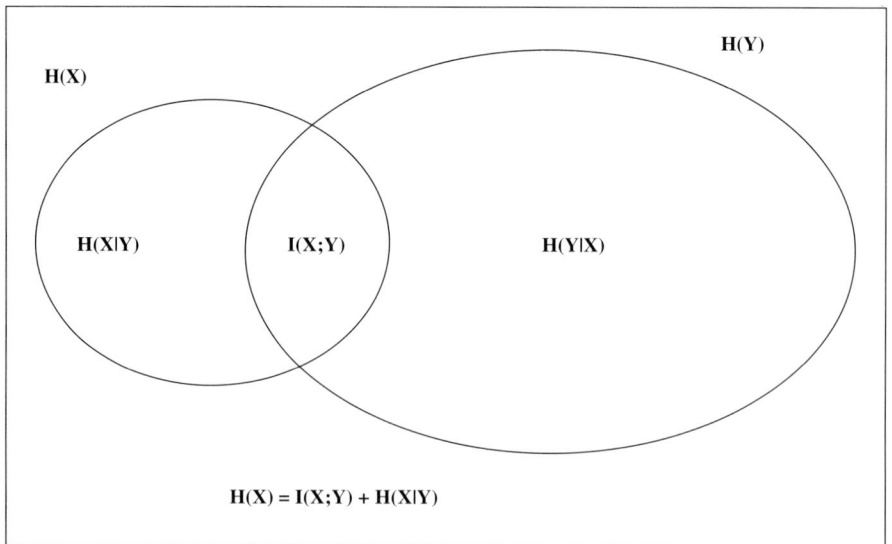

H(Y)

H(X)

H(X|Y)   I(X;Y)   H(Y|X)

H(X) = I(X;Y) + H(X|Y)

Fig. 1:   Components of information with two random vectors.

Note that, since the three-way information is symmetric in its arguments, **X**, **Y** and **Z**, there are three different but equivalent forms, obtained by permuting the arguments. It is possible to construct an information diagram (see Figure 2) for the case of three random components thus extending the diagram presented as Figure 1.

This decomposition of information can only make strict sense when the measure of three-way information is nonnegative; however, that cannot be guaranteed (Whittaker, 1990; Kay, 1994) and it is easy to construct simple examples to demonstrate this—for example, where two of the three random components are marginally independent, and thus have zero mutual information, while at the same time they are not conditionally independent given the third component, and so have non-zero conditional mutual information. It is important to stress, however, that this seeming pathology does not create problems in practical examples when three-way shared information does exist and when the computational goal is to maximise the three-way mutual information; in such cases the three-way mutual information is driven towards positivity during the learning process. More importantly, in real applications the three-way mutual information may be shown to be equal to a two-way mutual information and so is nonnegative (see Section 3.1); in these cases the three-way mutual information is a well-defined component of information. This potentially pathological behaviour is similar to the well-known Simpson's paradox in the analysis of categorical data. In the (usual) case in which the three-way mutual information is positive it is clear from

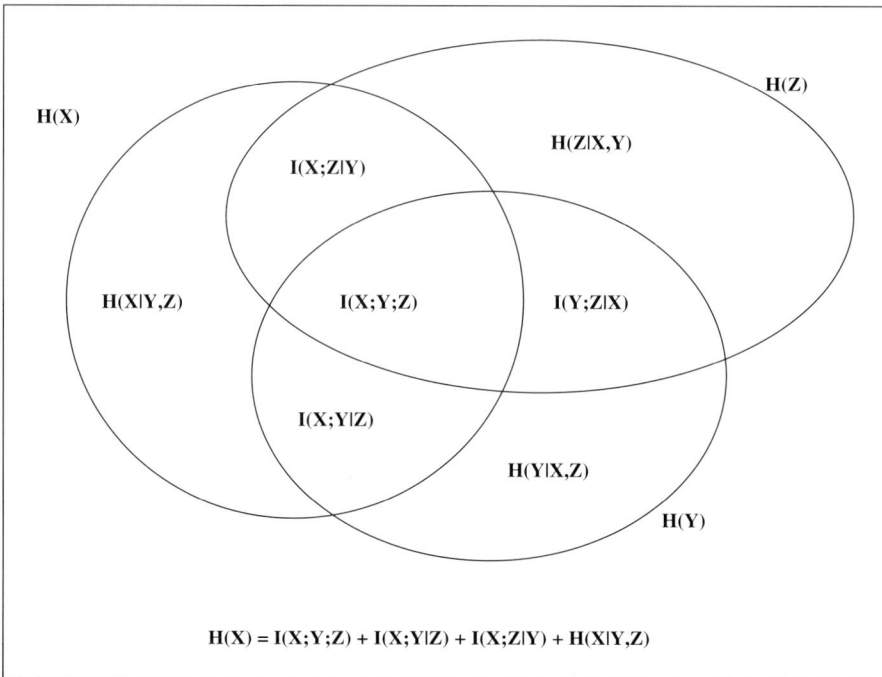

$$H(X) = I(X;Y;Z) + I(X;Y|Z) + I(X;Z|Y) + H(X|Y,Z)$$

Fig. 2:   Components of information with three random vectors.

Figure 2, and may be deduced from equations (2.1), (2.4) and (2.5), that the following decomposition holds:

$$H(\mathbf{Y}) = I(\mathbf{Y}; \mathbf{X}; \mathbf{Z}) + I(\mathbf{Y}; \mathbf{X}|\mathbf{Z}) + I(\mathbf{Y}; \mathbf{Z}|\mathbf{X}) + H(\mathbf{Y}|\mathbf{X}, \mathbf{Z}). \qquad (2.6)$$

Each of the four components of this equation will be of particular use in the general form of objective function considered in Section 3. We note for future reference some integral representations of Shannon entropy; in the case where the random variables are discrete, the integrals are replaced by summations, and the densities by probabilities:

$$H(\mathbf{Y}) = -\int p(\mathbf{y}) \log p(\mathbf{y}) d\mathbf{y} \qquad (2.7)$$

$$H(\mathbf{Y}|\mathbf{X}) = -\int \int p(\mathbf{y}|\mathbf{x}) \log p(\mathbf{y}|\mathbf{x}) d\mathbf{y} d\mathbf{x} \qquad (2.8)$$

$$H(\mathbf{Y}|\mathbf{X}, \mathbf{Z}) = -\int \int \int p(\mathbf{y}|\mathbf{x}, \mathbf{z}) \log p(\mathbf{y}|\mathbf{x}, \mathbf{z}) d\mathbf{y} d\mathbf{x} d\mathbf{z}. \qquad (2.9)$$

3   A single local processor: objective functions and learning rules

3.1   Objective functions

First we consider the case of a local processor which has a single output unit and whose inputs are separated into two distinct types, namely, receptive field (RF) inputs and contextual field (CF) inputs; see Figure 3. If we were to consider only a single processor then this distinction would be unnecessary; it is proposed, however, to consider multilayered, multistream networks built by connecting together such local processors. Hence it is envisaged that the contextual field will consist of units from neighbouring streams at the same layer of processing as well as, possibly, back-projections from higher layers. On the other hand, the receptive field will generally consist of units in the layers below the output unit. We use the random variable Y to denote the value of the output unit and the random vectors $\mathbf{X}$ and $\mathbf{Z}$ to denote, respectively, the RF inputs and the CF inputs.

In terms of such a local processor we may now interpret the four information components defined in equation (2.6) as follows. The three-way mutual information $I(Y; \mathbf{X}; \mathbf{Z})$ represents the information that is common to the output and to both RF and CF inputs; we wish to maximise this term so as to maximise the transmission of the information in the RF that is related to the current CF. The term $I(Y; \mathbf{X}|\mathbf{Z})$ denotes the information that the output shares with the RF inputs that is not contained in the CF units. It is sensible for this term to be allowed to increase because, while the information in the RF might not be relevant to the current context, it might nevertheless be relevant to some other contextual units in the system. The term $I(Y; \mathbf{Z}|\mathbf{X})$ denotes the information that is shared between the output unit and the CF units but not with

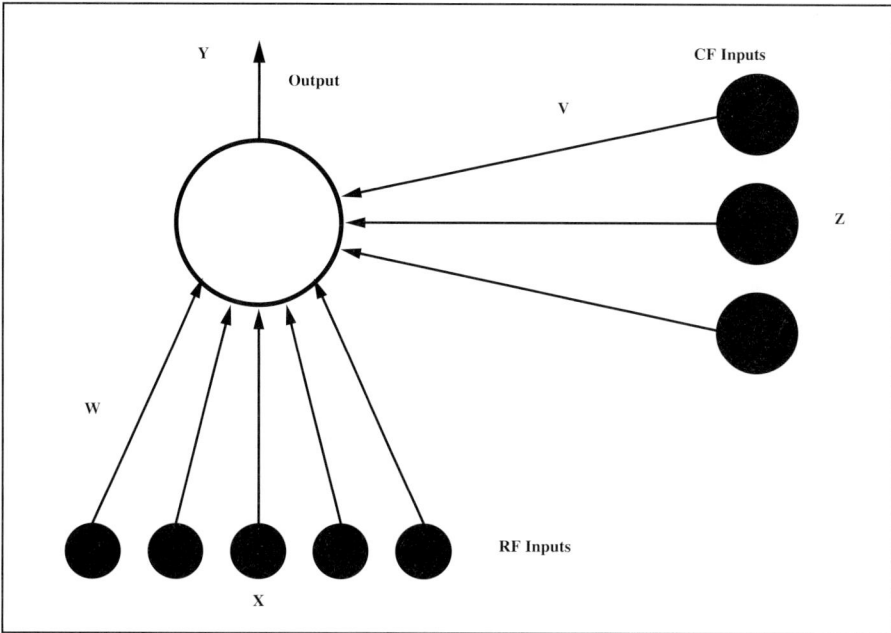

Fig. 3: A local processor with RF inputs **X**, CF inputs **Z**, and output Y. The vector **w** denotes the weights on the connections from the RF inputs into the output unit and **v** denotes the corresponding weights for the CF inputs.

the RF units. We don't necessarily wish this term to increase as that would allow the processing to be driven by the CF units rather than the RF units and we prefer the RF units to be the 'primary drivers' of the processor. Putting these information terms together gives the following general class of information-theoretic objective functions:

$$F = \phi_0 I(Y; \mathbf{X}; \mathbf{Z}) + \phi_1 I(Y; \mathbf{X}|\mathbf{Z}) + \phi_2 I(Y : \mathbf{Z}|\mathbf{X}) + \phi_3 H(Y|\mathbf{X}, \mathbf{Z}). \qquad (3.1)$$

We normally take $\phi_0 = 1$, so that ϕ_1, ϕ_2, and ϕ_3 express the relative importance of their respective components of information relative to the three-way information term. We allow the $\{\phi_i\}$ to take values in the interval $(-1, 1)$.

We now discuss some links between this class of objective functions and other work. Taking $\phi_1 = 1$, $\phi_2 = \phi_3 = 0$ gives formally

$$F = I(Y; \mathbf{X}), \qquad (3.2)$$

which is the objective function used by Linsker. This equivalence is formal, but to actually implement it within our more general framework it is required to cut the contextual connections.

Taking $\phi_1 = \phi_3 = 0$ and $\phi_2 = 1$ gives

$$F = I(Y; \mathbf{Z}), \qquad (3.3)$$

which is consistent with the approach of Becker and Hinton were similar architectures, connectivities and activation functions to be employed.

Taking $\phi_1 = \phi_2 = \phi_3 = 0$ gives

$$F = I(Y; \mathbf{X}; \mathbf{Z}). \tag{3.4}$$

This is the objective function on which we will mainly focus because it measures the information shared among the RF inputs, the CF inputs and the output; thus its maximisation enables the extraction of that information from the RF inputs that is coherently related to the information in the CF inputs, and this approach has been termed *coherent infomax*. It is important here to emphasise a special case of this objective function and one which arises in real applications. It is usually the case that, when working with real data, the RF inputs will form a set of distinct patterns. In this case, when the network is operated deterministically, there is only one CF input corresponding to each RF input at any particular time and the conditional mutual information between the output and the CF inputs, given the RF inputs, will be zero. It follows that the three-way mutual information in equation (2.5) reduces to the two-way mutual information between the output and the CF inputs given in equation (3.3). This is a nonnegative quantity and so in this case the objective function cannot possibly be negative (as alluded to in Section 2).

Taking $\phi_1 = 1 - \epsilon$, $\phi_2 = \epsilon$ and $\phi_3 = 0$ gives

$$F = \epsilon I(Y; \mathbf{Z}) + (1 - \epsilon)I(Y; \mathbf{X}) \tag{3.5}$$

which is an information-theoretic analogue of the objective function used by Schmidhuber and Prelinger (1993). Hence we see the generality of the proposed class of objective functions and also that important precursors to the approach described here may be viewed as special cases, especially those of Becker and Hinton and Linsker.

In the sequel we take a conditional approach to the modelling of the output given the RF and CF inputs and therefore it is convenient to use equations (2.1) and (2.4) to rewrite the objective function as

$$F = H(Y) - \psi_1 H(Y|\mathbf{X}) - \psi_2 H(Y|\mathbf{Z}) - \psi_3 H(Y|\mathbf{X}, \mathbf{Z}), \tag{3.6}$$

where $\psi_1 = 1 - \phi_2$, $\psi_2 = 1 - \phi_1$ and $\psi_3 = \phi_1 + \phi_2 - \phi_3 + 1$. Note from equation (3.6) that F contains the same entropic terms as does the three-way mutual information but that they are weighted differently. We now focus attention on the simple case of a single binary output.

3.2   Single binary output

We consider the case where there is a single output unit whose value is described by the random variable Y. We assume that all the units are stochastic and that the joint distribution of the RF inputs $\mathbf{X}$ and the CF inputs $\mathbf{Z}$ follows some general probability model which will remain unspecified; a useful advantage of the implementation here is that the empirical distribution of the inputs is used in learning and this makes the approach quite general by avoiding the need for explicit modelling of the distribution of the inputs.

We denote the connection weights between the RF inputs and the output by the vector $\mathbf{w}$ and the connections between the CF inputs and the output by $\mathbf{v}$, and adopt the familiar practice of treating biases by using additional units clamped at -1. The integrated RF and CF inputs are defined by

$$R = \sum_{i=1}^{m} w_i X_i - w_0$$

$$C = \sum_{j=1}^{n} v_j Z_j - v_0, \tag{3.7}$$

where w_0 and v_0 are the biases and the $\{X_i\}$ and the $\{Z_i\}$ are random variables representing the components of the random vectors $\mathbf{X}$ and $\mathbf{Z}$, respectively. We denote the activation function at the output unit by $A(r, c)$, where A is a general nonlinear function of r and c, the respective observed integrated RF and CF. We assume further that the output has a logistic nonlinearity with unit scale parameter. Hence the conditional probability that the output takes the value unity given the observed RF and CF inputs is given by

$$\Pr(Y = 1 | \mathbf{X} = \mathbf{x}, \mathbf{Z} = \mathbf{z}) = \frac{1}{1 + \exp(-A(r, c))} \tag{3.8}$$

and we denote this probability by $\theta \equiv \theta(\mathbf{w}, \mathbf{v}, \mathbf{x}, \mathbf{z})$; in the sequel, for simplicity, we won't indicate this explicit dependence of the output probability on the RF and CF weights and inputs.

We now list some basic mathematical results which will be used in the sequel. Since Y is a binary random variable, it follows from equations (2.7) and (3.8) that the conditional entropy of the output distribution is given by

$$H(Y | \mathbf{X}, \mathbf{Z}) = -\langle \theta \log \theta + (1 - \theta) \log(1 - \theta) \rangle_{\mathbf{x}, \mathbf{z}}. \tag{3.9}$$

The notation $\langle \ldots \rangle_{\mathbf{x}, \mathbf{z}}$ denotes the operation of taking the expectation with respect to the joint distribution of $\mathbf{X}$ and $\mathbf{Z}$.

We now state two simple results:

$$\frac{d}{du}[u \log u + (1 - u) \log(1 - u)] = \log \frac{u}{1 - u} \tag{3.10}$$

$$\frac{\partial}{\partial \mathbf{w}} \theta = \frac{\partial A}{\partial r} \theta(1 - \theta)\mathbf{x}$$

$$\frac{\partial}{\partial \mathbf{v}} \theta = \frac{\partial A}{\partial c} \theta(1 - \theta)\mathbf{z}. \tag{3.11}$$

3.3 Derivation of learning rules

Equation (3.9) provides an expression for the conditional entropy $H(Y|\mathbf{X}, \mathbf{Z})$ and we shall now use it to obtain expressions for the other entropies $H(Y)$, $H(Y|\mathbf{X})$ and $H(Y|\mathbf{Z})$. Hence we require to consider the distributions $p(y)$, $p(y|\mathbf{x})$ and $p(y|\mathbf{z})$. Now,

$$p(y) = \int \int p(y|\mathbf{x}, \mathbf{z}) p(\mathbf{x}, \mathbf{z}) d\mathbf{x} d\mathbf{z}$$
$$= \langle p(y|\mathbf{x}, \mathbf{z}) \rangle_{\mathbf{x}, \mathbf{z}}. \tag{3.12}$$

So,

$$\Pr(Y = 1) = \langle \Pr(Y = 1|\mathbf{X} = \mathbf{x}, \mathbf{Z} = \mathbf{z}) \rangle_{\mathbf{x}, \mathbf{z}}$$
$$= \langle \theta \rangle_{\mathbf{x}, \mathbf{z}}. \tag{3.13}$$

Hence the marginal probability of obtaining an output value of unity is simply the average of all the conditional output probabilities taken over all possible values of the RF and CF units; in practice, however, this average is taken empirically over all the actual inputs 'seen' by the processor. Note also that

$$p(y|\mathbf{x}) = \int p(y|\mathbf{x}, \mathbf{z}) p(\mathbf{z}|\mathbf{x}) d\mathbf{z} \tag{3.14}$$

and so

$$\Pr(Y = 1|\mathbf{X} = \mathbf{x}) = \langle \theta \rangle_{\mathbf{z}|\mathbf{x}}. \tag{3.15}$$

Similarly,

$$\Pr(Y = 1|\mathbf{Z} = \mathbf{z}) = \langle \theta \rangle_{\mathbf{x}|\mathbf{z}}. \tag{3.16}$$

Note that the calculation of the conditional expression (3.16) requires the averaging of the conditional output probabilities over all values of the RF inputs while keeping the CF units fixed at the particular value $\mathbf{z}$; that is, it requires a conditional average to be calculated, and stored, for each pattern taken on by the CF units. However, in practice, these averages are taken empirically over the actual RF inputs 'seen' by the processor for each particular CF pattern in the data. The expectation $\langle \theta \rangle_{\mathbf{z}|\mathbf{x}}$ has a similar interpretation with the roles of the RF and CF inputs being reversed. We denote these averages (expectations) by

$$E = \langle \theta \rangle_{\mathbf{x}, \mathbf{z}} \tag{3.17}$$
$$E_{\mathbf{z}} = \langle \theta \rangle_{\mathbf{x}|\mathbf{z}} \tag{3.18}$$
$$E_{\mathbf{x}} = \langle \theta \rangle_{\mathbf{z}|\mathbf{x}}. \tag{3.19}$$

Note also that

$$\langle E_{\mathbf{x}} \rangle_{\mathbf{x}} \equiv E \equiv \langle E_{\mathbf{z}} \rangle_{\mathbf{z}}. \tag{3.20}$$

We also derive expressions for the various entropic terms. Since the marginal distribution of Y is binary, with $\Pr(Y = 1) = E$, it follows from the argument used to derive equation (3.9) that

$$H(Y) = -[E \log E + (1 - E) \log(1 - E)]. \tag{3.21}$$

Since the conditional distribution of Y given that $\mathbf{X} = \mathbf{x}$ is binary with probability $\Pr(Y = 1 | \mathbf{X} = \mathbf{x}) = E_{\mathbf{x}}$, it follows that

$$H(Y|\mathbf{X}) = - \langle E_{\mathbf{x}} \log E_{\mathbf{x}} + (1 - E_{\mathbf{x}}) \log(1 - E_{\mathbf{x}}) \rangle_{\mathbf{x}}. \tag{3.22}$$

Similarly,

$$H(Y|\mathbf{Z}) = - \langle E_{\mathbf{z}} \log E_{\mathbf{z}} + (1 - E_{\mathbf{z}}) \log(1 - E_{\mathbf{z}}) \rangle_{\mathbf{z}}. \tag{3.23}$$

We now calculate the derivatives of each of the entropic terms in the objective function F, defined in equation (3.6), with respect to the connection weights $\mathbf{w}$ and $\mathbf{v}$ and the biases. First, we consider the derivatives of $H(Y)$. From equation (3.21), and using equation (3.10), we obtain

$$\frac{\partial H(Y)}{\partial \mathbf{w}} = - \log \frac{E}{(1 - E)} \frac{\partial E}{\partial \mathbf{w}}. \tag{3.24}$$

Now, using equation (3.11), we obtain

$$\frac{\partial E}{\partial \mathbf{w}} = \left\langle \frac{\partial \theta}{\partial \mathbf{w}} \right\rangle_{\mathbf{x},\mathbf{z}}$$

$$= \left\langle \frac{\partial A}{\partial r} \theta (1 - \theta) \mathbf{x} \right\rangle_{\mathbf{x},\mathbf{z}}. \tag{3.25}$$

Hence, combining equations (3.24) and (3.25) gives

$$\frac{\partial H(Y)}{\partial \mathbf{w}} = \left\langle \log \frac{E}{(1 - E)} \frac{\partial A}{\partial r} \theta (1 - \theta) \mathbf{x} \right\rangle_{\mathbf{x},\mathbf{z}}. \tag{3.26}$$

Arguing similarly, we deduce the following results for the other entropic terms:

$$\frac{\partial H(Y|\mathbf{X})}{\partial \mathbf{w}} = \left\langle \log \frac{E_{\mathbf{x}}}{(1 - E_{\mathbf{x}})} \frac{\partial A}{\partial r} \theta (1 - \theta) \mathbf{x} \right\rangle_{\mathbf{x},\mathbf{z}} \tag{3.27}$$

$$\frac{\partial H(Y|\mathbf{Z})}{\partial \mathbf{w}} = \left\langle \log \frac{E_{\mathbf{z}}}{(1 - E_{\mathbf{z}})} \frac{\partial A}{\partial r} \theta (1 - \theta) \mathbf{x} \right\rangle_{\mathbf{x},\mathbf{z}} \tag{3.28}$$

$$\frac{\partial H(Y|\mathbf{X},\mathbf{Z})}{\partial \mathbf{w}} = \left\langle \log \frac{\theta}{(1 - \theta)} \frac{\partial A}{\partial r} \theta (1 - \theta) \mathbf{x} \right\rangle_{\mathbf{x},\mathbf{z}}. \tag{3.29}$$

Combining equations (3.27)–(3.29) we obtain

$$\frac{\partial F}{\partial \mathbf{w}} = \left\langle (\psi_3 A - \bar{O}) \frac{\partial A}{\partial r} \theta (1 - \theta) \mathbf{x} \right\rangle_{\mathbf{x},\mathbf{z}}. \tag{3.30}$$

Applying a similar approach gives the following derivatives for the weights connecting the CF inputs to the output unit:

$$\frac{\partial F}{\partial \mathbf{v}} = \left\langle (\psi_3 A - \bar{O}) \frac{\partial A}{\partial c} \theta (1 - \theta) \mathbf{z} \right\rangle_{\mathbf{x}, \mathbf{z}}. \tag{3.31}$$

The term $\bar{O}$ is a nonlinear floating average given by

$$\bar{O} = \log \frac{E}{(1 - E)} - \psi_1 \log \frac{E_{\mathbf{x}}}{(1 - E_{\mathbf{x}})} - \psi_2 \log \frac{E_{\mathbf{z}}}{(1 - E_{\mathbf{z}})}. \tag{3.32}$$

The derivatives for the biases are given by

$$\frac{\partial F}{\partial w_0} = \left\langle (\psi_3 A - \bar{O}) \frac{\partial A}{\partial c} \theta (1 - \theta)(-1) \right\rangle_{\mathbf{x}, \mathbf{z}} \tag{3.33}$$

$$\frac{\partial F}{\partial v_0} = \left\langle (\psi_3 A - \bar{O}) \frac{\partial A}{\partial c} \theta (1 - \theta)(-1) \right\rangle_{\mathbf{x}, \mathbf{z}}. \tag{3.34}$$

Equations (3.30)–(3.31) and (3.33)–(3.34) provide the derivatives of F required for incremental gradient-ascent learning.

3.4 Discussion of the learning rules

As we wish to learn the weights and biases in order to maximise the objective function F, the gradient-ascent learning rules take the weight changes at each step as being proportional to these derivatives. If necessary, different learning rates could be used for the RF and the CF weights. In particular, it may be advantageous to fix the CF weights at zero for the first few epochs in order to counteract the possibility that initial RF input noise will be modulated rather than signal; normally, however, the same learning rate is used. The learning rules based on the derivatives in equations (3.30)–(3.31) and (3.33)–(3.34) are written for batch learning, but it is possible to perform on-line learning by removing the averaging brackets and by recursively building up the required averages of output probability; see Kay (1994).

The terms $\partial A / \partial r$ and $\partial A / \partial c$ give the rate of change of the pre-synaptic activation with respect to the integrated RF and CF fields. The term $\theta (1 - \theta)$ provides intrinsic weight stabilisation provided that the activation term A grows large when the weights grow in magnitude. The presence of the term $\psi_3 A - \bar{O}$ ensures that the weight change is nonmonotonically related to the pre-synaptic activation A. This property of these learning rules is therefore similar to the type of nonmonotonicity present in the behaviour of the BCM and ABS learning rules (Artola et al., 1990; Intrator and Cooper, 1992), which have been shown to enjoy some biological plausibility. The rules derived here are distinctive, however, particularly with regard to the floating average $\bar{O}$; here the average depends on the current integrated RF and CF and, in particular, it is context-sensitive. In order to further elucidate this nonmonotonicity we consider the term $\psi_3 A - \bar{O}$ which may be written as the single logarithm

$$\log \frac{(\theta/(1-\theta))^{\psi_3}(E_{\mathbf{x}}/(1-E_{\mathbf{x}}))^{\psi_1}(E_{\mathbf{z}}/(1-E_{\mathbf{z}}))^{\psi_2}}{E/(1-E)}. \tag{3.35}$$

Hence, in the case where ψ_3 is positive, this term will be positive provided that the current conditional output probability θ is greater than the threshold $t/(1+t)$, where

$$t = \exp(\bar{O}/\psi_3) \tag{3.36}$$

and otherwise nonpositive. On the other hand, when ψ_3 is negative, the term is positive when θ is less than the threshold and otherwise nonpositive.

3.5 The activation function

The activation function A may be any differentiable function of the integrated RF and CF. Here we discuss the activation function which was introduced by Kay and Phillips (1994, 1997) and discussed in Kay (1994). This particular form of activation function was derived from a set of requirements for the operation of a local processor, described as follows.

(1) If the integrated RF input is zero then the activation is zero.
(2) If the integrated CF input is zero then the activation should be the integrated RF input.
(3) If the integrated RF and CF inputs have the same sign, then the activation should be greater than when it is based on the RF input alone. On the other hand, if the integrated RF and CF inputs disagree, i.e. are of opposite sign, then the activation should be less than it would be if based on the RF input alone.
(4) The sign of the activation should be that of the integrated RF so that the context cannot affect the direction of the output decision.

Requirements (1) and (2) say that $A(r, 0) = r$ and $A(0, c) = 0$. These, taken together with requirement (4), suggest that $A(r, c)$ take the form $rA1(r, c)$, where $A1$ is a positive function with $A1(r, 0) = 1$. Requirements (3) and (4) suggest now that $A1(r, c)$ has the form $A2(r, c)$, where $A2$ is a function of a single variable and is greater than unity when its argument is positive and less than unity when its argument is negative, and also $A2(0) = 1$. The exponential function has these properties, and so we now consider the form

$$A(r, c) = r \exp(k_2 rc). \tag{3.37}$$

This form for $A(r, c)$ has the property that when rc is very negative it approaches 0. This means that in cases of very strong disagreement between the integrated RF and CF inputs the output is then placed in its most uncertain state. However, because we wish the RF inputs to be primary drivers, we don't necessarily desire that the RF information in such a combination is entirely ignored. Hence we introduce a constant k_1 which plays the role of a depression factor; that is, in the extreme case when $rc \longrightarrow -\infty$, the activation will be $k_1 r$. These requirements lead naturally to the following class of activation functions:

$$A(r, c) = r[k_1 + (1 - k_1)\exp(k_2 rc)], \tag{3.38}$$

with $k_2 > 0$ and $0 \leq k_1 < 1$.

In practical examples we normally take $k_1 = 0.5$ and $k_2 = 2$. This class of activation functions is sufficient to meet the stated requirements; it is clearly not unique and other nonlinear functions could be suggested which satisfy the requirements. What matters is that the family derived here is sufficient for our purposes. Some discussion of other activation functions is given by Kay (1994) and some simple experiments reported in Kay and Phillips (1994) demonstrate the necessity and sufficiency of this form of activation function. Finally we present the derivatives of the activation with respect to the integrated RF and CF inputs.

$$\frac{\partial A}{\partial r} = k_1 + (1 - k_1)(1 + k_2 rc)\exp(k_2 rc)$$

$$\frac{\partial A}{\partial c} = (1 - k_1)k_2 r^2 \exp(k_2 rc). \tag{3.39}$$

4   More complex architectures

4.1   Mutual contextual guidance

We now consider a simple network which has two channels and which is composed of two interfaced local processors of the type discussed in Section 3. In this case there are two RFs, whose inputs we denote by the random vectors $\mathbf{X}_1$ and $\mathbf{X}_2$, and two outputs represented by the random variables Y_1 and Y_2. The output Y_2 provides the contextual field for the output unit, with output Y_1, and vice versa. There are two objective functions F_1 and F_2, one for each of the local processors. F_1 has the form of the objective function defined in equation (3.6) with $\mathbf{X}_1$ as receptive field, Y_2 as contextual input and Y_1 as output. Similarly, F_2 has $\mathbf{X}_2$ as receptive field, Y_1 as contextual input and Y_2 as output. This network is illustrated in Figure 4.

We take the general objective function to be

$$F = F_1 + F_2. \tag{4.1}$$

In seeking to maximise F we are stipulating only that each of the local processors should be operated so as to maximise their respective objective functions. It is possible to perform these maximisations separately because, while the activation at output Y_1 depends implicitly on the weights $\mathbf{u}_2$ via the output Y_2, we shall ignore this dependence in the derivation of the learning rules. This simplification makes the computation more local because the weight-change rules for the learning of a particular set of receptive

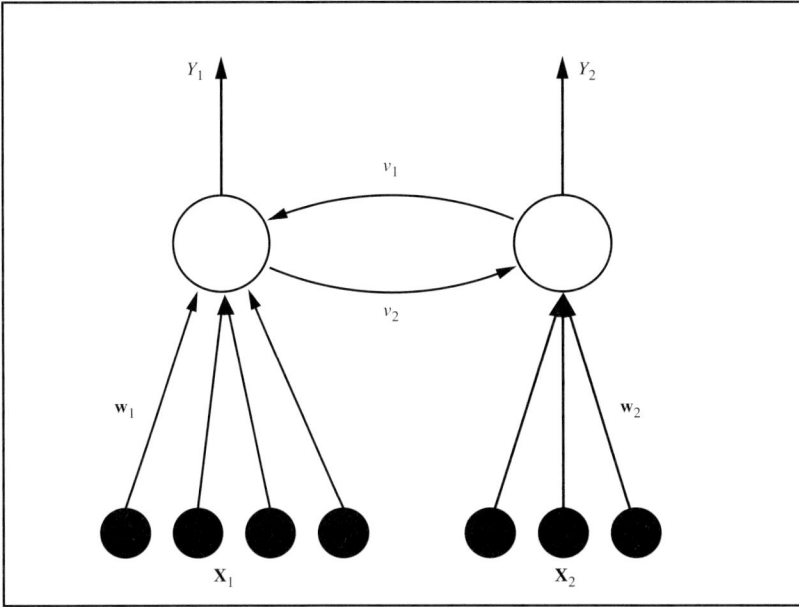

Fig. 4: Two local processors are interfaced and provide mutual contextual guidance to each other. There are two RFs ($\mathbf{X_1}$ and $\mathbf{X_2}$) and the outputs (Y_1 and Y_2) provide contextual guidance to each other.

field weights does not then involve the partial derivatives of the objective function associated with the neighbouring channel. This is tantamount to assuming that, given $\mathbf{X_2}$, Y_1 is conditionally independent of Y_2 and also that, given $\mathbf{X_1}$, Y_2 is conditionally independent of Y_1.

Note that the network displayed in Figure 4 differs from that used by Becker and Hinton in that here there are explicit connections between the output units. We now state explicitly the derivatives required for the learning rules. The integrated RF and CF inputs are defined as

$$r_1 = \mathbf{w}_1^T \mathbf{x}_1 - w_{10}$$
$$c_1 = v_1 y_2 - v_{10}$$
$$r_2 = \mathbf{w}_2^T \mathbf{x}_2 - w_{20}$$
$$c_2 = v_2 y_1 - v_{20}. \qquad (4.2)$$

From the additive form of the objective function defined in equation (4.1), and the assumptions of conditional independence, we know that $\partial F/\partial \mathbf{w}_i = \partial F_i/\partial \mathbf{w}_i$ ($i = 1, 2$), and similarly for the other weights. From equations (3.30)–(3.31) and (3.33)–(3.34) we obtain the following derivatives. In what is written, the same values for the parameters $\{\psi_i\}$ have been used in both channels; clearly, however, it is possible to use different values of these parameters in different channels, depending on the local purposes of learning and processing.

$$\frac{\partial F}{\partial \mathbf{w}_1} = \left\langle (\psi_3 A_1 - \bar{O}_1) \frac{\partial A_1}{\partial r_1} \theta_1 (1 - \theta_1) \mathbf{x_1} \right\rangle_{\mathbf{x}_1, y_2} \tag{4.3}$$

$$\frac{\partial F}{\partial v_1} = \left\langle (\psi_3 A_1 - \bar{O}_1) \frac{\partial A_1}{\partial c_1} \theta_1 (1 - \theta_1) y_2 \right\rangle_{\mathbf{x}_1, y_2} \tag{4.4}$$

$$\frac{\partial F}{\partial \mathbf{w}_2} = \left\langle (\psi_3 A_2 - \bar{O}_2) \frac{\partial A_2}{\partial r_2} \theta_2 (1 - \theta_2) \mathbf{x_2} \right\rangle_{\mathbf{x}_2, y_1} \tag{4.5}$$

$$\frac{\partial F}{\partial v_2} = \left\langle (\psi_3 A_2 - \bar{O}_2) \frac{\partial A_2}{\partial c_2} \theta_2 (1 - \theta_2) y_1 \right\rangle_{\mathbf{x}_2, y_1}. \tag{4.6}$$

To obtain the derivatives for the learning of the biases, $\mathbf{x}_1, \mathbf{x}_2, y_1$ and y_2 are replaced within the angled brackets in equations (4.3)–(4.6) by -1, cf. equations (3.33)–(3.34). In equations (4.3)–(4.6)

$$A_i \equiv A(r_i, c_i) = r_i[k_1 + (1 - k_1) \exp(k_2 r c)] \tag{4.7}$$

and

$$\bar{O}_i = \log \frac{E_i}{(1 - E_i)} - \psi_1 \log \frac{E_{i\mathbf{x}}}{(1 - E_{i\mathbf{x}})} - \psi_2 \log \frac{E_{i\mathbf{z}}}{(1 - E_{i\mathbf{z}})}, \tag{4.8}$$

where $i = 1, 2$ and the subscript i in the averages indicates which objective function is under consideration.

So far we have considered the case where there are only two channels. In general there might be any number of parallel channels and the local learning and processing illustrated here may be extended, but keeping the connectivity of the outputs local in a relatively sparse arrangement. With more streams, the contextual field of each output unit would consist, at least, of several of the outputs from neighbouring channels and the methodology extended in the obvious way, including the assumptions of local conditional independence which help to ensure local learning and processing. We develop this further in Section 4.3.

4.2 Multiple-output processors: competitive learning

It will be necessary when considering the case of multiple-output units to consider units possessing an internal structure of a cluster of cells. In this section we consider the simple case of local-coding, even though the multinomial specification considered here could be developed to allow more than one cell to fire. We describe the output of such a unit by a random vector which follows a multinomial probability model. The formulation developed here could be used to code multivalued units and this has been discussed by Becker (1992). In the particular case when only one of the cells is allowed to fire we term this a *multinomial winner-take-all unit*. Under the multinomial assumption, the probability distribution of the random output vector $\mathbf{Y}$ has the form

$$p(\mathbf{y}) \equiv \Pr(Y_1 = y_1, Y_2 = y_2, \ldots, Y_k = y_k) \propto \prod_{i=1}^{k} \theta_i^{y_i} \tag{4.9}$$

where $\sum_{i=1}^{k} \theta_i = 1$, $\theta_i = \Pr(Y_i = 1)$, $\sum_{i=1}^{k} y_i = n$, and the $\{y_i\}$ are integers.

In general more than one of the cells may signal an integer as long as the total is n. In the winner-take-all (WTA) case in which $n = 1$ only one of the cells can output a 1 with the rest being 0 (assuming that 0/1 units are being employed; if $-1/+1$ units are being used then the outputs are $2Y_i - 1$). Note that in a multinomial model only $k - 1$ cells are required to represent k outputs. However, due to the biological implausibility of one particular cell being the 'fill-up' cell, we use the symmetric formulation with k cells.

The Shannon entropy associated with a multinomial unit may be easily calculated as follows:

$$H(\mathbf{Y}) = -I\!E(\log p(\mathbf{Y}))$$

$$= -I\!E\left[\sum_{i=1}^{k} Y_i \log \theta_i\right]$$

$$= -\sum_{i=1}^{k} \theta_i \log \theta_i, \tag{4.10}$$

since Y_i has expectation θ_i. Here the notation $I\!E$ denotes the operation of theoretical expectation.

Before proceeding with the derivation of the learning rules, we mention a possible link with anti-Hebbian learning. In the multinomial model, the covariance between any pair of outputs Y_i and Y_j is $-n\theta_i\theta_j$ and so the estimated covariance between the ith and jth outputs is given by $-n\hat{\theta}_i\hat{\theta}_j$ which is negatively proportional to the product of the respective outputs. So if all of the outputs were pairwise-connected, as in a lateral inhibition network, and if a covariance rule was used to learn the weights on these connections, we would obtain anti-Hebbian learning; see, for example, Foldiak (1990). Actually we shall see that the learning rules have an element of anti-Hebbianism about them.

We now consider the probabilistic modelling of the vector of outputs. We use the multiple-logistic distribution, which dates back to Anderson (1972) at least, in order to model the conditional distribution of the outputs $\mathbf{Y}$ given the values of all inputs whether they are from the receptive field (usually the layer below) or from the contextual field, which includes neighbouring output vectors as well as possible backprojections from higher layers.

Consider the sth cell in the multiple-output unit ($s = 1, 2, \ldots, k$) and let $A_s(r_s, c_s)$ denote the activation of the sth cell, given the integrated inputs r_s and c_s from the receptive and contextual fields, respectively. Note that r_s and c_s will typically be calculated from the same inputs for all k cells, but they will take different values because they involve different connection weights and because the normalisation of output probabilities will tend to force the outputs apart. In general, however, the connectivity could be sparse and nonhomogeneous, e.g. by using RFs which overlap only partially. While Anderson (1972) used the multiple-logistic distribution in its nonredundant form we use the symmetric version which has been named the *Softmax* approach by Bridle (1990). We take

$$\theta_s = \frac{\exp(A_s(r_s, c_s))}{\sum_{j=1}^{k} \exp(A_j(r_j, c_j))} \qquad (s = 1, 2, \dots, k). \qquad (4.11)$$

Hence by substituting (4.11) into (4.10) we obtain an expression for the conditional entropy of $\mathbf{Y}$ given the receptive and contextual field inputs, $\mathbf{X}$ and $\mathbf{Z}$, respectively:

$$H(\mathbf{Y}|\mathbf{X}, \mathbf{Z}) = -\left\langle \sum_{i=1}^{k} \theta_i A_i(r_i, c_i) - \log\left[\sum_{j=1}^{k} \exp(A_j(r_j, c_j)) \right] \right\rangle_{\mathbf{X}, \mathbf{Z}} \qquad (4.12)$$

where the integrated RF and CF inputs are given by

$$r_i = \mathbf{w}_i^T \mathbf{x} - w_{0i}$$
$$c_i = \mathbf{v}_i^T \mathbf{z} - v_{0i}. \qquad (4.13)$$

We now require to derive expressions for the other entropic terms $H(\mathbf{Y})$, $H(\mathbf{Y}|\mathbf{X})$ and $H(\mathbf{Y}|\mathbf{Z})$. Recall that the multinomial model is the conditional distribution of $\mathbf{Y}$ given $\mathbf{X}$ and $\mathbf{Z}$. Therefore the conditional output probabilities $\{\theta_s\}$ may be written as

$$\theta_s = \Pr(Y_s = 1 | \mathbf{X} = \mathbf{x}, \mathbf{Z} = \mathbf{z}) \qquad (s = 1, 2, \dots, k). \qquad (4.14)$$

Hence, as in equation (3.17),

$$\Pr(Y_s = 1) = \langle \theta_s \rangle_{\mathbf{x}, \mathbf{z}} \qquad (4.15)$$

where we have chosen not to make explicit the dependence of θ_s on the RF and CF inputs, $\mathbf{x}$ and $\mathbf{z}$, and the connection weights, $\mathbf{w}_s$ and $\mathbf{v}_s$. Similarly, as in equations (3.18)–(3.19),

$$\Pr(Y_s = 1 | \mathbf{x}) = \langle \theta_s \rangle_{\mathbf{z}|\mathbf{x}} \qquad (4.16)$$

and

$$\Pr(Y_s = 1 | \mathbf{z}) = \langle \theta_s \rangle_{\mathbf{x}|\mathbf{z}}. \qquad (4.17)$$

We denote the averages in equations (4.15)–(4.17), respectively, by $E^{(s)}$, $E_{\mathbf{X}}^{(s)}$, and $E_{\mathbf{Z}}^{(s)}$.

In the WTA case the output unit codes k categories which are represented by k-tuples; for example, category s is represented by the vector having a 1 in the sth position and zeros elsewhere. We may also view this k-cell unit as coding a multi-valued discrete variable (Becker, 1992). In this WTA case the marginal probability distribution of the output vector $\mathbf{Y}$ is also multinomial with $n = 1$ and probabilities $\{E^{(s)}\}$. Hence it follows from equation (4.10) that the marginal output entropy is

$$H(\mathbf{Y}) = -\sum_{s=1}^{k} E^{(s)} \log E^{(s)} \tag{4.18}$$

where $\sum_{s=1}^{k} E^{(s)} = 1$. Similarly,

$$H(\mathbf{Y}|\mathbf{X}) = -\left\langle \sum_{s=1}^{k} E_{\mathbf{x}}^{(s)} \log E_{\mathbf{x}}^{(s)} \right\rangle_{\mathbf{x}} \tag{4.19}$$

where $\sum_{s=1}^{k} E_{\mathbf{x}}^{(s)} = 1$, and

$$H(\mathbf{Y}|\mathbf{Z}) = -\left\langle \sum_{s=1}^{k} E_{\mathbf{z}}^{(s)} \log E_{\mathbf{z}}^{(s)} \right\rangle_{\mathbf{z}} \tag{4.20}$$

where $\sum_{s=1}^{k} E_{\mathbf{z}}^{(s)} = 1$. Also, we have from equations (4.10) and (4.11) that

$$H(\mathbf{Y}|\mathbf{X}, \mathbf{Z}) = \left\langle \sum_{i=1}^{k} \theta_s \log \theta_s \right\rangle_{\mathbf{x}, \mathbf{z}}. \tag{4.21}$$

We now repeat the analyses of Section 3.2 in order to derive learning rules for the case of multinomial WTA output units. For simplicity of notation we assume that each output cell receives input from every RF and CF input; this need not be the case, but the derivations can be easily applied when the connectivities are quite general and unbalanced by making appropriate modifications. We consider the objective function defined in equation (3.6) where now the random variable Y is replaced by the random vector $\mathbf{Y}$.

We now calculate the derivatives of the entropies in equations (4.18)–(4.21) with respect to $\mathbf{w}_s$ and $\mathbf{v}_s$ and also the biases w_{0s} and v_{0s}. Recall that, even though the dependence is not expressed, the conditional output probability θ_s is a function of $\mathbf{w}_s$, $\mathbf{v}_s$, $\mathbf{x}$ and $\mathbf{z}$. We also write the activation term $A_s(r_s, c_s)$ as A_s for notational simplicity.

We begin with the conditional output entropy $H(\mathbf{Y}|\mathbf{X}, \mathbf{Z})$ and consider the term inside the averaging brackets in equation (4.12):

$$\frac{\partial}{\partial \mathbf{w}_s} \left[\sum_{i=1}^{k} \theta_i A_i - \log \left(\sum_{j=1}^{k} \exp(A_j) \right) \right] = \sum_{i=1}^{k} \frac{\partial \theta_i}{\partial \mathbf{w}_s} A_i. \tag{4.22}$$

Now,

$$\frac{\partial \theta_i}{\partial \mathbf{w}_s} = \begin{cases} \frac{\partial A_s}{\partial r_s} \theta_s (1 - \theta_s) \mathbf{x} & (i = s) \\ \frac{\partial A_s}{\partial r_s} \theta_s \theta_i \mathbf{x} & (i \neq s). \end{cases} \tag{4.23}$$

Combining equations (4.21)–(4.23) we obtain

$$\frac{\partial}{\partial \mathbf{w}_s} H(\mathbf{Y}|\mathbf{X}, \mathbf{Z}) = -\left\langle \frac{\partial A_s}{\partial r_s} \theta_s \left[A_s - \sum_{i=1}^{k} \theta_i A_i \right] \mathbf{x} \right\rangle_{\mathbf{X}, \mathbf{Z}}. \tag{4.24}$$

We consider now the other entropy terms. The partial derivative of $H(\mathbf{Y})$ with respect to $\mathbf{w}_s$ may be expressed as follows using equation (4.22):

$$\frac{\partial}{\partial \mathbf{w}_s} H(\mathbf{Y}) = -\sum_{i=1}^{k} \log E^{(i)} \frac{\partial E^{(i)}}{\partial \mathbf{w}_s}. \tag{4.25}$$

Now,

$$\frac{\partial E^{(i)}}{\partial \mathbf{w}_s} = \langle \frac{\partial \theta_i}{\partial \mathbf{w}_s} \rangle_{\mathbf{X}, \mathbf{Z}} \tag{4.26}$$

and so using equation (4.23) we obtain

$$\frac{\partial}{\partial \mathbf{w}_s} H(\mathbf{Y}) = \left\langle \frac{\partial A_s}{\partial r_s} \theta_s \left[E^{(s)} - \sum_{i=1}^{k} \theta_i E^{(i)} \right] \mathbf{x} \right\rangle_{\mathbf{X}, \mathbf{Z}}. \tag{4.27}$$

By using a similar argument it follows from equations (4.19)–(4.20) that

$$\frac{\partial}{\partial \mathbf{w}_s} H(\mathbf{Y}|\mathbf{X}) = \left\langle \frac{\partial A_s}{\partial r_s} \theta_s \left[E_{\mathbf{x}}^{(s)} - \sum_{i=1}^{k} \theta_i E_{\mathbf{x}}^{(i)} \right] \mathbf{x} \right\rangle_{\mathbf{X}, \mathbf{Z}} \tag{4.28}$$

and

$$\frac{\partial}{\partial \mathbf{w}_s} H(\mathbf{Y}|\mathbf{Z}) = \left\langle \frac{\partial A_s}{\partial r_s} \theta_s \left[E_{\mathbf{z}}^{(s)} - \sum_{i=1}^{k} \theta_i E_{\mathbf{z}}^{(i)} \right] \mathbf{x} \right\rangle_{\mathbf{X}, \mathbf{Z}}. \tag{4.29}$$

By combining equations (4.24) and (4.27)–(4.29) we obtain

$$\frac{\partial F}{\partial \mathbf{w}_s} = \left\langle \frac{\partial A_s}{\partial r_s} \theta_s \left[(\psi_3 - \bar{O}_s) - \sum_{i=1}^{k} \theta_i [\psi_3 A_i - \bar{O}_i] \right] \mathbf{x} \right\rangle_{\mathbf{X}, \mathbf{Z}} \tag{4.30}$$

where F is the objective function

$$F = H(\mathbf{Y}) - \psi_1 H(\mathbf{Y}|\mathbf{X}) - \psi_2 H(\mathbf{Y}|\mathbf{Z}) - \psi_3 H(\mathbf{Y}|\mathbf{X}, \mathbf{Z}) \tag{4.31}$$

and for $i = 1, 2, \ldots, k$

$$\bar{O}_i = \log E^{(i)} - \psi_1 \log E_{\mathbf{x}}^{(i)} - \psi_2 \log E_{\mathbf{z}}^{(i)}. \tag{4.32}$$

The partial derivatives required for changing the CF weights are obtained from equation (4.30) by replacing r_s by c_s and $\mathbf{x}$ by $\mathbf{z}$, yielding

$$\frac{\partial F}{\partial \mathbf{v}_s} = \left\langle \frac{\partial A_s}{\partial c_s} \theta_s \left[(\psi_3 - \bar{O}_s) - \sum_{i=1}^{k} \theta_i [\psi_3 A_i - \bar{O}_i] \right] \mathbf{z} \right\rangle_{\mathbf{x}, \mathbf{z}}. \qquad (4.33)$$

The derivatives for the biases are obtained from equations (4.30) and (4.33) by replacing $\mathbf{x}$ and $\mathbf{z}$, respectively, with -1. We take each of the activation functions $\{A_s\}$ to have the form defined in equation (3.38) and derivatives in equation (3.39), where the expressions there are indexed by the cell index s.

Since we are using multinomial WTA units the inputs will be categorised into one of k mutually exclusive classes under contextual guidance. Hence we may view this processor as performing contextually modulated unsupervised feature discovery—a form of contextually modulated competitive learning. The learning rules based on the derivatives in equations (4.30) and (4.33) have the same structure, and this is the same as that of the learning rules in the case of a single binary output unit, apart from a term that is common to all the rules. It is possible to show that these rules enjoy intrinsic weight stabilisation provided that the activation terms grow large as the weights grow large. The term that is common to the equations is

$$\sum_{i=1}^{k} \theta_i (\psi_3 A_i - \bar{O}_i) \qquad (4.34)$$

and it is the average excess $\psi_3 A_i - \bar{O}_i$ of the activation $\psi_3 A_i$ over the 'dynamic average' of the ith cell $\bar{O}_i$. Given positive input, and assuming that the partial derivatives $\partial A_i / \partial r_i$ and $\partial A_i / \partial c_i$ are positive, the weights connecting into the ith output cell will be increased if the excess for the ith cell is greater than the average excess for the whole multiple unit. These learning rules are local, however, as may be seen by expressing the computation of the learning rules as an anti-Hebbian network which connects together all the cells in the multiple-output unit. Let us denote the excess at the sth cell by D_s; then the central term in the learning rules is

$$\theta_s \left(D_s - \sum_{i=1}^{k} \theta_i D_i \right), \qquad (4.35)$$

which may be written as

$$\theta_s (1 - \theta_s) D_s - \sum_{i \neq s} \theta_s \theta_i D_i. \qquad (4.36)$$

This term can be computed via a fully pairwise-connected lateral network in which the weight on the connection linking the sth and ith cells is given by $-\theta_s \theta_i$. This is an anti-Hebbian network. A point of note is that the weights on the pairwise connections are precisely the covariances between the output units. Hence the computation of the learning rules may be organised locally within a network connecting all cells in the multiple-output unit; this network provides lateral inhibition and the weights are adapted using an exact covariance rule. It is of interest that this information-theoretic approach produces in this case a form of contextually modulated competitive learning which flows from a coherent probabilistic approach; for another approach, see Becker (1997).

This network will produce locally coded outputs and, while this is a useful thing for a multiple-output unit to do, it is also important to consider the production of coarsely coded outputs. These allow some redundancy amongst the output patterns and their advantages have been discussed by Churchland and Sejnowski (1992). Within this information-theoretic approach we require to model such outputs probabilistically and we now consider the case where more than one unit in a local processor may fire simultaneously given a particular input pattern.

4.3   Multiple-output processors: multivariate binary outputs

We now suppose that a local processor has p outputs which we represent by the collection of random variables $\mathbf{Y} = \{Y_1, Y_2, \ldots, Y_p\}$, where each Y_i is a binary random variable. As before, each processor receives information from two distinct sources, namely (a) its receptive field inputs and (b) its contextual field inputs. We also represent these inputs as random variables, with $\mathbf{X} = \{X_1, X_2, \ldots, X_m\}$ denoting the receptive field (RF) inputs and $\mathbf{Z} = \{Z_1, Z_2, \ldots, Z_n\}$ the contextual field (CF) inputs, respectively. To formally indicate the possibility of incomplete connectivity, we define connection neighbourhoods for each output unit Y_i within a local processor. Let $\partial i(x)$, $\partial i(z)$ and $\partial i(y)$ denote, respectively, the set of indices of the RF input units, the CF input units and the output units that are connected to the ith output unit Y_i. The corresponding random variables are denoted, respectively, by $\mathbf{X}_{\partial i}$, $\mathbf{Z}_{\partial i}$ and $\mathbf{Y}_{\partial i}$. We term these links the RF connections, the CF connections and the WP (within-processor) connections, respectively. The set of all components of $\mathbf{Y}$, excluding the ith component, is denoted by $\mathbf{Y}_{-i}$. The weights on the connections into the ith output unit are given by w_{ij}, v_{ij} and u_{ij} for the jth RF input, the jth CF input and the jth output unit, respectively, and we assume that the weights connecting each pair of output units to each other are symmetric ($u_{ij} = u_{ji}$) and also that these units are not self-connected ($u_{ii} = 0$). We now define the integrated fields in relation to the ith output:

$$S_i(x) = \sum_{j \in \partial i(x)} w_{ij} X_j - w_{i0} \tag{4.37}$$

is the random variable representing the integrated receptive field input to the ith output;

$$S_i(z) = \sum_{j \in \partial i(z)} v_{ij} Z_j - v_{i0} \tag{4.38}$$

is the random variable representing the integrated contextual field input to the ith output;

$$S_i(y) = \sum_{j \in \partial i(y)} u_{ij} Y_j \qquad (4.39)$$

is the random variable representing the within-processor integrated field input to the ith output.

We take a conditional approach to the modelling of the outputs $\mathbf{Y}$ given the RF and CF inputs $\mathbf{X}$ and $\mathbf{Z}$ and assume that $\mathbf{Y}$ follows a multivariate binary probability model (Besag, 1974), given the realised values of the RF and CF inputs, with probability mass function

$$\Pr(\mathbf{Y} = \mathbf{y} | \mathbf{X} = \mathbf{x}, \mathbf{Z} = \mathbf{z}) = \frac{1}{Z(\mathbf{a}, \mathbf{u})} \exp\left(\sum_{i=1}^{p} a_i y_i + \frac{1}{2} \sum_{i=1}^{p} \sum_{j \in \partial i(y)} u_{ij} y_i y_j\right),$$
$$(4.40)$$

where $Z(\mathbf{a}, \mathbf{u})$ is the normalisation constant (i.e. not a function of $\mathbf{y}$) required to ensure that the probabilities sum to unity. We assume that the output binary levels are 0 and 1; in applications in which bipolar units are used it is then necessary to adjust output probabilities using the mapping $p \mapsto 2p - 1$. The terms $\{a_i\}$ and $\{u_{ij}\}$ are parameters and in general may be functions of $\mathbf{x}$ and $\mathbf{z}$. In this article we shall shortly define the $\{a_i\}$ as a function of the RF and CF inputs but will take the $\{u_{ij}\}$ to be independent of these inputs, although there are other possibilities.

Equation (4.40) defines a regression model in two distinct senses. First, via the terms $\{a_i\}$ which may be taken to be general nonlinear functions of the RF and CF inputs, it is a nonlinear regression of the outputs as a function of all of the inputs. Secondly, when written in conditional form in terms of the distribution of the ith output given its neighbours, as in equation (4.41) below, it expresses, for given inputs, an autoregression for each output unit in terms of the other output units in its neighbourhood. It is an example of what has become known as a generalised linear model and the marginal probability model for the outputs is a generalised linear mixed model, the linearity referring to the parameters $\{a_i\}$ and $\{u_{ij}\}$ (McCullagh and Nelder, 1989; Clayton, 1996).

This model, in unconditional form, was used by Ising (1925) in statistical mechanics and developed as a model for spatial data by Besag (1974); however the general methodology works also with more general graph structures (Geman and Geman, 1984). The Ising model has been employed in a different way in neural networks by Hopfield (1982) and many others within the area of recurrent networks. The formulation described in the above model has the advantage of interfacing a feedforward network between layers with a recurrent network structure within a layer in a single coherent probabilistic framework. Not only that, but it is possible to connect the multiple output local processors themselves in a multilayered and multistream network structure in a probabilistically coherent manner.

It is natural to consider the local conditional distributions for each output unit given its RF, CF and WP inputs. Under the assumed restrictions on the WP connection weights, described above, the Hammersley–Clifford theorem (Besag, 1974) ensures that working locally with the conditional distributions is equivalent to assuming a coherent global model for all the output units. If the WP units are fully connected, then it is unnecessary to invoke this theorem as the local distributions then automatically have the form given in equation (4.41). The local conditional distribution for the ith output, given the values of its RF, CF and WP inputs, is Bernoulli with success probability

$$\theta_i \equiv \Pr(Y_i = 1 | \mathbf{X}_{\partial i} = \mathbf{x}_{\partial i}, \mathbf{Z}_{\partial i} = \mathbf{z}_{\partial i}, \mathbf{Y}_{\partial i} = \mathbf{y}_{\partial i}) = 1/(1 + \exp(-A_i)). \quad (4.41)$$

Here $A_i = a_i + s_i(y)$, where a_i may be taken to be any differentiable function of the integrated RF and CF fields and $s_i(y)$ is defined in equation (4.39). It then remains to specify the activation function for each output unit.

The activation function at the ith output unit is now a function of three integrated fields—how should they be combined? It will be assumed that the general function decomposes into the following sum, although there are other possibilities:

$$A_i = A(s_i(y), s_i(z)) + s_i(y) \equiv a_i + s_i(y). \quad (4.42)$$

The learning rules will be derived for the general case, however. This form in which the output integrated field is bound additively to a general function of the integrated RF and CF fields makes the specification of the activation function consistent with the multivariate probability model for the outputs defined above in equation (4.40). In this paper, the function A_i is chosen so that the CF input can modulate the response of the ith output given the RF input and in the next section we define the particular form used in this paper.

5   Global objective functions and learning rules

We now consider a global objective function based on the joint distribution of all outputs, RF inputs and CF inputs. In the case of multivariate outputs, we consider the general version of the objective function considered in Section 3.1, namely

$$F = I(\mathbf{Y}; \mathbf{X}; \mathbf{Z}) + \phi_1 I(\mathbf{Y}; \mathbf{X}|\mathbf{Z}) + \phi_2 I(\mathbf{Y}; \mathbf{Z}|\mathbf{X}) + \phi_3 H(\mathbf{Y}|\mathbf{X}, \mathbf{Z}), \quad (5.1)$$

and its conditional form

$$F = H(\mathbf{Y}) - \psi_1 H(\mathbf{Y}|\mathbf{X}) - \psi_2 H(\mathbf{Y}|\mathbf{Z}) - \psi_3 H(\mathbf{Y}|\mathbf{X}, \mathbf{Z}), \quad (5.2)$$

where $\psi_1 = 1 - \phi_2$, $\psi_2 = 1 - \phi_1$, and $\psi_3 = \phi_1 + \phi_2 - \phi_3 - 1$. We now present expressions for the entropic term of equation (5.2) and also develop the learning rules based on the derivatives of F with respect to the weights.

First, a simple calculation based on equation (4.40) shows that the conditional entropy $H(\mathbf{Y}|\mathbf{X}, \mathbf{Z})$ may be written as

$$H(\mathbf{Y}|\mathbf{X}, \mathbf{Z}) = \left\langle \log Z(\mathbf{a}, \mathbf{u}) - \sum_{i=1}^{p} a_i e_i - \frac{1}{2} \sum_{i=1}^{p} \sum_{j \in \partial i(y)} u_{ij} f_{ij} \right\rangle_{\mathbf{x}, \mathbf{z}} \tag{5.3}$$

where

$$e_i = E(Y_i|\mathbf{x}, \mathbf{z}) \tag{5.4}$$

and

$$f_{ij} = E(Y_i Y_j|\mathbf{x}, \mathbf{z}) \tag{5.5}$$

denote, respectively, the conditional means of Y_i and $Y_i Y_j$ given that $\mathbf{X} = \mathbf{x}$ and $\mathbf{Z} = \mathbf{z}$ ($j \in \partial i(y)$; $i = 1, \dots, p$).

Now we require the derivatives of the entropic terms. Let α be a generic weight, that is, one of the $\{w_{ij}\}$, $\{v_{ij}\}$ or $\{u_{ij}\}$. After some algebra we find that these derivatives are as follows:

$$\frac{\partial H(\mathbf{Y}|\mathbf{X}, \mathbf{Z})}{\partial \alpha} = \left\langle -\sum_{i=1}^{p} a_i \frac{\partial e_i}{\partial \alpha} - \frac{1}{2} \sum_{i=1}^{p} \sum_{j \in \partial i(y)} u_{ij} \frac{\partial f_{ij}}{\partial \alpha} \right\rangle_{\mathbf{x}, \mathbf{z}} \tag{5.6}$$

where

$$\frac{\partial e_i}{\partial \alpha} = \sum_{k=1}^{p} \frac{\partial a_k}{\partial \alpha} \mathrm{cov}(Y_i, Y_k|\mathbf{x}, \mathbf{z}) \tag{5.7}$$

and

$$\frac{\partial f_{ij}}{\partial \alpha} = \sum_{k=1}^{p} \frac{\partial a_k}{\partial \alpha} \mathrm{cov}(Y_i Y_j, Y_k|\mathbf{x}, \mathbf{z}). \tag{5.8}$$

Here $\mathrm{cov}(Y, Z|\mathbf{x}, \mathbf{z})$ denotes the conditional covariance between the random variables Y and Z given that $\mathbf{X} = \mathbf{x}$ and $\mathbf{Z} = \mathbf{z}$. Now

$$\frac{\partial H(\mathbf{Y})}{\partial \alpha} = -\left\langle \sum_{k=1}^{p} \frac{\partial a_k}{\partial \alpha} \mathrm{cov}(Y_k, \log p(\mathbf{Y})|\mathbf{x}, \mathbf{z}) \right\rangle_{\mathbf{x}, \mathbf{z}} \tag{5.9}$$

$$\frac{\partial H(\mathbf{Y}|\mathbf{X})}{\partial \alpha} = -\left\langle \sum_{k=1}^{p} \frac{\partial a_k}{\partial \alpha} \mathrm{cov}(Y_k, \log p(\mathbf{Y}|\mathbf{x})|\mathbf{x}, \mathbf{z}) \right\rangle_{\mathbf{x}, \mathbf{z}} \tag{5.10}$$

$$\frac{\partial H(\mathbf{Y}|\mathbf{Z})}{\partial \alpha} = -\left\langle \sum_{i=1}^{p} \frac{\partial a_k}{\partial \alpha} \mathrm{cov}(Y_k, \log p(\mathbf{Y}|\mathbf{z})|\mathbf{x}, \mathbf{z}) \right\rangle_{\mathbf{x}, \mathbf{z}}. \tag{5.11}$$

Collecting together terms from equations (3.6), (5.6) and (5.9)–(5.11) gives the derivatives of F as

$$\frac{\partial F}{\partial \alpha} = \left\langle \sum_{i=1}^{p} (\psi_3 a_i \frac{\partial e_i}{\partial \alpha} - \bar{O}_i) + \frac{1}{2} \sum_{i=1}^{p} \sum_{j \in \partial i(y)} u_{ij} \frac{\partial f_{ij}}{\partial \alpha} \right\rangle_{\mathbf{x,z}} \tag{5.12}$$

where

$$\bar{O}_i = \frac{\partial a_i}{\partial \alpha} (\text{cov}(Y_i, \log p(\mathbf{Y})|\mathbf{x}, \mathbf{z}) - \psi_1 \text{cov}(Y_i, \log p(\mathbf{Y}|\mathbf{x})|\mathbf{x}, \mathbf{z})$$
$$- \psi_2 \text{cov}(Y_i, \log p(\mathbf{Y}|\mathbf{z})|\mathbf{x}, \mathbf{z})). \tag{5.13}$$

It remains to specify the partial derivatives of a_i with respect to the weights. Simple calculation gives that

$$\frac{\partial a_i}{\partial w_{ij}} = \frac{\partial A_i}{\partial s_i(x)} x_j \tag{5.14}$$

$$\frac{\partial a_i}{\partial v_{ij}} = \frac{\partial A_i}{\partial s_i(z)} z_j \tag{5.15}$$

$$\frac{\partial a_i}{\partial u_{ij}} = \frac{\partial A_i}{\partial s_i(y)} y_j. \tag{5.16}$$

In the learning rules given above, we employ for each output unit the following activation function:

$$A_i = a_i + s_i(y), \tag{5.17}$$

where

$$a_i = \tfrac{1}{2} s_i(x)(1 + \exp(2s_i(x)s_i(z)) \tag{5.18}$$

is a member of the class of activation functions considered in Section 3.5. The parital derivatives are given by

$$\frac{\partial A_i}{\partial s_i(x)} = \tfrac{1}{2} + (\tfrac{1}{2} + s_i(x)s_i(z)) \exp(2s_i(x)s_i(z)) \tag{5.19}$$

$$\frac{\partial A_i}{\partial s_i(z)} = s_i(x)^2 \exp(2s_i(x)s_i(z)) \tag{5.20}$$

$$\frac{\partial A_i}{\partial s_i(y)} = 1. \tag{5.21}$$

The learning rules derived in this section, particularly the terms $\{\bar{O}_i\}$, are quite complicated; they are global at the level of the processor in that the weights connected to all the units are considered simultaneously. The computation of some of the average terms in equation (5.13) is particularly cumbersome as they involve entire conditional distributions of $\mathbf{Y}$. It is possible, however, to develop approximations of the zeroth order, but we don't present the details here. Despite this, these rules are computable when the number of units in the processor is limited, but approximations will be required if the number of output units is large; see Section 8. Primarily out of a wish to develop learning rules that are more biologically plausible and simpler to operate, we now turn our attention to local approximations of the global objective function considered above, and we shall see that these lead to local learning rules at the level of the units within the processor.

6   Local objective functions with local learning rules

In a processor with multiple output units, it is natural to consider processing in a local manner with each output unit using the information available to it from its RF, CF and WP neighbourhood connections. This suggests that we might focus on the joint distribution of each output and its RF and CF inputs, conditionally on its WP inputs. It then seems natural to consider the conditional three-way mutual information that is shared mutually by the ith output and its RF and CF inputs but not shared with its WP output units, defined by

$$I(Y_i; \mathbf{X}_{\partial i}; \mathbf{Z}_{\partial i}|\mathbf{Y}_{\partial i}) = I(Y_i; \mathbf{X}_{\partial i}|\mathbf{Y}_{\partial i}) - I(Y_i; \mathbf{X}_{\partial i}|\mathbf{Z}_{\partial i}, \mathbf{Y}_{\partial i}). \qquad (6.1)$$

It is possible, in general, to decompose the global three-way mutual information as follows:

$$I(\mathbf{Y}; \mathbf{X}; \mathbf{Z}) = I(Y_i; \mathbf{X}; \mathbf{Z}|\mathbf{Y}_{-i}) + I(\mathbf{Y}_{-i}; \mathbf{X}; \mathbf{Z}), \qquad (6.2)$$

which, given the connection structure, may be expressed as

$$I(\mathbf{Y}; \mathbf{X}; \mathbf{Z}) = I(Y_i; \mathbf{X}_{\partial i}; \mathbf{Z}_{\partial i}|\mathbf{Y}_{\partial i}) + I(\mathbf{Y}_{-i}; \mathbf{X}; \mathbf{Z}). \qquad (6.3)$$

This decomposition may be repeated recursively and is of particular relevance when the output units represent some directional structure such as a time series or a predecessor/successor hierarchy; then the well-known general factorization of joint probability into a product of marginal and conditional distributions allows the general three-way mutual information to be written as a sum of local conditional three-way mutual information terms. Such simplicity is not possible here, although the first-step decomposition, given in equation (6.3), shows that the conditional three-way information is a part of the global three-way information in a well-defined sense. Note that, under these local objective functions defined in equation (6.1), each output unit within the multiple-output local processor is attempting to transmit the information shared with its local RF and CF that is not being transmitted by the other units within the multiple-output processor. Hence one would expect these within-processor units to transmit slightly different aspects of the available information; but note that the outputs of these within-processor units can be correlated and hence this is not enforcing a winner-take-all scenario. There are other possible ways of defining local objective functions, but they will be discussed elsewhere.

The same conditioning idea may be applied to the other components of information within the objective function F and this leads to the specification of a local objective function for the ith output unit defined by

$$F_i = I(Y_i; \mathbf{X}_{\partial i}; \mathbf{Z}_{\partial i}|\mathbf{Y}_{\partial i}) + \phi_1 I(Y_i; \mathbf{X}_{\partial i}|\mathbf{Z}_{\partial i}, \mathbf{Y}_{\partial i})$$
$$+ \phi_2 I(Y_i; \mathbf{Z}_{\partial i}|\mathbf{X}_{\partial i}, \mathbf{Y}_{\partial i}) + \phi_3 H(Y_i|\mathbf{X}_{\partial i}, \mathbf{Z}_{\partial i}, \mathbf{Y}_{\partial i}), \qquad (6.4)$$

and, given our conditional approach to the modelling, we express $\{F_i\}$ in the more useful form

$$F_i = H(Y_i|\mathbf{Y}_{\partial i}) - \psi_1 H(Y_i|\mathbf{X}_{\partial i}, \mathbf{Y}_{\partial i})$$
$$- \psi_2 H(Y_i|\mathbf{Z}_{\partial i}, \mathbf{Y}_{\partial i}) - \psi_3 H(Y_i|\mathbf{X}_{\partial i}, \mathbf{Z}_{\partial i}, \mathbf{Y}_{\partial i}). \qquad (6.5)$$

This means that we envisage each output unit within a local processor working to maximise F_i and, because of the fact that mutually distinct sets of weights connect into each of the outputs, this is equivalent to maximising the sum of the F_is. We view this sum as a locally based approximation to the global objective function F defined in equation (5.1). In the extreme case where the outputs are conditionally independent this sum is equivalent to F. Obviously, the approximation will be better the smaller the sizes of the output neighbourhood sets relative to the number of outputs. We consider two particular forms for the $\{F_i\}$. If we take $\phi_1 = \phi_2 = \phi_3 = 0$, then equation (6.5) gives the conditional three-way mutual information shared among the output of the ith unit and its RF and CF inputs given the outputs of the neighbours of the ith output unit. Maximisation of this objective function means that each unit in each processor is being adapted to maximise the coherent information it shares with its RF and CF inputs conditional on the information flowing from its neighbouring outputs within the processor to which it belongs. This objective function generalises the *coherent infomax* goal proposed by Kay and Phillips (1994, 1997) and we term it the *local conditional coherent infomax* criterion. Another possibility is the choice $\phi_1 = 1$ and $\phi_2 = \phi_3 = 0$. In this case equation (6.5) gives the conditional mutual information shared between the output of the ith unit and its RF inputs given the information flowing from its neighbouring outputs within the processor. This generalises the infomax objective function proposed by Linsker (1988) and we term it the *local conditional infomax* criterion.

We now provide formulae for the local entropic terms and the components of local information for the ith output unit. The conditional distribution of Y_i given its RF, CF and WP inputs is a Bernoulli distribution with success probability given by equation (4.41). It is easy to show that the corresponding entropy term is

$$H(Y_i|\mathbf{X}_{\partial i}, \mathbf{Z}_{\partial i}, \mathbf{Y}_{\partial i}) = -\langle \theta_i \log \theta_i + (1 - \theta_i) \log(1 - \theta_i) \rangle_{\mathbf{x}_{\partial i}, \mathbf{z}_{\partial i}, \mathbf{y}_{\partial i}}. \qquad (6.6)$$

It is also easy to show that the conditional distribution of Y_i given its RF and WP inputs is also Bernoulli with success probability given by

$$E^{(i)}_{\mathbf{x}_{\partial i}, \mathbf{y}_{\partial i}} = \langle \theta_i \rangle_{\mathbf{z}_{\partial i}|\mathbf{x}_{\partial i}, \mathbf{y}_{\partial i}}. \qquad (6.7)$$

Hence a similar argument gives the following entropy term:

$$H(Y_i | \mathbf{X}_{\partial i}, \mathbf{Y}_{\partial i}) = - \left\langle E^{(i)}_{\mathbf{x}_{\partial i}, \mathbf{y}_{\partial i}} \log E^{(i)}_{\mathbf{x}_{\partial i}, \mathbf{y}_{\partial i}} + (1 - E^{(i)}_{\mathbf{x}_{\partial i}, \mathbf{y}_{\partial i}}) \log(1 - E^{(i)}_{\mathbf{x}_{\partial i}, \mathbf{y}_{\partial i}}) \right\rangle_{\mathbf{x}_{\partial i}, \mathbf{y}_{\partial i}}.$$
(6.8)

Similarly the conditional distribution of Y_i given its CF and WP inputs is Bernoulli with probability defined by

$$E^{(i)}_{\mathbf{z}_{\partial i}, \mathbf{y}_{\partial i}} = \langle \theta_i \rangle_{\mathbf{x}_{\partial i} | \mathbf{z}_{\partial i}, \mathbf{y}_{\partial i}},$$
(6.9)

and the corresponding entropy is

$$H(Y_i | \mathbf{Z}_{\partial i}, \mathbf{Y}_{\partial i}) = - \left\langle E^{(i)}_{\mathbf{z}_{\partial i}, \mathbf{y}_{\partial i}} \log E^{(i)}_{\mathbf{z}_{\partial i}, \mathbf{y}_{\partial i}} + (1 - E^{(i)}_{\mathbf{z}_{\partial i}, \mathbf{y}_{\partial i}}) \log(1 - E^{(i)}_{\mathbf{z}_{\partial i}, \mathbf{y}_{\partial i}}) \right\rangle_{\mathbf{z}_{\partial i}, \mathbf{y}_{\partial i}}.$$
(6.10)

Also the conditional distribution of Y_i given its WP inputs is Bernoulli with probability

$$E^{(i)}_{\mathbf{y}_{\partial i}} = \langle \theta_i \rangle_{\mathbf{x}_{\partial i}, \mathbf{z}_{\partial i} | \mathbf{y}_{\partial i}},$$
(6.11)

and entropy term

$$H(Y_i | \mathbf{Y}_{\partial i}) = - \left\langle E^{(i)}_{\mathbf{y}_{\partial i}} \log E^{(i)}_{\mathbf{y}_{\partial i}} + (1 - E^{(i)}_{\mathbf{y}_{\partial i}}) \log(1 - E^{(i)}_{\mathbf{y}_{\partial i}}) \right\rangle_{\mathbf{y}_{\partial i}}.$$
(6.12)

It follows that the components of local information at the ith output unit, given in terms of equations (6.6), (6.8), (6.10) and (6.12), are as follows:

$$I(Y_i; \mathbf{X}_{\partial i}; \mathbf{Z}_{\partial i} | \mathbf{Y}_{\partial i}) = (6.12) - (6.10) - (6.8) + (6.6)$$
(6.13)

$$I(Y_i; \mathbf{X}_{\partial i} | \mathbf{Z}_{\partial i}, \mathbf{Y}_{\partial i}) = (6.10) - (6.6)$$
(6.14)

$$I(Y_i; \mathbf{Z}_{\partial i} | \mathbf{X}_{\partial i}, \mathbf{Y}_{\partial i}) = (6.8) - (6.6)$$
(6.15)

$$H(Y_i | \mathbf{X}_{\partial i}, \mathbf{Z}_{\partial i}, \mathbf{Y}_{\partial i}) = (6.6).$$
(6.16)

We now present the learning rules which have the same general structure as those discussed in Section 3.2. We now give the gradient-ascent learning rules in relation to the ith output unit.

- **Receptive field connection weights**

$$\frac{\partial F_i}{w_{is}} = \left\langle (\psi_3 A_i - \bar{O}_i) \theta_i (1 - \theta_i) \frac{\partial A_i}{\partial s_i(x)} x_s \right\rangle_{\mathbf{x}_{\partial i}, \mathbf{z}_{\partial i}, \mathbf{y}_{\partial i}},$$
(6.17)

for each RF input s which connects into the ith output unit. For RF inputs s that do not connect into the ith output, $\partial F_i / \partial w_{is} = 0$.

- **Contextual field connection weights**

$$\frac{\partial F_i}{v_{is}} = \left\langle (\psi_3 A_i - \bar{O}_i) \theta_i (1 - \theta_i) \frac{\partial A_i}{\partial s_i(z)} z_s \right\rangle_{\mathbf{x}_{\partial i}, \mathbf{z}_{\partial i}, \mathbf{y}_{\partial i}},$$
(6.18)

for each CF input s which connects into the ith output. For CF inputs s which do not connect to the ith output, $\partial F_i / \partial v_{is} = 0$.

- **Within-processor connection weights**

$$\frac{\partial F_i}{\partial u_{is}} = \left\langle (\psi_3 A_i - \bar{O}_i)\theta_i(1 - \theta_i)\frac{\partial A_i}{\partial s_i(y)}y_s \right\rangle_{\mathbf{x}_{\partial i},\mathbf{z}_{\partial i},\mathbf{y}_{\partial i}}, \tag{6.19}$$

for each output unit s which connects into the ith output. For output units s which do not connect to the ith output, $\partial F_i/\partial u_{is} = 0$.

The partial derivatives of A_i are defined above in equations (5.19)–(5.21). Note that these learning rules are **local** and this results from the decision to separately maximise the local objective functions $\{F_i\}$ (or equivalently to maximise the sum of the $\{F_i\}$). They involve another level of averaging taken over the neighbouring output units. The dynamic average for the ith output unit is

$$\bar{O}_i = \log\frac{E_{\mathbf{y}_{\partial i}}^{(i)}}{(1 - E_{\mathbf{y}_{\partial i}}^{(i)})} - \psi_1\log\frac{E_{\mathbf{x}_{\partial i},\mathbf{y}_{\partial i}}^{(i)}}{(1 - E_{\mathbf{x}_{\partial i},\mathbf{y}_{\partial i}}^{(i)})} - \psi_2\log\frac{E_{\mathbf{z}_{\partial i},\mathbf{y}_{\partial i}}^{(i)}}{(1 - E_{\mathbf{z}_{\partial i},\mathbf{y}_{\partial i}}^{(i)})}. \tag{6.20}$$

The dynamic averages are more complicated than in the single-unit output case and their calculation involves storing the average probability at the ith output unit for each pattern of the other outputs that connect into the ith output unit, for the combination of each of the neighbouring output and RF input patterns and for the combination of each of the neighbouring output and CF input patterns. The computational implications resulting from the need to store these conditional averages means that it is feasible to apply these learning rules only to processors of limited size. However, as noted above, connecting together many such local processors would enable a complex architecture to be constructed in which the computation required in each local processor is manageable. This issue is discussed further in Section 8.

7   Two illustrations

We now briefly consider two illustrations of the methodology developed in Sections 4.3 and 6; for further details, see Kay *et al.* (1998).

7.1   First experiment

The architecture used in this experiment is shown in Figure 5. The input patterns for each stream were lines presented at four different orientations: horizontal, vertical, and two opposite diagonals. The lines were presented at the centre of a 3×3 square matrix. Each of the four lines could take a negative or a positive value: a negative line was composed of -1's against a background of $+1$'s, and a positive one was just the opposite. Each of the eight patterns was presented with equal probability to each stream, and the orientation and sign of each line were perfectly correlated across streams. Using the methodology described in Section 6, several simulations with networks of three, four, five and six units per processor were run on this training set.

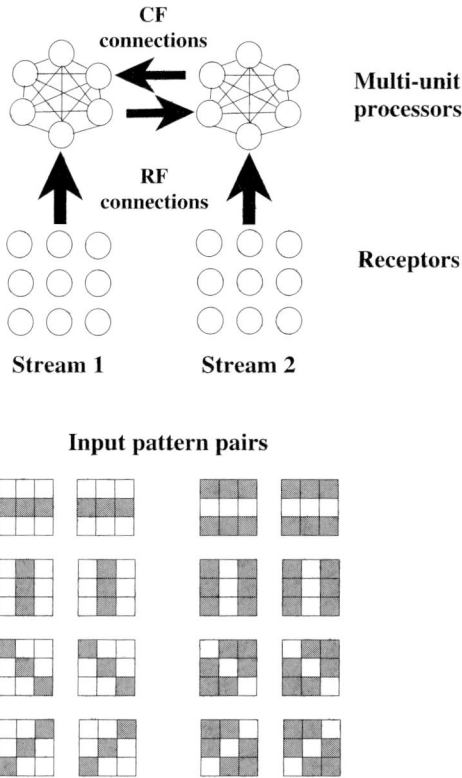

Fig. 5: **Top**: Network architecture; the number of units per processor was varied in different experiments (3, 4, 5 and 6). **Bottom**: Pairs of input patterns presented to the network during training. Each input pattern to a stream is arranged as a square matrix to visualise it as an oriented line with a negative or positive sign; negative lines are formed by -1's (grey squares) on a $+1$'s (white squares) background, and positive lines are formed of $+1$'s on a -1's background. The sign and orientation of each line is correlated across streams.

We first report the results of the experiments with the three-unit processors. The local conditional coherent infomax criterion was quickly maximised by all three units in both streams of the network. Learning took place at approximately the same time for all six units and the results were not sensitive to the value of the learning-rate parameter. With eight equiprobable input patterns and only three binary output units each unit was bound to take part in all the codes, so instead of allocating a particular unit to a particular whole input pattern that could occur in either of two signs, each unit was used as part of a distributed code by signalling a specific 'micro-structure' that occurred in each of the input patterns, these micro-structures being indicated by the RF weight structures. The CFs learned the appropriate cross-stream predictions, even though they were not necessary to discover the relevant within-stream features. Since each unit took part in the representation of all the input patterns all the CF connections were strengthened. In order to make sure that the CF weights were properly developed so as to predict the information transmitted by the RF weights, we checked that the integrated CF input to each unit matched in sign the integrated RF input for each pattern. This was always true, except for a few cases where a weak RF integrated input of opposite sign was compensated by a relatively strong WP integrated input to that unit. The final WP weights tended to approach zero, except for the few situations where they compensated the slight mismatch between RF and CF integrated inputs.

Similar results were obtained in experiments with more than three units per processor. However, since three units were sufficient for transmitting the entire information contained in eight patterns, one might wonder what use the algorithm made of the additional units. One way of assessing whether redundant units are used effectively consists in measuring the loss of information transmitted after systematic pruning of the trained units. After training a four-unit processor network to convergence, we systematically pruned out single units, tested the network on all the input patterns, and measured the loss of information. It turned out that pruning out any one unit reduced the information transmitted. This meant that all units were used in the distributed code developed for discriminating the eight patterns. We repeated the same procedure with a five-unit processor network. This time we found two units that, if pruned out (not together), did not affect the amount of information transmitted. Although these units were 'redundant', they were in fact taking part in the distributed representation. Finally, the same procedure was applied to a six-unit processor network. Here we found that we could prune out *any* one unit at a time without losing information. There was no information loss also when some combinations of two units were eliminated. However, the distributed code developed by the network was not a trivial duplication of the code developed in the three-unit case described above. When the number of units per processor was larger than the minimum number necessary to transmit the coherent input information, not all units could be fully decorrelated and, therefore, the WP weights remained strong. When several units and several input features were allowed a distributed representation was obtained and each unit specialised in detecting specific micro-features of the input patterns.

When four or more units were used per processor it would have been possible to transmit the relevant information by allocating one unit to each line orientation, as in the very simple conditions of the preceding experiment. We have not found the algorithm to develop this form of local coding.

7.2   Second experiment

The architecture used in the second experiment is shown in Figure 6. The networks used were composed of 25 streams arranged as a 5 × 5 matrix. Each stream had a single processor composed of a number of units fully interlinked by symmetric WP connections. Each unit in the network received RF input from all the receptors in its own stream. Each stream had nine receptors which could be visualised as a 3 × 3 matrix. Nearest-neighbour processors were fully interlinked via CF connections. The

Architecture

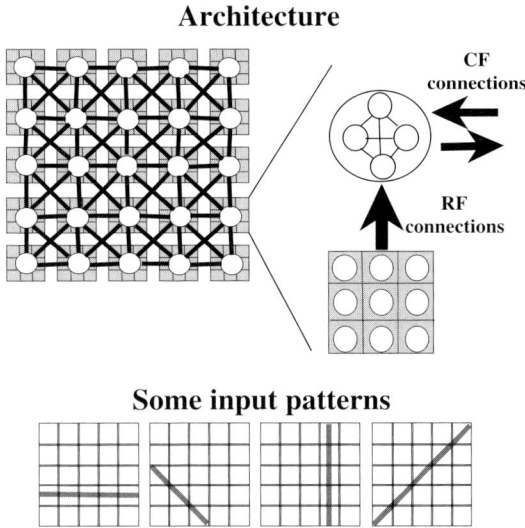

Fig. 6:   **Top**: Architecture and magnification of a single stream with four units per processor. The thick lines between processors show the CF connectivity between nearest-neighbour processors. Each unit receives RF input from all the receptors in its own stream, CF connections from all the units in the nearest-neighbour processors and WP connections from all the units within its own processor. **Bottom**: Some examples of the 56 input lines used during learning; each line is composed of an alignment of small lines with identical sign and orientation. This makes the RF input to each individual processor identical to that used in the previous experiments.

input patterns were long lines with four different orientations (horizontal, vertical, and 45° diagonals) and two opposite signs shifted across the whole input surface (the 5×5 stream surface was considered as a patch of a larger input surface). Some input patterns are shown in the bottom half of Figure 6. Each line was composed of an alignment of identical small lines and these were shown at all possible positions of the input array. Altogether, there were 56 input lines presented across the 15×15 receptor array—half positive and half negative, five horizontal, five vertical, nine diagonal in one direction and nine diagonal in the other direction. Hence, the RF input patterns to each processor were identical to those employed in the experiment discussed in Section 7.1.

We ran different versions of the experiment, varying the number of units per processor (three and four), the strength of the training patterns (four strengths were used: $\{+1, -1\}$, $\{+0.75, -0.75\}$, $\{+0.5, -0.5\}$, and $\{+0.25, -0.25\}$) and the values applied to the 'background', i.e. to the receptors in the streams where no line was presented (all 0's, all $+1$'s, all -1's or equally probable $+1$'s and -1's). After learning, the networks were tested on a variety of input data characterised by a weak signal and different types of noise to test the effects of learned contextual input on generalisation to patterns corrupted by noise.

All the processors in the network developed a distributed representation of the local eight patterns with the same properties already described in Section 7.1.

The learning dynamics were little affected by the number of units per processor, the strength of the input or the values applied to the background streams, so we describe in detail just the results obtained with the three-unit processors. The CF connections learned to correctly predict the RF features signalled by the units to which they projected. Post-training tests with weak RF inputs showed that contextual signals from neighbouring processors which were presented with the same RF inputs boosted the unit outputs in the right direction, whereas contextual signals from the same processors presented with opposite RF inputs further dampened the unit outputs. The role of the learned CF connection weights was also shown by tests with noisy weak inputs. In this case the network displayed double generalisation. The RF weights generalised to the noisy input by getting the unit signs correct and the CF weights generalised to the varying output strengths by boosting all the unit activations in the right direction. Although the relevant RF variables in the patterns used during training could have been discovered using the *infomax* approach, the CF connections learned using the *coherent infomax* approach provide the network with the additional ability of exploiting contextual information when the primary RF evidence is uncertain.

8   Computational complexity

8.1   Another look at the modelling

In Section 3 the probabilistic modelling was developed in terms of the actual RF inputs, CF inputs and the outputs. In that formulation it was necessary to store conditional averages for each RF and CF input. Hence as the dimensionality of the RF and CF vectors grows large the required amount of storage grows exponentially fast for each local processor in the network. This presents a serious limitation to the scalability of the approach described above. In order to overcome these computational difficulties, we now exploit a special property of our empirical approach to the computation and rethink the probabilistic modelling. Rather than model the outputs conditionally in terms of the *actual* RF and CF inputs via $p(\mathbf{y}|\mathbf{x}, \mathbf{z})$, we now develop the modelling in terms of the integrated RF and CF fields, R and C, via the probability density function $p(\mathbf{y}|r, c)$. The immediate advantages of this approach are as follows.

(1) It is required to store conditional averages for each (r,c) combination and so it is only necessary to deal with a two-dimensional space indexed by r and c.

(2) As the variables r and c are better thought of as quantitative variables, it now does not matter whether the RF and CF inputs are categorical (properly coded), discrete or continuous.

(3) The dimensionalities of the conditional averages is now independent of the dimensions of the RF and CF inputs.

(4) The required conditional averages may now be computed from a single function $\theta(r, c)$—the output probability when the integrated RF and CF inputs are r and c, respectively.

This new approach to the modelling resolves the scalability problem by working instead with a single two-dimensional function of the integrated fields. This would appear to present a different type of difficulty, namely, the specification of the form of this function. However, here we adopt an empirical approach to the modelling; that is, no explicit distributional assumptions are made concerning the integrated RF and CF inputs and we use the joint empirical distribution of R and C which is derived from the empirical distribution of the input patterns themselves. This leads to ways of learning approximations of the function based on the observed patterns. So this proposed remodelling does have advantages, but doesn't this change the objective function and the learning rules?

The objective function defined in equation (3.6) now becomes

$$F = \phi_0 I(Y; R; C) + \phi_1 I(Y; R|C) + \phi_2 I(Y; C|R) + \phi_3 H(Y|R, C). \qquad (8.1)$$

In the derivatives required for the learning rules defined in equations (3.30)–(3.34) we replace $\mathbf{y}$ and $\mathbf{z}$ with r and c, respectively, in the calculation of the averages. For example, the learning rule based on equation (3.30) becomes

$$\frac{\partial F}{\partial \mathbf{w}} = \left\langle (\psi_3 A - \bar{O}) \frac{\partial A}{\partial r} \theta (1 - \theta) \mathbf{x} \right\rangle_{r,c} \qquad (8.2)$$

and the conditional averages are

$$E = \langle \theta \rangle_{r,c} \qquad (8.3)$$
$$E_c = \langle \theta \rangle_{r|c} \qquad (8.4)$$
$$E_r = \langle \theta \rangle_{c|r}. \qquad (8.5)$$

Note that in all the equations in this new approach the terms within the angled brackets remain unaltered; the only difference now is that the averaging is being taken with respect to the joint distribution of R and C or the conditional distributions of R given that $C = c$ and C given that $R = r$. So the expressions are indeed different and in general are not the same as those obtained under the previous approach to the modelling taken in Section 3. However, with the empirical approach being employed here, in which all expectations are taken with respect to the empirical distributions of the input patterns, it turns out that all equations give identical results to those derived before. The reason for this is based on the fact that to each of the p primary input patterns $\{(\mathbf{x}_i, \mathbf{z}_i) : i = 1, \ldots, p\}$ there corresponds a single (r,c) pattern, and so working empirically with the actual patterns seen by the net there is a one-to-one correspondence between input patterns and integrated fields i.e. $(\mathbf{x}_i, \mathbf{z}_i) \leftrightarrow (r_i, c_i)$ $(i = 1, \ldots, p)$. Hence we may continue to calculate the components of the objective function and the learning rules as before, with the advantage now that the conditional averages may be obtained from a single two-dimensional function. As a result the batch and on-line learning algorithms may be used as before together with an approximation strategy for the conditional averages.

This remodelling may also be used in the case of the multivariate processor considered in Sections 4.3 and 6. In this case there are three integrated fields, $S_i(X)$, $S_i(Z)$ and $S_i(Y)$, associated with the ith output unit of each processor. Hence the output probability is a function of these three integrated fields. The argument given above applies, with the complication that for each unit a three-dimensional function must be learned. We now consider some approximations to the averages of output probability.

8.2   Approximations

In Kay (1994) two possible approaches are discussed in the case of a univariate processor. First, it seems reasonable to assume, via a central limit theorem, that the joint distribution of R and C may be approximated by a bivariate Gaussian probability model. Then the required computation consists of computing averages of a nonlinear function—the output probability function—with respect to this bivariate distribution and also the requisite conditional distributions. There is no closed-form expression for these averages, but two simple approaches are as follows. We could approximate these averages based on a zeroth-order approximation. This provides a neat approach, if perhaps quite crude, in that the computation of the averages reduces to the learning of only five parameters in the univariate case (and nine parameters per unit in the multivariate binary case); furthermore, these parameters may be updated on-line using simple recursive formulae. In the second approach the output probability function may be learned adaptively using nonparametric techniques; this is also a simple approach and the required storage per unit depends on the 'bin-size' used in the nonparametric estimation. For further details, see Kay (1994), where other simplifications are presented.

9   Acknowledgements

It is a pleasure to acknowledge that much of this work was stimulated by collaboration with Bill Phillips. Dario Floreano and Darragh Smyth provided excellent computational support. I also thank Sue Becker for helpful communication.

10   References

Aitchison, J. and Kay, J.W. (1975). Principles, practice and performance in decision making in clinical medicine. In *The Role and Effectiveness of Theories of Decision in Practice*, Eds. D.J. White and K.C. Bowen, pp. 252–272. Hodder and Stoughton, London.

Anderson, J.A. (1972). Separate sample logistic discrimination. *Biometrika* **59**, 19–35.

Artola, A., Brocher, S. and Singer, W. (1990). Different voltage-dependent thresholds for the induction of long-term depression and long-term potentiation in slices of rat visual cortex. *Nature* **347**, 69–72.

Atick, J.J. (1992). Could information theory provide an ecological theory of sensory processing? *Network: Computation in Neural Systems* **3**, 213–251.

Attneave, F. (1959). *Applications of Information Theory to Psychology*. Holt, Rinehart and Winston, New York.

Becker, S. (1992). An information-theoretic unsupervised learning algorithm for neural networks. Ph.D. Thesis. University of Toronto.

Becker, S. (1993). Learning to categorise objects using temporal coherence. In *Advances in Neural Information Processing Systems* **5**, Eds. S.J. Hanson, J.D. Cowan and C.L. Giles, pp. 361–368. Morgan Kaufmann, San Mateo, CA.

Becker, S. (1995). JPMAX: learning to recognise moving objects as a model-fitting problem. In *Advances in Neural Information Processing Systems* **7**, Eds. G. Tesauro, D.S. Touretzky and T.K. Leen, pp. 933–940. MIT Press, Cambridge, MA.

Becker, S. (1996). Mutual information maximization: models of cortical self-organization. *Network: Computation in Neural Systems* **7**, 7–31.

Becker, S. (1999). Implicit learning in 3D object recognition: the importance of temporal context. *Neural Computation* **11**, 347–374.

Becker, S. and Hinton, G.E. (1992). Self-organizing neural network that discovers surfaces in random-dot stereograms. *Nature* **355**, 161–163.

Becker, S. and Hinton, G.E. (1995). Spatial coherence as an internal teacher for a neural network. In *Backpropagation: Theory, Architectures and Applications*. Eds. Y. Chauvin and D. Rumelhart, pp. 313–349. Lawrence Erlbaum Associates, Hillsdale, NJ.

Bell, A.J. and Sejnowski, T.J. (1995). An information maximisation approach to blind separation and blind deconvolution. *Neural Computation* **7**, 1129–1159.

Besag, J. (1974). Spatial interaction and the statistical analysis of lattice systems (with discussion). *J. R. Statist. Soc.* B **36**, 192–236.

Bridle, J.S. (1990). Training stochastic model recognition algorithms as networks can lead to maximum mutual information estimation of parameters. In *Neural Information Processing Systems* **2**, Ed. D.S. Touretzky, pp. 211–217. Morgan Kaufman, San Mateo, CA.

Churchland, P.S. and Sejnowski, T.J. (1992). *The Computational Brain*. MIT Press, Cambridge, MA.

Clayton, D.G. (1996). Generalised linear mixed models. In *Markov Chain Monte Carlo in Practice,* Eds. W.R. Gilks, S. Richardson and D.J. Spiegelhalter, pp. 275–301. Chapman and Hall, London.

DeWeese, M. (1996). Optimisation principles for the neural code. *Network: Computation in Neural Systems* **7**, 325–331.

Foldiak, P. (1990). Forming sparse representations by anti-hebbian learning. *Biological Cybernetics* **64**, 165–170.

Geman, S. and Geman, D. (1984). Stochastic relaxation, Gibbs distributions, and the Bayesian restoration of images. *IEEE Trans. Pattern Anal. Machine Intell.* **6**, 721–741.

Gokhale, D.V. and Kullback, S. (1978). *The Information in Contingency Tables*. M. Dekker, New York.

Hamming, R.W. (1980). *Coding and Information Theory*. Prentice-Hall, Englewood Cliffs, NJ.

Hopfield, J. (1982). Neural networks and physical systems with emergent collective computational abilities. *Proceedings of the National Academy of Sciences* **79**, 2553–2558.

Intrator, N. and Cooper, L.N. (1992). Objective function formulation of the BCM theory of visual cortical plasticity: statistical connections, stability conditions. *Neural Networks* **5**, 3–17.

Intrator, N. and Cooper, L.N. (1995). Information theory of visual plasticity. In *The Handbook of Brain Theory and Neural Networks,* Ed. M.A. Arbib, pp. 484–487. MIT Press, Cambridge, MA.

Ising, E. (1925). Beitrag zur theorie des ferromagnetismus. *Zeitschrift Physik* **31**, 253.

Kay, J. (1994). Information-theoretic neural networks for unsupervised learning: Mathematical and statistical considerations. Technical Report. University of Stirling.

Kay, J., Floreano, D. and Phillips, W.A. (1998). Contextually guided unsupervised learning using local multivariate binary processors. *Neural Networks* **11**, 117–140.

Kay, J. and Phillips, W.A. (1994). Activation functions, computational goals and learning rules for local processors with contextual guidance. Technical Report CCCN-15. Centre for Cognitive and Computational Science, University of Stirling.

Kay, J. and Phillips, W.A. (1997). Activation functions, computational goals and learning rules for local processors with contextual guidance. *Neural Computation* **9**, 895–910.

Kullback, S. (1959). *Information Theory and Statistics.* Wiley, New York.

Lindley, D.V. (1956). On a measure of information provided by an experiment. *Ann. Math. Statist.* **27**, 986–1005.

Linsker, R. (1988). Self-organization in a perceptual network. *Computer* **21**, 105–117.

MacKay, D.J.C. (1992). *Bayesian methods for adaptive methods.* Ph.D. Thesis. California Institute of Technology.

McCullagh, P. and Nelder, J.A. (1989). *Generalized Linear Models.* Chapman and Hall, London.

McGill, W.J. (1954). Multivariate information transmission. *Psychometrika* **19**, 97–116.

Optican, L.M., Gawne, T.J., Richmond, B.J. and Joseph, P.J. (1991). Unbiased measures of transmitted information and channel capacity from multivariate neuronal data. *Biological Cybernetics* **65**, 305–310.

Phillips, W.A., Kay, J. and Smyth, D. (1995). The discovery of structure by multistream networks of local processors with contextual guidance. *Network: Computation in Neural Systems* **6**, 225–246.

Phillips, W.A. and Singer, W. (1997). In search of common foundations for cortical computation. *Behavioural and Brain Sciences* **20**, 657–722.

Redlich, A.N. (1993). Redundancy reduction as a strategy for unsupervised learning. *Neural Computation* **5**, 289–304.

Rieke, F., Warland, D., de Ruyter van Steveninck, R. and Bialek, W. (1997). *Spikes.* MIT Press, Cambridge, MA.

Sanger, T.D. (1997). A probability interpretation of neural population coding for movement. In *Self-Organisation, Computational Maps and Motor Control,* Eds. P. Morasso and V. Sanguineti, pp. 75–116. Elsevier, Amsterdam.

Schmidhuber, J. and Prelinger, D. (1993). Discovering predictable classifications. *Neural Computation* **5**, 625–635.

Shannon, C.E. and Weaver, W. (1949). *The Mathematical Theory of Communication.* Univ. of Illinois Press, Chicago.

Taylor, J.G. and Plumbley, M.D. (1993). Information theory and neural networks. In *Mathematical Approaches to Neural Networks,* Ed. J.G. Taylor, pp. 307–340. Elsevier, Amsterdam.

Tononi, G., Sporns, O. and Edelman, G.M. (1994). A measure for brain complexity: relating functional segregation and integration in the nervous system. *Proc. Nat. Acad. Sci., USA* **91**, 5033–5037.

Treves, A. and Panzeri, S. (1995). The upward bias in measures of information derived from limited data samples. *Neural Computation* **7**, 399–407.

Whittaker, J. (1990). *Graphical Models in Applied Multivariate Statistics*. Wiley, Chichester.

Zador, A. (1998). Impact of synaptic unreliability on the information transmitted by spiking neurons. *J. Neurophysiol.* **79**, 1219–1229.

3
Radial Basis Function Networks and Statistics

David Lowe

Neural Computing Research Group
Aston University, Aston Triangle, Birmingham B4 7ET, UK
`d.lowe@aston.ac.uk`

Abstract

The class of neural network architectures known as radial basis functions has a structure and history which has produced strong links with various areas in the statistical sciences. This chapter reviews some of the more salient aspects, which include the links with conditional and unconditional density estimation (such as Gaussian mixture models), regression problems and links with kernel estimators and linear smooths, supervised feature extraction and discriminant analysis, and topographic feature extraction and links to multidimensional scaling, Sammon mappings and nonlinear principal component analysis.

1 Introduction

1.1   The basic RBF structure

The radial basis function is a single hidden layer feedforward network with linear output transfer functions and nonlinear transfer functions, $\phi_j(\dots)$, on the hidden-layer nodes. Many types of nonlinearity may be used. There is also typically a bias weight on each output node. The main adjustable parameters are the final layer weights, $\{\lambda_{jk}\}$, connecting the jth hidden node to the kth output node. There are also weights $\{\mu_{ij}\}$ connecting the ith input node with the jth hidden node and occasionally 'smoothing' factor matrices, $\{\Sigma_j\}$. The mathematical embodiment of the radial basis function with h hidden nodes takes the following form. The kth component of the output vector $\boldsymbol{o}_p$ corresponding to the pth input pattern $\boldsymbol{x}_p$ is expressed as

$$[\boldsymbol{o}(\boldsymbol{x}_p)]_k = \sum_{j=0}^{h} \lambda_{jk}\phi_j(\boldsymbol{x}_p - \boldsymbol{\mu}_j; \Sigma_j), \tag{1.1}$$

where $\phi_j(\dots)$ denotes the nonlinear transfer function of hidden node j, $(\phi_0(\dots) \equiv 1$ is the bias node), and the possible dependence on a 'smoothing' matrix is left explicit.

The most common example of the smoothing factor is in the use of a general Gaussian transfer function, i.e. $\phi(z; \Sigma) \approx \exp(-[z^T \Sigma^{-1} z])$. Since the general expression in (1.1) is an analytic function of the parameters corresponding to the basis function positions and smoothing factors, it is possible to estimate them from training data by a full nonlinear least squares process if required (Lowe, 1989; Moody and Darken, 1989). This is usually not necessary. As can be seen from equation (1.1), the main difference from a multilayer perceptron is that the output of the hidden node j, H_j, is given as a *radial* function of the distance between each pattern vector and each hidden node weight vector, $H_j = \phi_j(x - \mu_j)$, rather than in terms of a scalar product: $H_j^{\text{MLP}} = \phi_j(x.\mu_j)$.

One of the advantages of the radial basis function network is that the first-layer weights $\{\mu_j, \Sigma_j; j = 1, \ldots, h\}$ may often be determined or specified by a judicious use of prior knowledge, or estimated by simple techniques. This is discussed later. Once the weights associated with the first layer have been specified, the major problem in 'training' a radial basis function network is focussed upon the determination of the final-layer weights, $\{\lambda_{jk}\}$. Since the radial basis function network is typically employed to perform a *supervised* discrimination or prediction task such as time series forecasting, this training usually takes the form of the optimisation of a cost function requiring the outputs of the network to approximate a set of known target values. It is common to attempt to minimise a standard residual sum-of-squares cost function, though other cost functions may be employed. Since this is a linear optimisation process (the parameters $\{\lambda_{jk}\}$ occur linearly when minimising the residual sum squared error measure), the radial basis function network is computationally more attractive in applications than a multilayer perceptron, even though they are both computationally universal architectures (Park and Sandberg, 1991).

1.2   Historical overview

The radial basis function network architecture was originally introduced in order to provide an interpretation of the concepts of 'learning' and 'generalisation' in artificial neural networks (Broomhead and Lowe, 1988). The primary motivation was the ongoing fundamental mathematical work into the theory of function approximation by a process of strict interpolation, principally by Powell (1985, 1987, 1990) and Micchelli (1986). This work was extended to map over the basic concepts into a neural network description. The embodiment of function approximation techniques as a neural network permitted the interpretation of 'learning' in neural networks as a special form of 'curve fitting' (actually curve fitting to the *generator* of the data as opposed to fitting the data themselves), and consequently 'generalisation' could then be interpreted as 'interpolation' along the fitting surface. As a result of the computational universality of radial basis function networks and more recent stronger results on the uniform approximation of functions on a compact subset by a radial basis function expansion (Powell, 1990; Park and Sandberg, 1991; Hartman and Keeler, 1990), there exists a radial basis function network that arbitrarily closely approximates any multilayer perceptron. In essence the interpretation embodied in radial basis function networks maps over to an interpretation of learning and generalisation in multilayer perceptrons as well. The function approximation background of radial basis function networks is not discussed extensively in this article. For details of this see Broomhead and Lowe (1988) and Powell (1990).

Despite its introduction through the route of mathematical function approximation, the radial basis function network has much stronger appeal when related to various other statistical descriptions of information. The primary aim of this chapter is to develop some of these links between the radial basis function network and statistics. As a consequence of the links between radial basis function networks and a variety of information processing domains, the radial basis function network has found successful and generic applications in forecasting, regression, classification and feature extraction problems, though we are unable to discuss applications in this chapter.

Alternative motivations for radial basis function architectures have been proposed. For example, Moody and Darken (1989) introduced the same network structure but based on the biologically captivating idea of 'effective' neurons in certain parts of the nervous system with response characteristics which are *locally tuned* or exhibit a coarse selectivity over a range of input values. They illustrated that this network structure had convenient training properties, as one could now combine linear and non-linear optimisation approaches as well as combining supervised and 'self-organising' methods.

A less biologically motivated metaphor was employed by Poggio and Girosi (1990a, b). Whereas the original metamorphosis of approximation theory into a neural network context imposed constraints through model complexity limitations and a least squares approach, the perspective pursued by Poggio and Girosi considered the solution of ill-posed problems through *regularisation* (Tikhonov and Arsenin, 1977). Essentially, explicit smoothness constraints are built into the exact optimisation process by the addition of a 'regulariser', a smoothing term which penalises overly complex fitting surfaces. This regulariser emerges as a consequence of prior knowledge about the problem. A natural radial basis function architecture may be derived by seeking the optimum functional form which would minimise the total optimisation criterion. The precise form of the basis functions is determined by the particular regulariser used. This approach is discussed in Section 3.1. The above approaches discussed the radial basis function architecture in the context of function approximation, where the underlying data generator was deterministically approximated. This allowed radial basis function networks to be viewed as a tool in dynamical systems theory, allowing, for example, applications in time series prediction as well as multivariate regression. Alternatively, the mathematical utility of the radial basis function structure has been investigated as an effective method for density estimation. In the classical statistical literature, its closest analogue is the Parzen window kernel-based estimation technique (Parzen, 1962; Silverman, 1986), also developed as the potential function method primarily by the Russian school (Bashkerov *et al.*, 1964). Related work which also emerged from this literature during the 1960s include Specht's polynomial discriminant networks and Nilsson's 'Phi machines' (Nilsson, 1965). Specht recently reintroduced his work as probabilistic neural networks (Specht, 1990), which are related to a restricted form of radial basis function networks essentially with as many cluster centres as data points and where the final layer weights are prescribed. The links to statistical density estimation are discussed in Section 2.

There have also been recent innovations in the development of radial basis function network techniques for feature extraction and data visualisation (Lowe and Tipping, 1996; Webb, 1995, 1996) in such statistical areas as nonlinear principal component analysis and multidimensional scaling. This work is briefly discussed in Sections 4.2 and 4.3. However, we begin our foray into the relationships between radial basis function networks and statistics by exploring the more traditional areas of density modelling.

2   Density estimation

2.1   Introduction

In this section we exploit the link between the specific case of a *Gaussian* radial basis function network and mixture density models (McLachlan and Basford, 1988). In particular we obtain an *unsupervised* clustering mechanism for the first layer of the radial basis function network motivated by density modelling.

There are many unsupervised clustering techniques that can be employed to position the centres of the radial basis function network, and to adapt the 'smoothing' parameters corresponding to the first layer of the network. These range from simple clustering techniques to fully optimised adaptive methods. A simple heuristic technique is to select the position of the centre randomly according to the distribution of the data; this is actually quite effective in practice provided that the distribution of the training data is representative of the problem in general, since the centres are chosen to reflect the density of the information. Graph-theoretic methods include hierarchical clustering and the use of a minimal spanning tree, and Leader clustering. An example of an adaptive method based on the minimisation of a specified objective function, also known as the partition algorithm, is k-means clustering, employed by Moody and Darken (1989), based on the minimisation of the sum of squared Euclidean distances between patterns and cluster centres, which has also proven effective even though the algorithm converges to a local minimum solution. Once the data have been partitioned into clusters, each cluster may be modelled by its own explicit basis function, such as a Gaussian parameterised by a mean and a covariance matrix which may be estimated from the data clusters. This is equivalent to a Gaussian classifier algorithm which, for the two-class discrimination problem, reduces to a quadratic discriminant function.

However, in this section we choose to consider a 'principled' approach to clustering motivated by the links between the first layer of the radial basis function network and kernel-based unconditional density estimation by mixture distributions (Tråvèn, 1991; Lowe, 1991). It will be observed that a close link may be made to Kohonen's algorithm for producing a self-organising feature map.

2.2 Semiparametric estimation of density functions

The basic nonparametric model for estimating a probability density function is to construct a multidimensional histogram of the data. However, the number of discretisation cells grows exponentially with the dimensionality of the data. Hence Parzen (1962) introduced a window function centred on each data point, constructing an approximation to the underlying density function by combining the values from each window function. This windowing approach of Parzen has been employed by Specht (1990) in a 'probabilistic neural network'. Although superficially similar, the difference between the radial basis function approach and these alternatives is that the radial basis function network is *adapted* to the training set and scales with the complexity of the problem, rather than scaling with the size of the training set used. These alternatives based on Parzen windowing are more equivalent to a k-nearest neighbour approach, in that there is no 'training' process, merely a gathering and storing of training data which are subsequently used as a 'smoothed' look-up technique. The typically high storage and unfavourable data scaling behaviour are characteristic disadvantages of such nonparametric approaches.

The radial basis function model has been termed a 'semiparametric' method (Tråvèn, 1991; Lowe, 1991) in that, although it is a flexible network structure, the number of parameters does not scale with the dimensionality or the number of training data points. As the complexity of the model scales with the complexity of the underlying data generator, the adjustable model parameters in the radial basis function have to be adapted.

We may motivate the radial basis function structure from the perspective of kernel-based density estimation (Hand, 1982; Lowe and Webb, 1991; Tråvèn, 1991) as follows.

2.2.1 *Supervised probabilistic classification*

In supervised probabilistic classification tasks we are primarily interested in the 'posterior' probability, $p(c|x)$, the probability that class c is present given the observation x. Given an estimate of the conditional posterior probability for each of C classes of patterns, the Bayes decision rule for minimising the overall error rate is to select that class with the largest posterior. Note that this assumes equal decision costs. The more general case where we have an arbitrary vector of costs incurred for assigned patterns to classes is discussed by Lowe and Webb (1991).

In a radial basis function network optimised to minimise the residual error, it is now well accepted that the optimum network output is the 'conditional risk' vector (Lowe and Webb, 1991), $\rho(x) = \sum_{c=1}^{C} s_c p(c|x)$, where s_c is the vector of costs incurred in assigning an observation to class c. For a 1-from-C target coding scheme, equivalent to assuming equal gains in assigning patterns to classes, this optimum solution reduces to the class conditional probabilities. Hence an appropriately optimised model with appropriate data coding directly approximates the posterior probabilities, irrespective of the form of the modelling function.

However, it is often easier to model other related aspects of the data, such as the unconditional distribution of the data $p(x)$ and $p(x|c)$, which is the conditional probability function of the data, given that it came from a specific class c. We can use these quantities to 'derive' the structure of a radial basis function model in the context of probabilistic classification.

By Bayes' theorem, $p(c|x) = p(c)p(x|c)/p(x)$. The distribution of the data is modelled as if it were generated by a mixture distribution, i.e. a linear combination of parameterised state or basis functions such as Gaussians. Since individual data clusters for each class are not likely to be approximated by a single Gaussian distribution, we need several basis functions per cluster. It is assumed that $p(x|c)$ and the unconditional distribution can both be modelled by the same set of distributions, but with different mixing coefficients, i.e. $p(x) = \sum_s p(s)q(x|s)$ and $p(x|c) = \sum_s p(s|c)q(x|s)$. The basis functions need not be probability density functions themselves, but they are often taken to be so in kernel-based density estimation. Note, however, that this assumption incurs a penalty; see the discussion at the end of this chapter. Then the quantity we are interested in, $p(c|x) = p(c)p(x|c)/p(x)$, is given by

$$p(c|x) = \sum_s \frac{p(c)p(s|c)}{p(s)} \frac{p(s)q(x|s)}{\sum_{s'} p(s')q(x|s')}$$

$$\equiv \sum_s \lambda_{cs}\phi(x|s),$$

where $\lambda_{cs} = p(c)p(s|c)/p(s)$ relates the overall significance of state s to class c, and $\phi(x|s)$ is a normalised basis function, namely $p(s)q(x|s)/\sum_j p(j)q(x|j)$.

The above expression is recognised as a radial basis function architecture. For a total of h state functions used to approximate the likelihood and the unconditional density, there are h hidden nodes corresponding to the normalised basis functions, and the final layer weights relate the significance of the hidden nodes to the C output class nodes, providing the class-conditional information. Of course, the positions and possibly also the ranges of influence of each of these basis functions need to be specified/adapted to allow an adequate model of each data cluster. This is the specific topic considered next.

2.2.2   Unsupervised probabilistic clustering

From the previous motivation of interpreting the first layer of the radial basis function network as approximating the unconditional distribution of the data, we can now derive a simple adaptive algorithm for positioning the centres and selecting the 'widths' of the basis functions. This algorithm is essentially the expectation-maximisation algorithm (Baum et al., 1970).

It is assumed that the data samples are unconditionally probabilistically generated according to a weighted summation over n_0 parameterised 'substates'; i.e. $p(x) = \sum_{s=1}^{n_0} p(s)p(x|s) \equiv \sum_{s=1}^{n_0} p(s)\phi(x; \theta_s)$ where $\phi(x; \theta_s)$ corresponds to the transfer function of one of the hidden nodes of the radial basis function network which has a set of adjustable parameters $\boldsymbol{\theta}_s$. Therefore, if we assume that the data samples are independently generated, the *likelihood* that the whole dataset was generated according to the model is just $Q(\{\boldsymbol{x}_p\}) \propto \prod_p \sum_{s=1}^{n_0} p(\boldsymbol{x}_p|s)$, where we have set the priors on the 'substates' to be uniform. By maximising the relative entropy of this likelihood with respect to the actual density function of the data (by adjusting the kernel parameters) one arrives at the expectation-maximisation (EM) (Dempster *et al.*, 1977) iterative algorithm. For the explicit case of employing normalised Gaussians with a spherical covariance matrix for each hidden node in the radial basis function network the update equations for the means and variances are

$$\boldsymbol{\mu}_s = \frac{\sum_p \boldsymbol{x}_p q(s|\boldsymbol{x}_p)}{\sum_p q(s|\boldsymbol{x}_p)} \tag{2.1}$$

and

$$d\sigma_s^2 = \frac{\sum_p \|\boldsymbol{x}_p - \boldsymbol{\mu}_s\|^2 q(s|\boldsymbol{x}_p)}{\sum_p q(s|\boldsymbol{x}_p)}, \tag{2.2}$$

where d is the dimensionality of the data and $q(s|\boldsymbol{x}) \equiv p(\boldsymbol{x}|s)/\sum_s p(\boldsymbol{x}|s)$ is the probability of state s given the single data pattern $\boldsymbol{x}$. This process amounts to setting the first layer of the network by an unsupervised fuzzy clustering process which maximises the likelihood of the data being generated by the assumed functional form; for simplicity the relative importances of each state, the $\{p(s)\}$, have been assumed constant, though one may adapt these iteratively as well.

Of course, maximum likelihood estimators of mixture models can suffer from singularity problems, in that an infinite likelihood is obtained if one of the component densities captures a single data point by collapsing the variance to zero. This potential problem is avoided by either deleting such states if the variance becomes too small, or restricting the minimum size of the variance allowed.

Another way to circumvent the problem is based on a stochastic gradient descent approach, as suggested by Tråvèn (1991) and Lowe (1991). The stochastic update approach reformulates the previous iterative procedure for estimating the positions and smoothing factors of the basis functions as a sequential stochastic update method equivalent to a Robbins–Monro process (Robbins and Monro, 1951). For example, the update equation for the mean evaluated from $p + 1$ data points may be expressed in terms of the mean evaluated using the previous p data points combined with one new data point in the form

$$\boldsymbol{\mu}_s(p+1) = \boldsymbol{\mu}_s(p) - \eta_{p+1}\{\boldsymbol{\mu}_s(p) - \boldsymbol{x}_{p+1}\}. \tag{2.3}$$

The corresponding expression for sequentially updating the variances of each state is

$$d\sigma_s^2(p+1) = d\sigma_s^2(p) - \eta_{p+1}\{d\sigma_s^2(p) - \|\boldsymbol{x}_{p+1} - \boldsymbol{\mu}_s(p+1)\|^2\}. \tag{2.4}$$

In these sequential update equations, η_{p+1} is an important parameter. Typically in stochastic update expressions, there are conditions on the rate of decay of η_{p+1} to ensure convergence and consistency as $p \to \infty$. However, for this derivation η_{p+1} is determined dynamically from the distribution of the data itself. Indeed η_{p+1} obeys its own sequential update equation. Explicitly, $\eta_{p+1} = q(s|\boldsymbol{x}_{p+1})/\sum_{t=1}^{p+1} q(s|\boldsymbol{x}_t)$ and hence

$$\frac{1}{\eta_{p+1}} = \frac{1}{\eta_p}\frac{q(s|\boldsymbol{x}_p)}{q(s|\boldsymbol{x}_{p+1})} + 1. \tag{2.5}$$

On average, each new data point will influence the re-estimated quantities less and less, as the scaling parameter may be expected in general circumstances to decrease harmonically. The main problem with this algorithm is that it can suffer from slow convergence. Note that the re-estimation equation (2.3) is reminiscent of Kohonen's update procedure as part of creating a self-organising feature map. In the terminology of the radial basis function network, in Kohonen's algorithm the position $\boldsymbol{\mu}_j$ of centre j is moved iteratively and incrementally towards each data point in turn according to

$$\boldsymbol{\mu}_j[\text{new}] = \boldsymbol{\mu}_j[\text{old}] + \alpha(\boldsymbol{x}[\text{new}] - \boldsymbol{\mu}_j[\text{old}]),$$

where $0 < \alpha \leq 1$ is usually a monotonically decreasing function of iteration time which governs the step size change of the centre position and involves the neighbourhood relation in 'feature' space. Of course, in the radial basis function case we do not usually consider an explicit neighbourhood relation. The equivalent case is equation (2.4), which gives the formula for the contraction of the basis function widths over 'time'. However, the basis functions themselves are not 'tied' together to impose any topographic constraint on the basis functions. Nevertheless, the radial basis function network as a whole is capable of preserving topographic relationships, as we discuss later. Although Kohonen's algorithm does not reflect the underlying probability distribution of the data correctly, except in simple circumstances, it has proven to be an effective clustering technique which in some sense preserves local neighbourhood topology. It is also interesting that Kohonen's re-estimation algorithm was heuristically motivated, whereas the radial basis function clustering process just discussed has been motivated from a probabilistic likelihood-based approach. According to this latter approach, the 'update parameter' is specified according to the current estimates of the model; it is not a free parameter of the training process. In this sense it is similar to the differences between a steepest-descent backpropagation algorithm and a conjugate-gradient method, in which the free parameters of the model, step size and momentum term, become data driven, adapted according to the current state of the model.

To summarise, by exploiting the interpretation of the first layer of a radial basis function network as representing an unconditional density estimator according to an adaptive kernel-based mixture model, we are led to an unsupervised clustering process. Once the first-layer parameters have been optimised according to this process, the combined outputs of the hidden layer represent the statistical generator of the *unconditional* data, according to $p(\boldsymbol{x}) = \sum_{s=1}^{n_0} p(s)\phi(\boldsymbol{x};\boldsymbol{\theta}_s)$. The final layer weights may then be adapted according to a supervised process (e.g. linear least squares) to map the unconditional density expansion, $p(\mathbf{x})$, into conditional densities, such that the optimum network output approaches the probability of a 'class' given the data, $p(c|\mathbf{x})$, for example.

As a final comment in this section, we have considered the positions of the basis functions, the final layer weights and the smoothing widths of the basis functions to be an equivalent set of adjustable parameters. This is not correct. One consequence of this assumption was the introduction of the 'anomaly' of the possible singular solutions as the width collapsed around a data point. Generally, setting the widths of the radial basis function network according to a criterion such as maximum likelihood, or by minimising the error (Lowe, 1989), produces smoothing parameters which are too 'narrow'. Since the widths are the dominant features which restrict the smoothness of the network mapping, they should be considered as 'hyperparameters' and therefore *not* adapted according to the data at all, but determined by a higher level of prior knowledge.

3   Regression

In this section we consider the links between radial basis function networks and multivariate regression. In particular, we illustrate the relationships to *regularisation* methods on the one hand, and *kernel smooths* and *kriging* on the other.

3.1   Equivalence to regularisation

The original radial basis function approach was in terms of function approximation in which over-fitting was mediated against by restricting the degrees of freedom of the network. In the regularisation approach to function approximation, the over-fitting problem is addressed differently in that explicit smoothness constraints are folded into the optimisation process. In this approach (Poggio and Girosi, 1990a, b) the radial basis function network emerges by the following argument.

Consider a finite set of discrete data points, $S = \{(\mathbf{x}_p, y_p) \in \mathbb{R}^n \times \mathbb{R}, p = 1, 2, \ldots, P\}$; we only consider mappings into the real line, for simplicity. We wish to interpolate this dataset by an approximator, $s(\mathbf{x})$. According to the regularisation approach, of all possible approximators, we wish to choose the one that minimises the functional

$$H[s] = \sum_{p=1}^{P} \{y_p - s(\mathbf{x}_p)\}^2 + \eta \| \hat{O}s \|^2.$$

Here $\hat{O}$ is an operator which embodies the constraints of our prior knowledge, usually taking the form of a partial differential operator such as $\partial^2/\partial\mathbf{x}^2$, which constrains the curvature of the mapping, and $\| \ldots \|$ indicates a norm on the function space to which $\hat{O}s$ belongs; η is the regularisation parameter and usually embodies the degree to which the constraint should dominate the data. The value of η itself may be fixed by prior knowledge or may be chosen adaptively. Considering $\partial H[s]/\partial s(x) = 0$ leads to the operator equation

$$\hat{O}^\dagger \hat{O} s(x) = \frac{1}{\eta} \sum_{p=1}^{P} (y_p - s(x)) \delta(x - x_p),$$

where $\hat{O}^\dagger$ is the adjoint operator to $\hat{O}$. It is possible to obtain a formal solution to this operator equation by the method of Green's functions. Let us re-express the problem as follows. We wish to solve $\hat{L}|s\rangle = |f\rangle$ where $\hat{L} = \hat{O}^\dagger \hat{O}$, $|f\rangle$ represents the right-hand side of the operator equation and $|\ldots\rangle$ represents a column vector; $\langle\ldots|$ represents a row vector. In this notation, $\langle a|b\rangle$ is a scalar product between vectors $|a\rangle$ and $|b\rangle$, and $|a\rangle\langle b|$ is a tensor product, giving a matrix in this case. Provided there exists an operator $\hat{G}$ such that $\hat{G}\hat{L} = \hat{I}$, where $\hat{I}$ is the identity operator, then the solution of $\hat{L}|s\rangle = |f\rangle$ may be written as $|s\rangle = \hat{G}|f\rangle$. Writing $\hat{G}$ as an integral operator whose integral kernel is the Green's function $G(x', x'')$, $\hat{G} = \int\int dx'dx'' \, |x'\rangle w(x')G(x', x'')w(x'')\langle x''|$, in which $w(x)$ is a weight which will be taken to be unity, we find that the Green's function satisfies the differential equation $\hat{L}_{x'}G(x', x'') = \delta(x' - x'')/w(x')$. The Green's function satisfies the same homogeneous equation as $s(x)$. It should also satisfy the same boundary conditions. There are corresponding adjoint Green's functions $G^\dagger(x', x'')$ satisfying the adjoint equations. Now we can write down the solution of the original problem in terms of the (adjoint) Green's function corresponding to the projection operator $\hat{O}^\dagger \hat{O}$.

Converting the operator solution $|s\rangle = \hat{G}|f\rangle$ into the Green's function terminology gives $s(x) = \int dx' G^\dagger(x', x)f(x')w(x')$. Since $f(x) = \frac{1}{\eta}\sum_{p=1}^{P}(y_p - s(x))\delta(x - x_p)$, the solution may be expressed finally as

$$s(x) = \sum_{p=1}^{P} \lambda_p G^\dagger(x, x_p),$$

where $\lambda_p = (y_p - s(x_p))/\eta$. This indicates that the solution of the regularisation problem lies in a P-dimensional subspace, and a basis for this subspace is provided by the P Green's functions $\{G^\dagger(x, x_p)\}$, with a Green's function centred at each data point x_p. If the operator $\hat{O}^\dagger \hat{O}$ is rotationally and translationally invariant, then the Green's function is a function only of the radial differences of its arguments, i.e. $G(\|x - x_p\|)$. In this case the link to the previous interpretation of strict interpolation by radial basis function networks has been completed. As in the previous case, the weighting coefficients $\{\lambda_p\}$ may be determined from linear equations. Specifically, evaluating the function $s(x) = \sum_{p=1}^{P} G^\dagger(x, x_p)(y_p - s(x_p))/\eta$ at each of the P data points leads to P linear equations in P unknowns, i.e. in matrix form $[\eta I + G]\lambda = y$, where $[G]_{ij} = G^\dagger(x_i, x_j)$.

Gaussian basis functions of width σ emerge from the rather uninformative regularisation expression given by

$$\|\hat{O}s\|^2 \approx \sum_{l=0}^{\infty} \frac{\sigma^{2l}}{l!2^l} \int |\nabla^l s(x)|^2 dx,$$

where ∇ denotes the gradient operator.

Cubic spline basis functions (Wahba, 1990) emerge, when x is scalar, from regularisation expressions of the form

$$\|\hat{O}s\|^2 \approx \int \left(\frac{d^2 s(x)}{dx^2}\right)^2 dx.$$

Note that complete solution of the regularisation problem requires us to augment the Green's function expansion to allow for functions that lie in the null space of the operator $\hat{O}$. This null space depends on the specific Green's functions employed.

This straightforward regularisation approach is similar to the nonparametric methods of density modelling, as it requires as many basis functions as data points. However, if we relax the strict regularisation requirements, the regularisation network may employ fewer basis functions than data points and adapt the parameters of the Green's functions that are used.

Note that we have 'derived' the radial basis function architecture from two distinct perspectives, a probabilistic classification approach and a function-fitting regularisation approach. The ability of the radial basis function to draw together distinct strands of statistics is perhaps its main advantage over alternative network models.

3.2 Equivalence of kernel smooths and dual basis functions

We now explore additional relationships to statistical techniques in regression concerned with a specific but powerful class of smoothing models. A 'smoother' is a nonparametric approach to producing a summary of the trend of a response measurement driven by measurements of predictors (covariates) (Hastie and Tibshirani, 1990). In this sense a smoother may be considered as a regression fit to an observational set of data. There are different classes of smoothers, and the radial basis function has a particularly strong relationship to the class of statistical regression models known as *kernel smoothers*.

3.2.1 *Kernel smoothers*

In a kernel smoother the weight given to the pth point, x_p, in producing a regression estimate at some point x is given by

$$\Psi_p(x) = \frac{\alpha}{\sigma} \psi\left(\left\|\frac{x - x_p}{\sigma}\right\|\right),$$

where $\psi(\cdot)$ is an even, decreasing function, σ determines the width of the kernel function and α is usually determined to normalise the regression estimator such that the weights all sum to unity. Often the Gaussian function is chosen as the kernel function, though we would not recommend this in practice. Empirical evidence suggests that the precise choice of the kernel width is relatively unimportant.

In terms of these weights, the regression smooth obtained from a set of observations $\{y_p, p = 1, \ldots, P\}$ is given by the convolution

$$s(x) = \sum_{p=1}^{P} \Psi_p(x) y_p.$$

When the values of the convolution evaluated at the points x corresponding to the data are written as a vector, this gives a *linear smooth* $s = \Psi y$, in an obvious notation, as it is linear in the target values. We can introduce the concept of the *degrees of freedom* of a linear smooth as $df = \mathrm{tr}(\Psi)$, which is the sum of the eigenvalues of the smoother matrix Ψ and indicates the amount of smoothing produced by the smoother. This may be considered to be a generalisation of multivariate linear regression since this definition of the degrees of freedom reduces to the number of free parameters if we have a simple linear model. We will shortly use this definition of the degrees of freedom of a smooth to determine the appropriate model order complexity for radial basis function networks. There are other definitions of degrees of freedom which generalise from the linear case, in particular $\mathrm{tr}(\Psi \Psi^T)$ and $\mathrm{tr}(2\Psi - \Psi \Psi^T)$; see Hastie and Tibshirani (1990).

Other relevant properties of a linear smooth include the following.

- Pointwise estimates of confidence intervals can be obtained from

$$\mathrm{cov}(s) = \Psi \Psi^T \sigma^2,$$

  provided an estimate of the additive random noise component may be obtained.

- Constant-preserving: any reasonable smooth should be constant-preserving in the sense that $\Psi 1 = 1$; thus the row-sums of Ψ should equal unity.

3.2.2   Dual basis functions

To make the link between radial basis function networks and kernel smoothers we introduce the concept of *dual basis functions*; see also Girosi *et al.* (1995). If, referring to the basic definition of the radial basis function network, we solve for the weights of the network by minimising the residual training sum of squares, then the weights are determined from the pseudo-inverse of the matrix of hidden unit outputs multiplied by the target values. Hence the final-layer weights of the radial basis function network are linear in the target values. Mathematically, the output of the network, given input x, is

$$o_k(x) = \sum_{j=0}^{h} \phi_j(x) \lambda_{jk}.$$

Substituting in the minimum norm minimum error solution for the weights gives the least squares estimates of the $\{\lambda_{jk}\}$ as

$$\lambda_{jk} = \sum_{p=1}^{P} \Phi_j^+(p) y_k(p),$$

where $\Phi_j^+(p)$ contains the appropriate multipliers of the observations in the formulae for least squares estimates and $y_k(p)$ is the kth component of y_p. Substitution into the general output expression and rearranging terms gives

$$o_k(x) = \sum_{p=1}^{P} \left[\sum_{j=0}^{h} \phi_j(x) \Phi_j^+(p) \right] y_k(p)$$

$$= \sum_{p=1}^{P} \Psi_p(x) y_k(p),$$

(3.1)

where we have defined the 'dual basis functions' $\Psi(\cdot)$ as

$$\Psi_p(x) \equiv \sum_{j=0}^{h} \phi_j(x) \Phi_j^+(p).$$

(3.2)

Note that equation (3.1) redefines the radial basis function as equivalent to a kernel smoother, but with the restriction that the kernel functions are given by the dual basis functions (3.2). In the original interpretation, the radial basis function is a linear combination of nonlinear basis functions, where the basis functions are fixed and the weights need to be determined by an optimisation approach. Also, there are fewer basis functions than data points. In this alternative interpretation of a radial basis function there are as many dual basis functions as data points but the basis functions need to be obtained from an optimisation process and the weights are fixed, being simply the observed target values.

Determining the dual basis functions explicitly in this manner also constrains the behaviour of the smoothing matrix, Ψ, which is composed out of the values of these dual basis functions. In particular we note the following.

- Degrees of freedom. $df = \text{tr}(\Psi) = \text{rank}(\Phi)$. In a least squares optimised radial basis function the degrees of freedom is the same as the rank of the hidden-layer data matrix. We can use this link to determine the 'correct' model-order complexity as determined by the number of basis functions used to approximate a function. We demonstrate this in an example later. Note that this link is a justification for the use of the condition number of Φ, which leads to structure or 'kinks' in the singular spectrum of Φ, as indicated later, as a more robust measure of degrees of freedom, especially in noisy situations when the data matrix will tend to be of full rank anyway.

- Constant-preserving. The radial basis function smooth is naturally constant-preserving because of the properties of the pseudo-inverse. In particular we note that

$$\sum_p \Psi_p(x) \equiv \sum_j \phi_j(x) \sum_p \Phi_j^+(p) = 1.$$

Note also from (3.2) that the dual basis functions are generally all different. Also, we observe that the dual basis functions tend to be localised, even if the original basis functions are not. This reflects the characteristic that it is the network as a whole which needs to construct a localised representation, and this behaviour need not be reflected in the individual basis functions.

The contrast between the usual basis functions and the dual basis functions may be observed from the following simple example.

3.2.3 *Example: noisy sine wave*

Consider the simple one-dimensional example of data being generated according to a sine wave, with additive white noise.

Synthetic data were generated according to $y(x) = \sin(2\pi x) + n(x)$, where x is obviously scalar and $n(x)$ is uniform random noise with a maximum possible amplitude equal to 0.5: $n(x) \sim \mathrm{Un}(-0.5, 0.5)$. Also, the input variable x was sampled quadratically more densely nearer the origin. Figure 1 depicts the data samples and the underlying noise-free sine-wave generator function; the nonuniform data sampling is apparent. The radial basis function network was constructed using the nonlocal and

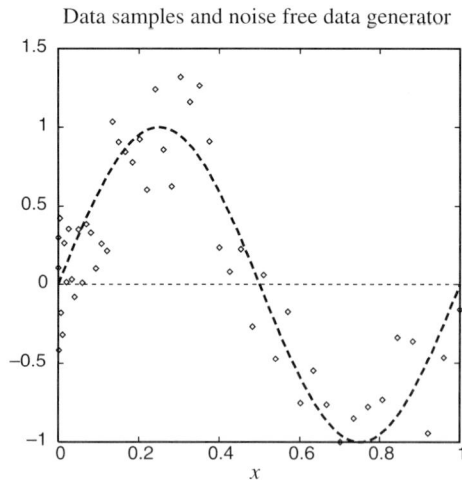

Data samples and noise free data generator

Fig. 1:   Raw data superimposed on the noise-free sine wave. Note that the sampling density of data points is nonuniform.

non-positive-definite basis functions $\phi(z) = z \log(z)$, where $z = \|x - \mu\|$ and the centres of the basis functions were distributed uniformly on $[0, 1]$.

Figure 2 shows the singular spectrum of the radial basis function data matrix using seven basis functions. The data matrix is the matrix of outputs of the hidden layer of the radial basis function network. From this figure we deduce that there are only three significant degrees of freedom in this problem, as characterised by the projection of the data into the hidden-unit space. We can corroborate this opinion by recourse to the links with linear smooths. We saw that the degrees of freedom in a linear smooth could be related to the eigenvalues of the 'smoother matrix', Ψ. Figure 3 depicts the eigenvalues of the smoother matrix. It is clear here that this is only a rank 3 problem.

On the basis of the evidence presented in Figures 2 and 3, Figure 4 demonstrates the fitting ability of the radial basis function network using approximations of rank 2, 3 and 21. This figure is a demonstration that the rank 3 level of description of the data is as reasonable as can be obtained, given the nature of the sampled data.

Finally, it is instructive to examine the dual basis functions used in this approximation. Figure 5 depicts just two of the dual basis functions located at different positions. The 'centres' of the basis functions are marked with a vertical line. Recall from Section 3.2.2 that there is one dual basis function for each data point in the training set. Figure 5 demonstrates that the dual basis functions are different, and, despite the fact that the original basis functions of the network were nonlocal, these dual basis functions are localised and have a dispersion related to the local density of data. In regions of high data density the dual basis functions are more tightly localised than in regions of low data density. Hence we see how the original linear combination of nonlocal basis functions may be combined in such a way as to create a network with an *overall localised* response.

Singular value spectrum of the RBF data matrix

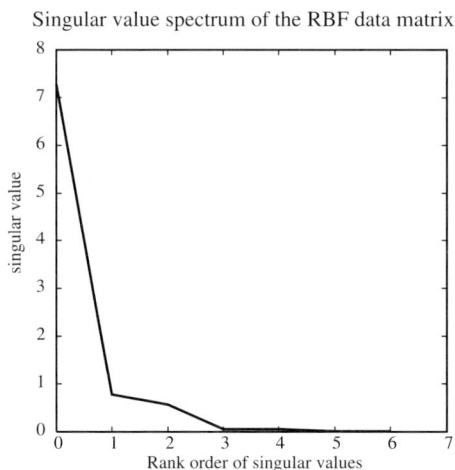

Fig. 2: The singular spectrum of the radial basis function data matrix. According to this figure a radial basis function with rank 3 data matrix should be adequate to resolve the complexity of the data.

Eigenvalues of the Smoother Matrix

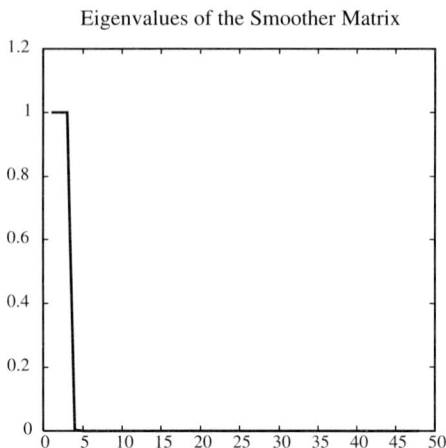

Fig. 3:   The eigenvalue spectrum of the smoother matrix. Note that there are only three eigenvalues of relevance, in accordance with the singular spectrum of the radial basis function data matrix.

The main conclusions of this simple experiment are as follows.

- Despite the use of nonlocal basis functions, the dual basis functions are localised, leading to an overall local response of the network.
- The data-dependent dual basis functions exhibit a narrowing in regions of high data density and an expansion in regions of low data density, as we might expect if we were naïvely using locally adaptive basis functions.
- The use of the singular spectrum of the hidden-unit data matrix serves as a robust indicator of appropriate model-order complexity.

In summary, in this section we have derived relationships between radial basis function network models and statistical methods in regression, namely regularisation and kernel smooths. Since most kernel smooths are equivalent to the solution of a penalised least squares problem with an appropriately defined regulariser, the explicit action of regression smoothing in penalising over-fitting is apparent. From the radial basis function perspective, the smoothing is governed by the properties of the hidden-layer data matrix and in particular its singular value spectrum. This is turn is determined by the model complexity and the form of the basis functions.

There are several obvious topics we have not had time to explore but which deserve a mention. One other method of relevance is *kriging*, popular in the geostatistics community. The kriging framework is related to smoothing splines, and hence to radial basis function networks, through the concept of dual basis functions. Kriging emphasises the stochastic approach to smoothing, in which the observation y is assumed to consist of a stochastic trend component $f(x)$ and an uncorrelated noise source ϵ. Kriging predicts the underlying model $f(x)$ by a linear combination of observations

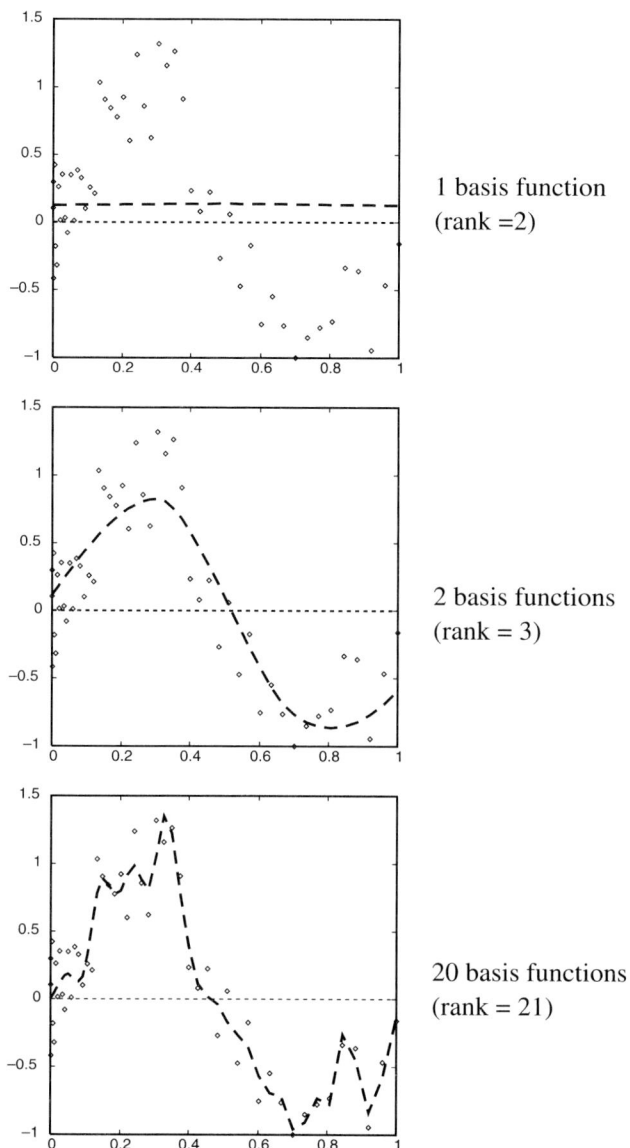

Fig. 4:   The radial basis function regression curves obtained from using rank 2, 3 and 21 approximations. The 'best' fit is that determined by using only three 'degrees of freedom'.

$g(x) = \alpha^T y$ by minimising $E\{g(x) - f(x)\}^2$. The optimum solution is exactly the same as a smoothing spline (Hastie and Tibshirani, 1990).

In addition, we have not explored the properties of confidence intervals of radial basis function networks as determined by these links with linear smooths, but it should

Two representative Dual Basis Functions

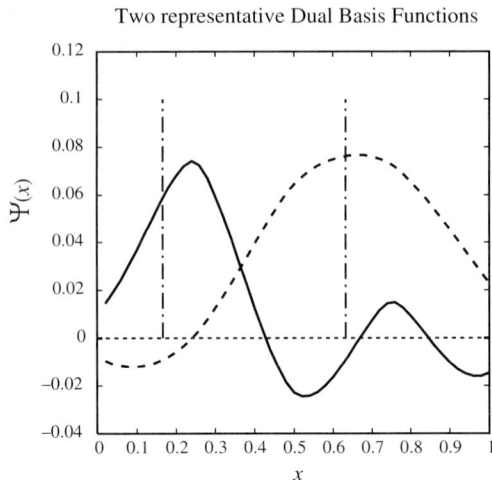

Fig. 5:   A diagram of just two of the dual basis functions. One is located in the higher density region, the other is centred in the lower density region of data. (Positions of the 'centres' of the dual basis functions are marked by vertical lines.) Note that both dual basis functions are localised, though oscillatory, and that the width of the dual basis functions is narrower in regions of higher data density.

be apparent from the previous discussion that confidence intervals may be estimated by using the properties of the radial basis function data matrix.

4   Feature extraction

The major attention of feedforward networks, including radial basis function models, has been diverted to problems such as classification and prediction. However, there are several developments of feedforward networks which render them as useful feature-extraction devices.

The hidden layer of all feedforward neural networks is crucial in that it is this part of the network that performs the feature-extraction stage. In the case of the radial basis function network, we can obtain explicit information on the nature of this transformation, because of the analytic tractability of the final, linear layer.

Feature extraction is the process by which the representation of data is transformed into another domain, usually involving dimension reduction, in such a way that essential features of the underlying problem are retained. For example, in a classification problem we have the extra information that patterns belonging to the same class should be grouped together, whilst patterns from different classes should be maximally separated if possible. This would allow the subsequent discrimination of the patterns to be made easier. Similarly, there are other problem domains where we may not have explicit class information, but we may require a transformation which preserves local or global neighbourhood properties. Using a feedforward network to effect topographic transformations is a recent innovation which relates to statistical work in multidimensional scaling. In this section we examine different types of feature extraction distinguished by the nature of the extra prior knowledge available in each problem domain. The extra knowledge is usually provided in the form of class labels, or desired targets, in which case we are interested in supervised feature extraction. Without class information or desired target values, we have to be content with unsupervised feature extraction. An in-between representation occurs when we have some extra knowledge. This may take the form of preferential, relative ordering of patterns or more explicit *topographic* or neighbourhood relationships. In this case the feature extraction we are interested in is more subtle. We now discuss the links between radial basis function networks and each of these different types of statistical feature extraction.

4.1   Supervised feature extraction: discriminant analysis

One important group of feature-extraction approaches is driven by classification tasks where the extra prior knowledge concerns the specific class label attached to each pattern. In this section we point out a link between adaptable radial basis function networks and traditional discriminant analysis.

In traditional discriminant analysis, a linear transformation is obtained into a space of reduced dimension such that the transformation maximises a specific discriminant criterion. For instance, Fisher's linear discriminant function applied to the two-class problem produces a projection into a one-dimensional space such that the transformed data maximises the ratio of the between-class to the within-class scatter (Fukunaga, 1972). This criterion has been extended to the multiclass case (Fukunaga, 1972; Gallinari *et al.*, 1988) in which, for a c-class problem, the optimum linear projection into a $(c - 1)$-dimensional space maximises the ratio of the *determinants* of the between-class to the within-class scatter matrices, $J_1 = |S_B|/|S_W|$. The ability of feedforward networks to perform static pattern discrimination stems from their ability to create a *specific* nonlinear transformation into a space spanned by the outputs of the layer of hidden units in which subsequent separation of classes is easier. Previous work on the *linear* multilayer perceptron (Gallinari *et al.*, 1988) demonstrated that the optimum hidden-layer projection also maximises the discriminant function J_1. A general result for nonlinear multilayer perceptrons is lacking, though empirical evidence (Asoh and Otsu, 1990) indicates that the full nonlinear multilayer perceptron maximises a more severe criterion, $J_2 = \text{tr}(S_B S_W^{-1})$, involving the trace of the between-class covariance matrix multiplied by the inverse of the within-class covariance matrix.

It has been noted that, for a radial basis function network to attempt to produce the expected value of the target given the input pattern, the network has to be able to form an internal representation of the generator of the data. We can interpret this representation in more detail, in terms of a supervised clustering process. The specific point that we wish to make is that minimising the error at the output of a network with linear output units is equivalent to maximising a *very specific* feature extraction/clustering criterion at the outputs of the hidden layer of the radial basis function network. A basic outline is as follows; details may be obtained from Webb and Lowe (1990) and Lowe and Webb (1990). We consider the minimisation of the weighted sum of squared errors function

$$\mathcal{E} = \sum_{p=1}^{P} d_p \|y_p - f(x_p)\|^2 \tag{4.1}$$

between the desired target values y_p and the radial basis function approximations, $f(x^p)$. We have also allowed for a differential weighting d_p for each pattern, often useful in equalising the priors in uneven classification problems.

First, minimising the error function (4.1) with respect to the bias vector gives the optimum solution for the biases as

$$\lambda_0 = \bar{y} - \Lambda \, m^H, \tag{4.2}$$

where Λ is the matrix of final layer weights, $\bar{y}$ is the weighted target mean, weighted by the 'significance' values $\sqrt{d}_p$, and m^H is the weighted mean pattern evaluated on the training set at the outputs of the hidden nodes. Therefore, the bias vector is important for compensating for the difference between the mean of what was wanted and the mean of what could be produced in the absence of a bias vector. We can use this result to replace the terms in the error expression with mean shifted versions of targets and hidden unit patterns. Minimising the modified error function with respect to the final-layer weights gives an expression for the minimum norm Moore–Penrose pseudo-inverse in terms of the hidden-layer patterns and the target patterns. Substituting the formal solution for the final-layer weights back into the error function and rearranging, we can finally show that minimising the error is equivalent to maximising the function

$$J_N = \text{tr}\left\{ S_B S_Y^+ \right\}, \tag{4.3}$$

where the matrices S_Y and S_B are specified entirely in terms of the targets and the distribution of patterns at the outputs of the hidden nodes:

$$S_Y \equiv \frac{1}{P} \hat{H} \, D^2 \hat{H}^*$$
$$\tag{4.4}$$
$$S_B \equiv \left(\frac{1}{P} \right)^2 \hat{H} D^2 \hat{Y}^* \hat{Y} D^2 \hat{H}^*,$$

where D is the diagonal matrix of the weighting factors, $\sqrt{d}_p$, $\hat{Y}$ is the mean-shifted target matrix and $\hat{H}$ is the matrix of mean-shifted hidden unit patterns.

These matrices, S_Y and S_B, may be interpreted as weighted total and between-class covariance matrices of the nonlinearly transformed input patterns, particularly

under the standard coding scheme of 1-from-C target values in a classification problem. This allows (4.3) to be interpreted as a feature extraction criterion for discriminant analysis, where the criterion is optimum for the subsequent linear classification
scheme to the output targets. The clustering process performed by the hidden layer of
the radial basis function network is such that the patterns are nonlinearly transformed
into a feature space such that patterns lying within one class are grouped together as
tightly as possible, whilst patterns in different classes are separated as far as possible. The specific feature-extraction criterion above is similar to, and has an equivalent
interpretation to, other suggested cost functions used in traditional discriminant analysis (Fukunaga, 1972). The difference is that this particular feature-extraction method
has been optimised to achieve the best performance consistent with a subsequent linear classification of the patterns. Thus, a fully optimised radial basis function network
may be interpreted as performing an optimised feature-extraction/clustering process
and classification task simultaneously, where the clustering process is 'supervised'
and optimal.

The significance of this relationship is the following: if the first layer of the
radial basis function network is adapted simultaneously with optimising the final-
layer weights to minimise the residual sum of squares, then the space spanned by the
outputs of the hidden layer is constructed in such a way that the projected patterns are
distributed so as to maximise the network discriminant function. This relationship is
not specific to radial basis function networks; in fact it is more relevant to multilayer
perceptron models with a linear final layer, since it is typically the case that the
first layer of the radial basis function network is not usually adapted (i.e. estimated)
according to the classification criterion. As noted in the previous discussion, it is
more common to fix the first layer or perhaps adapt the centres in order to capture the
unconditional distribution of the data.

4.2 Unsupervised feature extraction: nonlinear principal components

Recent work (Webb, 1996) has demonstrated the relationship and application of radial
basis function networks for extracting nonlinear principal components. In unsupervised dimension-reducing feature extraction we seek a transformation from data space
to feature space $\mathbb{R}^d \to \mathbb{R}^m$ ($m < d$) such that the 'variance' of the transformed data
is maximised. The radial basis function approach models this transformation as a
linearly weighted sum of kernel functions, which has certain advantages.

Linear principal components analysis may be considered as defining hyperplanes
of closest fit to the data such that each hyperplane projects to a point in the feature
space. A nonlinear principal components analysis produces curved surfaces of closest
fit to data. In this sense there is a similarity to the method of principal curves (Hastie
and Stuetzle, 1989). Principal curves constitute a nonparametric fit to data which aim
to provide a nonlinear summary of a data distribution. The specific type of nonlinear
summary is constrained by a 'unit speed' parameterisation in which the gradient of
the transformation along the principal curve is constrained to be unity.

In the radial basis function approach to nonlinear principal component analysis, we seek a transformation of the form $f : \mathbb{R}^d \to \mathbb{R}^m : f(x) = y$, where $f(x) = \sum_j \lambda_j \phi_j(x)$, such that a given component $f_k(x)$ of $f(x)$ takes the form $\sum_j \lambda_{jk} \phi_j(x)$. The parameters of this transformation are determined by seeking a solution which maximises the variance,

$$\text{var} = \langle (f - \bar{f})^2 \rangle \equiv \lambda^T C \lambda,$$

where C is the covariance matrix of the hidden-layer output space:

$$C = \langle (\phi - \bar{\phi})(\phi - \bar{\phi})^T \rangle.$$

To prevent unconstrained normalisation it is necessary to incorporate a constraint on the weights. The constraint chosen is to demand normalisation of the weights:

$$\lambda^T B \lambda = 1,$$

where B is a symmetric matrix. Webb (1996) selected the sensible choice

$$B = \left\langle \frac{\partial \phi_j}{\partial x} \frac{\partial \phi_j^T}{\partial x} \right\rangle.$$

This requirement, in distinction to the principal curves unit-speed parameterisation, imposes the constraint that the average squared magnitude of the gradient of the transformation is unity, which allows the principal component surface to distort and stretch whilst retaining normalisation. In the case of a linear transformation, B is the identity matrix and hence this constraint is equivalent to requiring that the transformed vectors be unit vectors.

Hence, because of the linearity of the radial basis function approach, the problem of extracting high-dimensional nonlinear principal components from data reduces to a simple linear problem. Since we wish to maximise the augmented variance function

$$\lambda^T C \lambda - \alpha (\lambda^T B \lambda - 1),$$

where α is the Lagrange multiplier, we find that we need to solve the generalised symmetric eigenvector equation

$$C\lambda = \alpha B \lambda.$$

Clearly, the variance obtained by using the eigenvectors of this equation is

$$\text{var} = \lambda^T C \lambda = \alpha.$$

Hence, the directions of maximum variance are those vectors with maximum eigenvalues. Obviously, if B is the identity matrix, then the solutions for the eigenvectors are just the usual linear principal components. This approach to extracting nonlinear principal components, and the reverse process of constructing embedded surfaces which provide a nonlinear summary of the original data, have been developed further by Webb (1996), where additional relationships to other statistical feature-extraction domains are considered.

4.3 Topographic feature extraction: multidimensional scaling and Sammon mappings

Recently (Mao and Jain, 1995; Lowe, 1993; Webb, 1995; Lowe and Tipping, 1996) an important class of topographic neural networks based on feedforward neural networks has been introduced. This new approach can be related to the traditional statistical methods of Sammon mappings (Sammon, 1969) and multidimensional scaling (Kruskal, 1964). These novel alternatives to Kohonen-like approaches for topographic feature extraction possess several interesting properties.

The approach to topographic feature extraction using radial basis function networks has been termed 'NEUROSCALE'. We now provide a brief introduction to the NEUROSCALE philosophy of nonlinear topographic feature extraction. Further details may be found in Lowe (1993), Webb (1995) and Lowe and Tipping (1996). We seek a dimension-reducing, *topographic* transformation of data for the purposes of visualisation and analysis. By 'topographic', it is implied that the geometric structure of the data is preserved in the transformation. The embodiment of this constraint is that 'distances' in the feature space should correspond as closely as possible to 'distances' in the data space. This is extra prior knowledge which is not as severe as demanding individually labelled target values, only a requirement that relative distances are preserved, if possible.

It is assumed that neighbourhood relations in the original data space are characterised by a user-specified 'dissimilarity matrix'. For instance, this could be defined by the relative interpoint Euclidean distances between data points, though it may be more general.

We seek a reduced-dimensional representation of the original data such that the relationship between the interpoint distances in the reduced feature space reflects the dissimilarity matrix. This aim is related to that of *Sammon mappings* (Sammon, 1969), in which a set of P data points in n dimensions is mapped on to P data points in two dimensions for data-visualisation purposes, such that the relative positions of the points in the two spaces are approximately preserved. In the Sammon-mapping case this is achieved by iteratively adjusting positions in two-dimensional space by gradient-descent techniques to try and minimise the error, or STRESS function

$$E = \sum_{p=1}^{P} \sum_{q<p} [d_n(p,q) - d_2(p,q)]^2, \tag{4.5}$$

where $d_n(p,q) = \|x_p - x_q\|$ is the distance between data points in the original space and $\{d_2(p,q) = \|y_p - y_q\|\}$ are the distances in the transformed space. The positions $\{y\}$ are iteratively adjusted to minimise the error.

The concept of generating a map from a matrix of distances is also central to the field of *multidimensional scaling* (Kruskal, 1964). This is a technique used widely in psychology and the social sciences, where it is desired to compute a configuration of points that reflect a set of *dissimilarities* between various stimuli. These measurements of dissimilarity are often derived from subjective human judgement, and the map is generated such that the interpoint distances in the map correspond to the measured dissimilarities. There are two main branches of multidimensional scaling. The first is *classical scaling*, and is equivalent to a principal component transformation if the dissimilarities correspond to a Euclidean distance matrix; this gives the optimal distance-preserving *linear* transformation. The second form is *nonmetric* or *ordinal* scaling, where, for reasons of psychological interpretation, the dissimilarities are an arbitrary monotonically increasing function of the distances in the map. In the latter method, map generation is again an iterative process, optimising a similar criterion to Sammon stress. The Sammon mapping is effectively a nonlinear metric scaling method, although its exact analogue does not exist in the multidimensional scaling field.

The main disadvantages of these techniques are that the required iterative optimisation process operates on the P coordinate vectors, $\{y_p\}$, involving $P \times m$ parameters, in m-dimensional feature space, and in addition it only provides a mapping of the 'training' data points; it does not provide a *transformation* from the input space to the feature space. In this sense the mapping techniques provide look-up tables without interpolative properties. There is no information indicating where a 'new' data point should transform to without having to optimise the full set of data points. The advantages of network methods include their *transformation* ability, such that new data may be projected into the newly derived feature space. The feedforward networks are used to try and find a *transformation* into a space in which the metric in the transformed space reflects the actual pairwise dissimilarities.

Therefore we seek a nonlinear transformation $\{f : \mathbb{R}^d \to \mathbb{R}^m : f(x) = y\}$ from the original data-configuration space that maps into the feature space. We choose the class of nonlinear parameterised transformations provided by radial basis function networks. This has the advantage that a *transformation* is now obtained, allowing interpolation to be considered, and in general a smaller number of parameters will have to be adapted. This is because, rather than iteratively adjusting the positions of the coordinates of all of the patterns in the training set directly in the transformation space, the positions are *indirectly* adjusted by modifying the parameters of the network. The network parameters are adjusted to minimise the global STRESS (4.5). The points y are generated by the radial basis function network, given the data points as input. That is, $y_q = f(x_q; \theta)$, where f is the nonlinear transformation effected by the radial basis function network with parameters, i.e. weights and any kernel smoothing factors, θ. The (squared) distances in the feature space may thus be given by

$$d_{qp}^2 = \| f(y_q) - f(y_p) \|^2$$
$$= \sum_{l=1}^{n} \left(\sum_{k} \lambda_{lk} \left[\phi_k(\|x_q - \mu_k\|) - \phi_k(\|x_p - \mu_k\|) \right] \right)^2.$$

The topographic nature of the transformation is imposed by the STRESS term which attempts to match the interpoint Euclidean distances in the feature space with those in the input space. This mapping is said to be *relatively supervised* because there is no specific target for each y_q; only a relative measure of target separation between each (y_q, y_p) pair is provided. In this form it does not take account of any additional information, such as class labels, that might be associated with the data points, but is determined strictly by their spatial distribution. However, the approach may be extended to incorporate the use of extra 'subjective' information which may be used to influence the transformation. Note that the objective function for the radial basis function network is no longer quadratic, and so a standard analytic matrix-inversion method for fixing the final layer weights cannot be employed.

Although any universal approximator may be exploited within this NEUROSCALE approach to topographic feature extraction, using a radial basis function network allows more theoretical analysis of the resulting behaviour, despite the fact that we have lost the usual linearity advantages of the radial basis function network. For instance, the NEUROSCALE architecture has the empirically observed property that the generalisation performance does not seem to depend critically on model-order complexity, contrary to intuition based upon knowledge of their supervised counterparts. This applies both to the number of centres used in the radial basis function network and the kernel smoothing factors, which themselves may be viewed as regularising hyperparameters in a feedforward supervised situation.

The robustness to model complexity is illustrated by Figure 6, which shows the training- and test-set performances on the standard IRIS dataset. The network had 5-45 basis functions, trained and tested on 45 separate patterns sampled from the dataset. To within acceptable deviations, the training- and test-set STRESS values are approximately constant. This behaviour is counterintuitive when compared with research on feedforward networks trained according to supervised approaches. We have observed this general trend on a variety of diverse real-world problems.

There are two fundamental causes of this observed behaviour. First, in contrast to supervised problems, in topographic feature extraction problems we have explicit

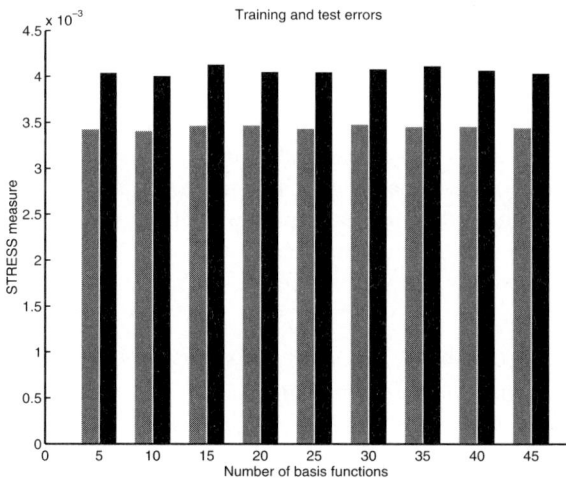

Fig. 6:   Training and test errors for NEUROSCALE radial basis functions as a function of the number of basis functions. Training errors are on the left, test errors are on the right.

prior knowledge that the transformation should be as smooth as possible. Hence there is an in-built regularising factor on the overall curvature of the model, as determined by the STRESS. Secondly, there is an appropriate regularising component implicitly incorporated in the training algorithm used to minimise the STRESS in the particular case of the radial basis function network. Details of this behaviour may be found elsewhere (Lowe and Tipping, 1997).

4.4   Summary

In this section we have considered the relevance of radial basis function networks for general *feature extraction*. The nature of the feature extraction process depends upon the amount of extra or prior knowledge available. For instance, in supervised classification problems it was observed that the optimum feature space was equivalent to a specific type of discriminant analysis which was a straightforward generalisation of results known in linear classification. In an unsupervised situation it was observed that the radial basis function network was an appropriate tool for extracting *nonlinear principal components*. In an intermediate 'relative supervision' situation, where constraints take the form of topographic requirements, the radial basis function was observed to be capable of extracting features in a manner reminiscent of multidimensional scaling.

5   Miscellaneous comments

Finally, we consider some interesting theoretical issues which might affect the application of radial basis function networks in real problems, especially in the context of the choice between local and nonlocal basis functions.

Conventional wisdom assumes that radial basis function network models are composed of basis functions with specific probabilistic qualities, i.e. that the basis functions themselves should be localised, normalisable and everywhere positive. In this section we raise a few doubts about this, and in particular indicate with some mathematical justification why it may be advantageous to consider basis functions which do not decay away from the centre, which are not normalisable, and which are not even necessarily positive definite, even in situations in which we are considering approximating probability density functions.

The first set of results is taken from the mathematical literature on function approximation.

Theorem 5.1   *(Powell, 1990) If the centres of an interpolating radial basis function network $g(x)$ with nonlinear basis functions $\phi(\|z\|)$ form an infinite regular grid then we can achieve $g(x) = f(x)$, where $f(x)$ is a low-order polynomial, only if*

$$\int \phi(\|z\|)dz$$

is unbounded.

The implication of this is that there are types of problem in which one should expect unbounded basis functions to have better approximation properties than bounded basis functions.

Also the following result, accredited to Brown (1980) and discussed in Powell (1990), indicates that nonlocalised basis functions have better approximation properties on bounded domains.

Theorem 5.2 *If $f : D \rightarrow \mathbb{R}$ is a continuous function which maps a closed bounded submanifold D of $\mathbb{R}^d$ into $\mathbb{R}$ then, irrespective of the form of $f(x)$, there exists a finite radial basis function network with unbounded basis functions that arbitrarily closely approximates $f(x)$:*

$$|f(x) - \sum_j \lambda_j \phi(\|x - c_j\|)| < \epsilon.$$

Note that this implies that, even using nonlocal basis functions, it is possible to choose the parameters of the network such that the network as a whole has a localised response on the bounded submanifold of interest. However, nothing is indicated about the behaviour outside this submanifold, where the network is effectively extrapolating beyond its region of validity. Another point of note in this theorem is that it is a rare result, in that it concerns the approximation behaviour of a *finite* radial basis function architecture.

There are also several results in the traditional density estimation literature (Hand, 1982) which are of relevance to radial basis function networks.

Consider first of all some results from kernel function models. Specifically, if we consider the finite sample bias properties of kernel estimators, then we are led to the conclusion that restricting the basis function to be positive definite has a consequence of restricting the estimator to be inevitably biased.

The next theorem indicates that if the estimator as a whole is globally positive then we are obliged, generally, to accept a certain level of bias of the estimator.

Theorem 5.3 *(Rosenblatt, 1956) Let $X_1, \ldots, X_n$ be i.i.d. random variables with continuous density $f(x)$. If $g(x)$ is an estimate of $f(x)$ symmetric in $X_1, \ldots, X_n$ and is jointly Borel measurable in $X_1, \ldots, X_n$ such that $g(x) > 0$, then $g(x; X_1, \ldots, X_n)$ is not an unbiased estimator of $f(x)$ for all x.*

In other words, there is no nonnegative estimator which is unbiased for all $f(x)$. This is a result concerning the ability to find unbiased estimators over the whole class of continuous functions.

Alternatively one could consider a *specific* class of estimators in which we assume that the individual kernel functions are themselves valid probability density functions, and examine the properties of the resulting estimator.

Theorem 5.4 *(Yamato, 1972) Let $X_1, \ldots, X_n$ be i.i.d. random variables with continuous density $f(x)$. Let $\phi(z)$ be a measurable function satisfying $\phi(z) > 0 \; \forall \, z$ and $\int \phi(z)dz = 1$. Then the kernel estimator $g(x) = \int \phi(x - y)f(y)dy$ is biased: $E[g(x)] \neq f(x)$ for at least some x.*

Therefore, for finite samples and restricting ourselves to positive definite kernel functions, the resulting density estimators are necessarily biased.

These are results based on finite sample approximations. They tell us nothing regarding how the bias might decrease as we tend towards asymptotic behaviour. However, results due to Shapiro (1969) and Nadaraya (1974) indicate that the rate of convergence of the bias is improved if we relax the requirement that the kernel function should be positive everywhere. In particular, the rate of convergence of the bias can be improved if we permit $\int y^r \phi(y) dy = 0$ for $r = 2, 3 \ldots$. Similarly the rate of convergence of the variance is restricted by $\lim_{n \to \infty} n \text{Var} g(x) = f(x) \int \phi(y)^2 dy$. For consistency, both the bias and variance should decrease asymptotically. The mean integrated square error combines both of these criteria, and it has been demonstrated by Nadaraya that the rate of convergence of the error is improved if we allow the kernel functions to take on negative values as well.

Hence if we combine the theoretical results from function approximation and kernel density estimation we conclude that the widespread practice of using Gaussian basis functions in radial basis function networks may not be such a good strategy. It has been known for some time that Gaussian basis functions are unsuitable for function approximation and empirically this is manifest in weaker generalisation performance when compared to, say, models based on nonlocal (and non-positive-definite) spline basis functions.

When performing density estimation, for example, it might be convenient to try and impose 'desirable' properties on the radial basis function estimator, such as having a model which produces normalisable and positive definite outputs. Imposition of these requirements on the individual basis functions themselves is one way to guarantee that the whole model retains these properties. However this apparently correct procedure is naïve, in that it is a constraint with a consequence: the estimator is doomed to be biased. This consequence does not seem to be appreciated in the neural network domain.

6 Conclusion

This chapter has outlined some theoretical relationships between radial basis function neural networks and traditional statistical methodologies. The fact that the radial basis function network relates to several diverse areas is encouraging in that the model can act as a focus to integrate knowledge from various domains. The neural network community should benefit by exploiting background knowledge derived from diverse statistical areas such as regression smoothers and density estimation. Similarly, the wide range of successful applications of the radial basis function network and the extensions to the traditional approaches that the network offers should act as a motivation to the statistics community to develop existing techniques.

7 References

Asoh, H. and Otsu, N. (1990). An approximation of nonlinear discriminant analysis by multilayer networks. *Proc. Int. Joint Conference on Neural Networks*, III, 211–216. IEEE, Piscataway, New Jersey.

Bashkerov, O.A., Braverman, E.M. and Muchnik, I.B. (1964). Potential function algorithms for pattern recognition learning machines. *Automation and Remote Control*. **25**, 629–631.

Baum, L.E., Petrie, T., Soules, G. and Weiss, N. (1970). A maximization technique in the statistical analysis of probabilistic functions of Markov chains. *Ann. Math. Statist.* **41**, 164–171.

Broomhead, D.S. and Lowe, D. (1988). Multi-variable functional interpolation and adaptive networks. *Complex Systems* **2**, 321–355.

Dempster, A.P., Laird, N.M. and Rubin, D.B. (1977). Maximum likelihood from incomplete data via the EM algorithm (with discussion). *J. R. Statist. Soc.* B **39**, 1–38.

Fukunaga, K. (1972). *Introduction to Statistical Pattern Recognition*. Academic Press, New York.

Gallinari, P., Thiria, S. and Fogelman Soulie, F. (1988). Multilayer perceptrons and data analysis. *IEEE Annual Conference on Artificial Neural Networks*, I, 391–399. SOS Printing. IEEE, New York.

Girosi, F., Jones, M. and Poggio, T. (1995). Regularization theory and neural networks architectures. *Neural Computation* **7**, 219–269.

Hand, D. (1982). *Kernel Discriminant Analysis*. Research Studies Press, John Wiley and Sons, Chrichester.

Hartman, E.J. and Keeler, J.D. (1990). Layered neural networks with Gaussian hidden units as universal approximations. *Neural Computation* **2**, 210–215.

Hastie, T.J. and Stuetzle, W. (1989). Principal curves. *J. Am. Statist. Assoc.* **84**, 502–516.

Hastie, T.J. and Tibshirani, R.J. (1990). *Generalized Additive Models*. Chapman and Hall, London.

Kruskal, J.B. (1964). Nonlinear multidimensional scaling: a numerical method. *Psychometrika* **29**, 115–130.

Lowe, D. (1989). Adaptive Radial Basis Function nonlinearities and the problem of generalisation. *1st IEE International Conference on Artificial Neural Networks, Conference Publication number 313*, pp. 171–175. Institute of Electrical Engineers, London.

Lowe, D. (1991). On the iterative inversion of RBF networks: a statistical interpretation. *2nd IEE International Conference on Artificial Neural Networks, Conference Publication number 349*, pp. 29–33. Institute of Electrical Engineers, London.

Lowe, D. (1993). Novel 'topographic' nonlinear feature extraction using Radial Basis Functions for concentration coding in the 'artificial nose'. *3rd IEE International Conference on Artificial Neural Networks, Conference Publication number 372*, pp. 95–99. Institute of Electrical Engineers, London.

Lowe, D. and Tipping, M.E. (1996). Feed-forward neural networks and topographic mappings for exploratory data analysis. *Neural Computing Applics.* **4**, 83–95.

Lowe, D. and Tipping, M.E. (1997). NeuroScale: novel topographic feature extraction using RBF networks. *Proc. Neural Information Processing Conference*, 1996, MIT Press, Cambridge, MA.

Lowe, D. and Webb, A.R. (1990). Exploiting prior knowledge in network optimisation: an illustration from medical prognosis. *Network: Computation in Neural Systems* **1**, 299–323.

Lowe, D. and Webb, A.R. (1991). Optimized feature extraction and the Bayes decision in feed-forward classifier networks. *IEEE Trans. Pattern Anal. Machine Intell.* **13**, 355–364.

Mao, J. and Jain, A.K. (1995). Artificial neural networks for feature extraction and multivariate data projection. *IEEE Trans. Neural Networks* **6**, 296–317.

McLachlan, G.J. and Basford, K.E. (1988). *Mixture Models: Inference and Applications to Clustering.* Marcel Dekker Inc., New York.

Micchelli, C.A. (1986). Interpolation of scattered data: distance matrices and conditionally positive definite functions. *Constructive Approximation* **2**, 11–22.

Moody, J. and Darken, C. (1989). Fast learning in networks of locally tuned processing units. *Neural Computation* **1**, 281–294.

Nadaraya, E. (1974). The integral mean square error of certain non-parametric estimates of a probability density. *Teor. Verojatnost. i Primenen* **19**, 131–139.

Nilsson, N.J. (1965). *Learning Machines.* McGraw-Hill Inc., New York.

Park, J. and Sandberg, I.W. (1991). Universal approximation using Radial Basis Function networks. *Neural Computation* **3**, 246–257.

Parzen, E. (1962). On estimation of a probability density function and mode. *Ann. Math. Statist.* **33**, 1065–1076.

Poggio, T. and Girosi, F. (1990a). Regularization algorithms for learning that are equivalent to multilayer networks. *Science* **247**, 978–982.

Poggio, T. and Girosi, F. (1990b). Networks for approximation and learning. *Proc. IEEE* **78**, 1481–1497.

Powell, M.J.D. (1985). Radial basis functions for multivariable interpolation: A review, *IMA Conference on Algorithms for the Approximation of Functions and Data, RMCS Shrivenham.*

Powell, M.J.D. (1987). Radial basis functions for multivariable interpolation: A review. In *Algorithms for Approximation*, Eds. J.C. Mason and M.G. Cox, pp. 143–167. Oxford University Press.

Powell, M.J.D. (1990). The theory of Radial Basis Function approximation in 1990. *Preprint 1990/NA11.* Dept. Applied Mathematics and Theoretical Physics, University of Cambridge.

Robbins, H. and Monro, S. (1951). A stochastic approximation method. *Ann. Math. Statist.* **22**, 400–407.

Rosenblatt, M. (1956). Remarks on some non-parametric estimates of a density function. *Ann. Math. Statist.* **27**, 832–837.

Sammon, Jr., J.R. (1969). A nonlinear mapping algorithm for data structure analysis. *IEEE Trans. Computers* **18**, 401–409.

Shapiro, J.S. (1969). *Smoothing and Approximation of Functions.* Van Nostrand Reinhold, New York.

Silverman, B.W. (1986). *Density Estimation for Statistics and Data Analysis.* Chapman and Hall, London.

Specht, D.F. (1990). Probabilistic neural networks. *Neural Networks* **3**, 109–118.

Tikhonov, A.N. and Arsenin, V.Y. (1977). *Solutions of Ill–posed Problems.* V.H. Winston and Sons, Washington D.C.

Tråvèn, H.G.C. (1991). A neural network approach to statistical pattern classification by 'semiparametric' estimation of probability density functions. *IEEE Trans. Neural Networks* **2**, 366–377.

Wahba, G. (1990). *Spline Models for Observational Data*. CBMS 59. SIAM, Philadelphia.

Webb, A.R. (1995). Multidimensional scaling by iterative majorization using Radial Basis Functions. *Pattern Recognition* **28**, 753–759.

Webb, A.R. (1996). An approach to non-linear principal component analysis using radially symmetric kernel functions. *Statist. Comp.* **6**, 159–168.

Webb, A.R. and Lowe, D. (1990). The optimised internal representation of multilayer classifier networks performs nonlinear discriminant analysis. *Neural Networks* **3**, 367–375.

Yamato, H. (1972). Some statistical properties of estimators of density and distribution functions. *Bull. Math. Statist.* **19**, 113–131.

4

Robust Prediction in Many-parameter Models

Nathan Intrator

School of Mathematical Sciences, University of Tel-Aviv

http://www.math.tau.ac.il/~nin

Abstract

When parameter estimation is based on a set of training patterns with size that is of the order of the number of free parameters, estimation may become very unreliable. Until recently, estimation in such cases had sounded unrealistic, but it has now been accepted that such estimation is possible under certain conditions. For example, linear discriminant analysis (Fisher, 1936) requires some adjustment when the input dimensionality is large (Buckheit and Donoho, 1995) to account for the added variability of the covariance matrices. In simple terms, innovative ways to reduce the variance portion of the error are required, as well as methods to impose (reasonable) bias.

One fundamental assumption concerns the 'true' dimensionality of the data. Although the data may be represented in a high-dimensional space, it is often the case that the actual dimensionality is much smaller. Clearly, a dimensionality reduction method which does not 'throw the baby out with the bath water', but retains important information in the data, leads to a smaller number of parameters for the required prediction and thus may be more robust. This chapter reviews several ways to robustify estimation and prediction in many-parameter models. We start with a short review of the bias/variance dilemma and then address separately ways to control the variance portion of the error and the bias. A central issue in variance control methods is that of ensemble averaging. We shall expand on this to present methods for extracting additional information from ensembles of experts by exploiting the variability in responses for estimating the confidence of a certain group of experts. The control or introduction of bias into the model leads naturally to dimensionality reduction. We make this connection and discuss methods for hybrid dimension reduction and classification, and we spend some time on dimension reduction methods. Throughout, we demonstrate the methods with real-world applications.

1   The variance–bias dilemma

The motivation of our approach follows from a key observation regarding the bias–variance decomposition, namely the fact that ensemble averaging does not affect the bias portion of the error, but reduces the variance, when the estimators on which averaging is done are independent.

The classification problem is to estimate a function $f_{\mathcal{D}}(x)$ of observed data characteristics x, for predicting a class label y, based on a given training set $\mathcal{D} = \{(x_1, y_1), \ldots, (x_L, y_L)\}$, using some measure of the estimation error on $\mathcal{D}$. A good estimator will perform well not only on the training set, but also on new *validation* sets which were not used during estimation.

Evaluation of the performance of the estimator is commonly done via the mean squared error distance (MSE) by taking the expectation with respect to the (unknown) probability distribution P of y:

$$E[(y - f_{\mathcal{D}}(x))^2 | x, \mathcal{D}].$$

This can be decomposed into

$$E[(y - f_{\mathcal{D}}(x))^2 | x, \mathcal{D}] = E[(y - E[y|x])^2 | x, \mathcal{D}] + E[(f_{\mathcal{D}}(x) - E[y|x])^2].$$

The first term typically depends on neither the training data $\mathcal{D}$ nor the estimator $f_{\mathcal{D}}(x)$; it measures the amount of noise or variability of y given x. Hence f can be evaluated using

$$E[(f_{\mathcal{D}}(x) - E[y|x])^2].$$

The empirical mean squared error of f is given by

$$E_{\mathcal{D}}[(f_{\mathcal{D}}(x) - E[y|x])^2],$$

where $E_{\mathcal{D}}$ represents expectation with respect to all possible training sets $\mathcal{D}$ of fixed size.

To investigate further the MSE performance we decompose the error into bias and variance components (Geman *et al.*, 1992) to obtain

$$E_{\mathcal{D}}[(f_{\mathcal{D}}(x) - E[y|x])^2] = (E_{\mathcal{D}}[f_{\mathcal{D}}(x)] - E[y|x])^2$$
$$+ E_{\mathcal{D}}[(f_{\mathcal{D}}(x) - E_{\mathcal{D}}[f_{\mathcal{D}}(x)])^2]. \qquad (1.1)$$

The first term on the right-hand side is called the bias term (strictly, the squared bias) of the estimator and the second term is called the variance term. When training on a fixed training set $\mathcal{D}$, reducing the bias with respect to this set may increase the variance of the estimator and contribute to poor generalisation performance. This is known as the trade-off between variance and bias. Typically, variance is reduced by smoothing, but this may introduce bias since, for example, it may blur sharp peaks. Bias is reduced by incorporating prior knowledge. When prior knowledge is used also for smoothing, it is likely to reduce the overall MSE of the estimator.

When training neural networks such as multilayer perceptrons, the variance arises from two terms. The first term comes from inherent data randomness and the second term arises from the nonidentifiability of the model, in that, for a given training dataset, there may be several local minima of the error surface.

Consider the ensemble average $\bar{f}$ of Q predictors, which in our case can be thought of as neural networks with different random initial weights which are trained on data with added Gaussian noise:

$$\bar{f}(x) = \frac{1}{Q} \sum_{i=1}^{Q} f_i(x).$$

These predictors are identically distributed, and thus the variance contribution to equation (1.1) becomes

$$E[(\bar{f} - E[\bar{f}])^2] = E\left[\left(\frac{1}{Q}\sum f_i - E\left[\frac{1}{Q}\sum f_i\right]\right)^2\right]$$

$$= E\left[\left(\frac{1}{Q}\sum f_i\right)^2\right] + \left(E\left[\frac{1}{Q}\sum f_i\right]\right)^2$$

$$- 2E\left[\frac{1}{Q}\sum f_i E\left[\frac{1}{Q}\sum f_i\right]\right]$$

$$= E\left[\left(\frac{1}{Q}\sum f_i\right)^2\right] - \left(E\left[\frac{1}{Q}\sum f_i\right]\right)^2; \qquad (1.2)$$

we omit mention of x and $\mathcal{D}$ for clarity. The first term in (1.2) can be rewritten as

$$E\left[\left(\frac{1}{Q}\sum f_i\right)^2\right] = \frac{1}{Q^2}\sum E\left[f_i^2\right] + \frac{2}{Q^2}\sum_{i<j} E[f_i f_j],$$

and the second term gives

$$\left(E\left[\frac{1}{Q}\sum f_i\right]\right)^2 = \frac{1}{Q^2}\sum(E[f_i])^2 + \frac{2}{Q^2}\sum_{i<j} E[f_i]E[f_j].$$

Plugging these equalities into (1.2) gives

$$E[(\bar{f} - E[\bar{f}])^2] = \frac{1}{Q^2}\sum\left\{E\left[f_i^2\right] - (E[f_i])^2\right\}$$

$$= \frac{2}{Q^2}\sum_{i<j}\{E[f_i f_j] - E[f_i]E[f_j]\}. \qquad (1.3)$$

Set

$$\gamma = \text{Var}(f_i) + (Q - 1)\max_{i,j}(E[f_i f_j] - E[f_i]E[f_j]).$$

It follows that[1]

$$\frac{1}{Q}\text{Var}(f_i) \leq \text{Var}(\bar{f}) \leq \frac{1}{Q}\gamma \leq \max_i \text{Var}(f_i). \qquad (1.4)$$

---

[1] We use the fact that $ab \leq (a^2 + b^2)/2$. Thus, $E[f_i f_j] - E[f_i]E[f_j] = E(\{f_i - E[f_i]\}\{f_j - E[f_j]\}) \leq \max_i \text{Var}(f_i)$.

This analysis suggests a simple extrapolation to large values of Q by giving an upper bound of $1/Q\gamma$ to the variance behaviour under large network ensembles from small-size ensembles (Naftaly *et al.*, 1997).

Note that

$$E[f_i f_j] - E[f_i]E[f_j] = E\Big(\{f_i - E[f_i]\}\{f_j - E[f_j]\}\Big).$$

Thus, the notion of independence can be understood as independence of the deviations of each predictor from the expected value of the predictor, which can be replaced, because of linearity, by

$$E\Big(\{f_i - E[\bar{f}]\}\{f_j - E[\bar{f}]\}\Big),$$

and is thus interpreted as an indication of the prediction variation around a common mean.

2 Variance control via various ensemble averaging methods

The success of ensemble averaging of neural networks in the past (Hansen and Salamon, 1990; Wolpert, 1992; Breiman, 1996; Perrone, 1993) is due to the fact that error surfaces used to estimate the weights (parameters) of neural networks have in general many local minima, and thus, even with the same training set, different local minima are found when starting from different random initial estimates. These different local minima lead to somewhat independent predictors, and thus the averaging can reduce the variance. When a larger set of independent networks is needed but no more data are available, data re-use methods can be of help. Bootstrapping (Breiman, 1996) has been very helpful, since, by resampling from the training data without replacement, the degree of independence among the training sets, and hence among the resulting sets of estimators, is increased, leading to improved ensemble results. The smoothed bootstrap (Efron and Tibshirani, 1993) is potentially more useful since a wider range of sets of independent training samples can be generated. The smoothed bootstrap approach amounts to generating larger datasets by simulating the *true* noise in the data.

The next section demonstrates what can be done with networks trained on the same data, and the following section demonstrates ways to increase the independence between networks and associated considerations.

2.1 Exhaustive training

The analysis presented in Section 1 suggests that training an ensemble of predictors should be done in a different way from training single predictors. This is because an ensemble of predictors has a lower overall bias, and thus the optimal trade-off between variance and bias should correspond to a lower level of bias so as to balance the contribution to error from the lower variance. In neural networks, trained by methods such as iterative gradient descent, this is achieved by stopping the training process at a later point, thus overfitting the individual training dataset somewhat, thereby leading to lower bias but higher variance. Naftaly *et al.* (1997) have recently demonstrated the effect of training on single and ensemble errors, using the well-known sunspots data (Murphy and Aha, 1992). After giving some technical details, we briefly review their results. For full details and comparison with other approaches see Naftaly *et al.* (1997). As in Weigend *et al.* (1990), feedforward, simple-recurrent (Elman and Zipser, 1988) networks with four sigmoidal hidden units were used with data from twelve consecutive time points as inputs. In other words, the neural network was used as a nonlinear predictor of the number of sunspots in a given month, using the data from the twelve previous months as covariates/predictors. The prediction error was measured according to the average relative variance (ARV). Training was done via the error backpropagation algorithm (Rumelhart *et al.*, 1986) with a fixed learning rate of 0.003. A validation set containing 35 randomly chosen points was left out during training to serve for performance validation. The only difference between the members of the ensemble of networks was that different sets of weights were used to initialise the training algorithm.

2.2 Estimating the variance reduction for large ensembles

Naftaly *et al.* (1997) presented a very simple way of estimating what the variance portion of the MSE will be for large ensemble sizes, based on the performance of small ensemble sizes: see Section 1. Figure 1 depicts results from Naftaly *et al.* (1997) based on the validation set (left) and on the test set (right). Values of ARV are shown as a function of the number of training epochs. The highest curve in each figure corresponds to $Q = 1$, i.e. the case of single networks. Below it appear the curves of $Q = 2, 4, 10, 20$, followed by the limiting case $Q \to \infty$. The extrapolation over ensemble size is demonstrated in Figure 2, where ARV values obtained for training times $t = 70$ and $t = 140$ kilo-epochs (KE) for the test set are depicted as a function of $1/Q$. It is quite clear that a linear extrapolation is very satisfactory.

The amount of variance due to initial conditions can be found by subtracting the $Q \to \infty$ result from that for $Q = 1$ (Figure 2). Although this variance is due to the choice of initialising estimates and not to different training sets, it does increase with time, as would be expected from the variance due to different training sets.

2.3 Noise injection

In the previous section we have emphasised the reduction in the variance portion of the error resulting from ensemble averaging. This reduction was a result of the independence between the errors made by different predictors. While the use of initial random weights may lead to some independence, training on different datasets improves the independence even further. Bootstrapping (Breiman, 1996) is most appropriate for

ARV as a function of ensemble size and training time

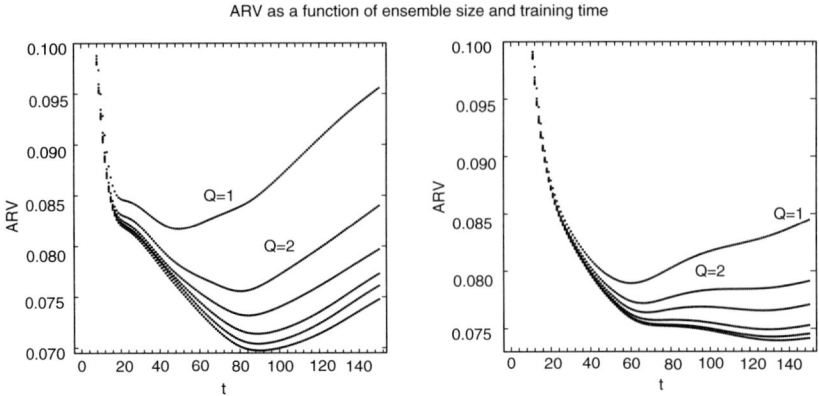

Fig. 1:   ARV vs. training time in kilo-epochs (KE) for the sunspots data (from Naftaly *et al.*
(1997)). The curves are shown for different choices of ensemble sizes: $Q = 1, 2, 4, 10, 20$
from top to bottom. The lowest curve is the extrapolation to $Q \to \infty$ (see Naftaly *et al.*
(1997).) *Left:* Results for a cross-validation set of 35 points. *Right:* Results on a test set.
Note the difference in minimum error between a single predictor and an ensemble and the
shallower curve as Q, the ensemble size, increases.

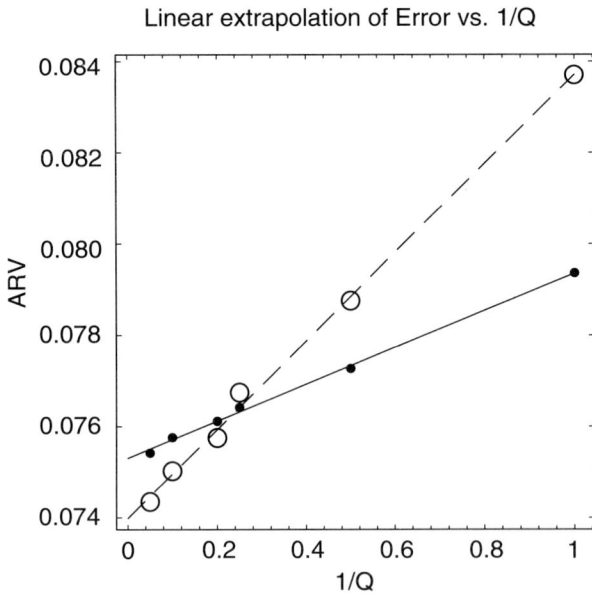

Fig. 2:   Extrapolation method used for extracting the $Q \to \infty$ prediction (from Naftaly
et al. (1997)). The results for different Q at two different training periods, $t = 70$ KE (top)
and 140 KE (bottom), can be extrapolated by a linear regression in $1/Q$.

small datasets since, by resampling with replacement from the training data, the independence between the training sets is increased, yet each predictor has more than a $1/Q$ fraction of the training set on which to train. The smoothed bootstrap (Efron and Tibshirani, 1993) is potentially more useful since larger sets of independent training samples can be generated. The smoothed bootstrap approach amounts to generating larger datasets by simulating the *true* noise in the data. In this section, we demonstrate that training with added noise that is more variable than the true data-noise, and which increases the variance of individual predictors, can still be helpful as it also increases the degree of independence between the training sets and, therefore, the degree of independence between predictors. A simple bootstrap procedure amounts to sampling with replacement from the training data and constructing several training sets, all of the same size as the original training set. Later, the variability between the sets of estimated parameters can be measured, and gives some indication about the true variability of the estimates of model parameters computed from the original data. Furthermore, variability or error-bars of the predictions can also be estimated in this way.

One version of the bootstrap involves estimation of a model of the form

$$y = f(x) + \epsilon \qquad (2.1)$$

for some parametric family to which f belongs, and a noise variable ϵ which is assumed to have small variance and zero mean. An estimate $\hat{f}$ of f is obtained from n training samples, leading to fitted residuals $\hat{\epsilon} = (\hat{\epsilon}_1, \ldots, \hat{\epsilon}_n)$. One can then sample n times with replacement from the ϵ_i, giving $\epsilon^* = (\epsilon_1^*, \ldots, \epsilon_n^*)$, and construct new samples of the form (x_i, y_i^*), in which ϵ_i is replaced by ϵ_i^* sampled from the above set. Clearly, this approach can be easily extended to a smoothed bootstrap version (Efron and Tibshirani, 1993). In such a case, one can increase the size of each bootstrap set, since because of the noise the different sets are sufficiently independent. It should be noted that, if $\hat{f}$ is biased, the noise variance may be over-estimated.

For classification problems, the form

$$y = f(x + \epsilon) \qquad (2.2)$$

may be more appropriate. In this case, applying noise injection to the inputs during training, one can improve the generalisation properties of the estimator (Sietsma and Dow, 1991).

Recently, Bishop has shown that training with small amounts of noise is locally equivalent to regularisation (Bishop, 1995). Here, we give a different interpretation of the addition of noise to the inputs during training, and view it as a regularising parameter that controls, in conjunction with ensemble averaging, the capacity and the smoothness of the estimator. The major role of this noise is to push different estimators to different local minima and thereby produce a more independent set of estimators. Best performance is then achieved by averaging the estimators. For this regularisation, the level of the noise may be larger than the 'true' level which can be indirectly estimated. Since we want to study the effect of bootstrapping with noise on

---

The BEN algorithm
- Let $\{(x_i, y_i)\}$ be a set of training patterns for $i = 1, \ldots, N$.
- Let $\epsilon = \{\epsilon_1, \ldots, \epsilon_J\}$.
- Let $\lambda = \{\lambda_1, \ldots, \lambda_I\}$.
- For a noise level ϵ_j estimate an optimal penalty term for weight decay λ_i:

    * Fix a size K for the bootstrap sample, such that $K \gg N$ (we used $K = 10N$).
    * Let $s_1, s_2, \ldots, s_K$ be a set of indices, each chosen independently at random from $\{1, \ldots, N\}$.
    * Create a noisy bootstrap resample of the training set inputs, of the form $\{x_{s_i} + \zeta_i\}_{i=1,\ldots,K}$ and the corresponding resampled outputs $\{y_{s_i}\}_{i=1,\ldots,K}$, where ζ_i is a vector whose components are $N(0, \epsilon_j^2)$.
    * Train several networks with the noisy samples using weight decay parameters $\lambda_1, \ldots, \lambda_I$.
    * Generate an ensemble average of the resulting set of networks.
    * Choose the optimal weight decay parameter λ via cross-validation or a test set.

- Repeat the process for a new choice of noise ϵ_j until there is no improvement in prediction.

---

the smoothness of the estimator, separately from the task of input noise estimation, we consider a highly nonlinear, noise-free classification problem, and show that, even in this extreme case, addition of noise during training improves results significantly.

We choose a problem that is very difficult for feedforward neural networks to deal with. It is difficult because of the highly nonlinear nature of the decision boundaries, and the fact that these nonlinearities are easier to represent in terms of local radially symmetric functions rather than ridge functions such as those given by feedforward sigmoidal functions. Since the training data are given with no noise, it seems unreasonable to train a network with noise, but we show that, even in this case, training with noise is a very effective approach for smoothing the estimator.

In the bootstrap ensemble with noise (BEN) (Raviv and Intrator, 1996), we push the idea of noise injection further. We observe that adding noise to the inputs increases the first term on the right-hand side of (1.3), in that it adds variance to each estimator, but it decreases the contribution of the second term as it increases the degree of independence between estimators. Instead of using the 'true' noise, estimated from the data, for the bootstrap, we seek an optimal noise level which gives smallest contribution to the error from the sum of the two components of the variance. It is impossible to calculate the optimal variance of the Gaussian noise without knowing f explicitly, and therefore the value of this variance remains a regularisation term, a parameter which has to be estimated so as to minimise the total contribution of the variance term to the total error measure. Furthermore, since the injection of noise increases the degree of independence between different training sets, we can use bootstrap samples that are larger than the original training set. This does not affect the bias, if the noise is symmetric around zero, but it can reduce the variance. Note that the bias contribution to the error is not affected by introducing the ensemble-average estimator, because of the linearity of the expectation operator.

It follows that the BEN approach has the potential to reduce the contribution of the variance term to the total error. We should therefore seek a different level of trade-off between the contributions of variance and bias. In other words, we are able to use large (unbiased) networks without being affected by the large variance associated with such networks. This observation implies that the estimation of optimal noise levels should not be based on the performance of a single estimator, but rather based on the ensemble performance. The large variance of each single network in the ensemble can be tempered with a regularisation term such as weight decay (Krogh and Hertz, 1992; Ripley, 1996), but again the estimation of the optimal regularisation factor should be based on the ensemble-averaged performance. Breiman (1996) and Ripley (1996) present compelling empirical evidence of the importance of weight decay as a single network stabiliser. Our results confirm this fact under the BEN model. When no noise is injected, it was found that the nonregularised ensembles perform better (Taniguchi and Tresp, 1997).

In the simplest version, the same noise level is used for each input dimension. This is suitable for problems in which all of the inputs are on the same scale, or, more precisely, when the noise distributions are similar in the different data dimensions. When all covariates have the same interpretation, e.g. similar measurements taken at different time steps, or when dealing with pixel data, such a noise assumption is adequate. However, when the noise is nonhomogeneous in space or has a nondiagonal covariance matrix, or when different dimensions represent completely different measurements, it is best to estimate the different noise levels in each dimension separately. When this is too costly, or there are insufficient data for robust estimation, a quick solution is to sphere the data, that is, transform the variables so that the sample covariance matrix is the identity matrix.

2.3.1   *The two-spirals problem*

The following *two-spirals* problem was chosen for the demonstration of the influence of added noise because (a) it is a hard problem for backpropagation networks because of its intrinsic high nonlinearity and radial symmetry, (b) it is a noise-free problem, and (c) the generalisation performance of different predictors can be easily visualised in two dimensions. This problem consists of a training set of 195 points, half of which are to produce an output value of 1 and half an output of 0. These training points are arranged in two interlocking spirals that go around the origin three times, as shown in Figure 3. The example was first proposed by Alexis Wieland of MITRE Corp. It is easy to see that the two sets of points in the spirals cannot be separated by a combination of a small number of linear separators. Lang and Witbrock (1988)

Original and noisy Two-Spirals data

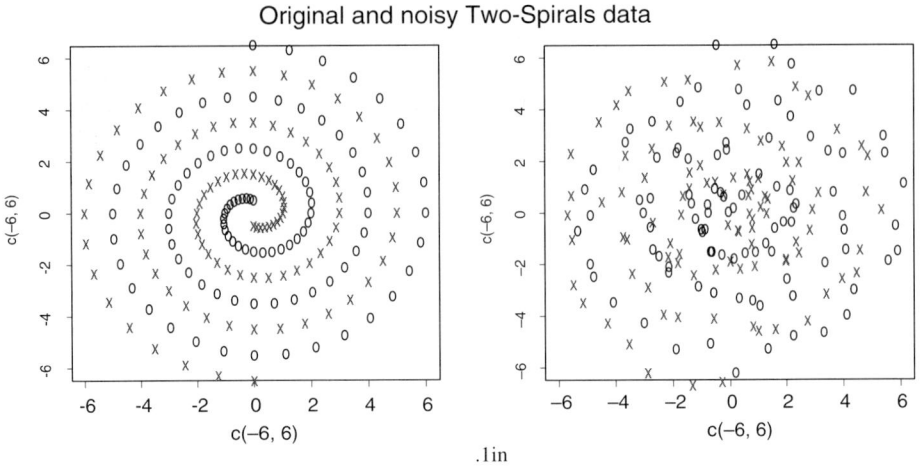

Fig. 3: The two-spirals training data (left); training points with noise, SD = 0.34 (right). As can be seen, the noise level that contaminates the data causes objects to cross the virtual boundary defined by the data, namely the noise leads to wrong class labelling for the training data. This reduces performance of single predictors, but the added independence between the predictors leads to improved ensemble performance.

proposed a $2 - 5 - 5 - 5 - 1$ multilayer network with logistic sigmoid activation functions and 'short-cuts' using 138 weights. They used a variant of the quickprop learning algorithm (Fahlman, 1988) with weight decay. They claimed that the problem could not be solved with simpler architecture involving fewer layers or without short-cuts. Their resulting predictor applied to the original dataset seems to give poor generalisation results. Baum and Lang (1991) demonstrated that there are many sets of weights that would cause a $2 - 50 - 1$ network to be consistent with the training set, but when such a network was trained with error backpropagation, starting with random initial weights, none of these desirable solutions was found. Fahlman and Lebiere (1990) used the cascade–correlation architecture for this problem, obtaining better results, but still little 'spiralness'. Recently, Deffuant (1995) suggested the 'perceptron membrane' method that uses piecewise linear surfaces as discriminators, and applied it to the spiral problem. He used 29 perceptrons but had difficulties capturing the structure of the spirals because of the piecewise linearity of his decision boundaries.

The minimisation criterion dictating the training process is mean squared error with weight decay regularisation:

$$E = \sum_p |t_p - y_p|^2 + \lambda \cdot \sum_{i,j} w_{i,j}^2, \qquad (2.3)$$

where t_p is the target and y_p the output for the pth example pattern. The $w_{i,j}$ are the weights and λ is a parameter that controls the amount of weight decay regularisation. This minimisation was achieved using Ripley's S-Plus 'nnet' package (Ripley, 1996), which implements backpropagation, along with our code for boosting, noise injection and ensemble averaging. The network had two inputs, 30 hidden units and one output. The first and last layers were fully connected to the hidden layer, giving a total of 121 weights. The activation function at the hidden and output units was the logistic sigmoidal function. The initialising weights were chosen randomly from $U(-0.7, 0.7)$. It should be noted here that, although we are training 5–40 networks, the effective number of parameters is not more, and is probably even less, than the number of parameters for a single network. This is because we do not have the flexibility to estimate an optimal combination of predictors, but rather take the simple average of them.

Baseline results were obtained by training 40 networks without any regularisation. We derived then an average predictor whose output is the mean of the outputs of the 40 nets (Figure 5, top left). The predictor had no smoothness constraint and therefore found relatively linear boundaries.

2.3.2 *Applying bootstrap to networks with weight decay*

The best results (Raviv and Intrator, 1996) were obtained when applying the bootstrap ensemble with noise (BEN) method to networks with optimal weight-decay regularisation ($\lambda = 3e^{-4}$). Figure 4 demonstrates the effect of bootstrap with noise on the performance of a five-net ensemble trained with optimal weight decay. Each image is a thresholded output of a five-net ensemble average predictor. Noise level goes from $\epsilon = 0$ in the upper left image through $\epsilon = 0.8$ in the lower right. The classification results are drawn on a uniform grid of 100×100 points (representing therefore a much larger test set) so as to get a clear view of the classification boundaries defined by the classifier. The effect of ensemble averaging over networks that were trained with different random initial conditions only is demonstrated in the top left image, which represents the case of adding no noise during training. It can be seen that, for small noise levels ϵ, the ensemble average predictor is unable to find any smooth structure in the data and merely over-fits the training data. For moderate levels of the noise, better structure can be found, and, for large levels of the noise, the data are so badly corrupted that again no structure can be found. The optimal noise standard deviation, σ, was around $\sigma = 0.35$. Optimal noise values are similar to those obtained when training with no weight decay, and are surprisingly high (see Figure 3 (right) for the result of adding noise to the data). Although the results look better than those obtained with no weight decay, in the sense that the boundaries look smoother, they can be further improved by averaging over a larger ensemble of 40 networks. This is demonstrated in Figure 5 which summarises the effect of averaging.

There was a clear improvement in all results with the large ensemble size as well as the addition of either noise or weight decay regularisation. Finally, the combination of weight decay, noise and a 40-net ensemble clearly gives the best results (Figure 5, bottom right).

Thus, while earlier work suggested that a single-layer feedforward network is not capable of capturing the structure in the spiral data, it is evident that a network

Different Levels of Gaussian Noise

Fig. 4: Ensembles of five networks with fixed weight decay and a varying degree of noise (top left is zero noise, bottom right is noise with SD = 0.8). The classification threshold is 0.5.

ensemble with strong control over its capacity (via weight decay) which is trained with heavy noise can discover the highly nonlinear structure of the problem.

2.3.3 *Properties of variance regularisation via noise*

The motivation of our approach comes from a key observation regarding the decomposition of prediction error into bias and variance components, namely the fact that ensemble averaging does not affect the bias portion of the error but reduces the variance, when the estimators on which averaging is done are independent. The level of noise affects the level of independence between the training sets, and thus the relative improvement of ensemble averaging. However, the level of noise also affects the quality of each predictor separately, increasing its variance by increasing the variability in the data. Thus, there should be an optimal level of the noise, which may not

Summary of 40-Net Ensemble-Average Results

Fig. 5: *Top left:* no constraints (no weight decay or noise); *top right:* optimal weight decay ($\lambda = 3e^{-4}$) and no noise; *bottom left:* optimal noise (Gaussian SD = 0.35) and zero weight decay; *bottom right:* optimal noise and optimal weight decay.

correspond to the true noise level and which leads to optimal ensemble performance. This performance can be further improved if the variance of individual networks can be tempered, e.g. with weight decay.

Raviv and Intrator (1996) demonstrated the effect of noise injection on prediction in three different cases. (i) Highly nonlinear (spiral) data, using a suboptimal model; the data are almost radially symmetric and the neural net is not. This required the use of an ensemble of high-capacity single predictors and thus made the regularisation task challenging. It was shown that the excess variance of high-capacity models could be effectively trimmed only by a combination of all three of weight decay, noise injection and ensemble averaging. (ii) Highly nonlinear (spiral) data with essentially the perfect model for it, namely a generalised additive model (GAM) with locally linear units (Hastie and Tibshirani, 1986). Even in this case, where regularisation provides the perfect degree of bias to the model, performance could be improved. (iii) A highly linear problem, where practically any network has excess capacity, that is, is over-parameterised. This case is a representative of a family of clinical datasets, in which (linear) variable selection was applied to a high-dimensional dataset and resulted in a highly linear low-dimensional data structure. It was thus challenging to be able to show that the BEN algorithm is useful in this case, and can lead to improved classification results. Performance was also evaluated using the receiver operating characteristic (ROC) measure (Hanley and McNeil, 1982), which is a standard model comparison tool for clinical data analysis.

Theoretical analysis suggests that it is best to start with a very flexible function approximation technique (e.g. a feedforward network with a large number of hidden units) and then control its capacity and smoothness using noise and averaging. These conclusions hold also for other models such as generalised additive models.

3   Integrating ensembles of estimators

After having established that ensembles of estimators are very useful for variance control and having discussed ways to increase the level of independence between ensemble members, as well as optimal stopping criteria for ensembles, we study the possible additional benefits of ensembles of experts. In particular, we would like to assign different ensembles to different regions in input space, or different representations of the input space, constructed by various types of preprocessing, in an attempt to exploit the independence between different ensembles. We start with a review of current related methods and demonstrate our ideas on a specific seismic classification problem. Full details of the algorithms and application to this problem are given in Shimshoni (1995) and Shimshoni and Intrator (1998).

The lack of a *priori* knowledge about the true underlying model of the data in the seismic classification problem, as in many other real-life problems, leads practitioners to examine various suboptimal classifiers. Different classifiers can exploit various types and sets of features, and hence combining multiple estimated classifiers might yield better performance than the best single candidate. In the neural network and machine learning literature there are several methods for combining estimators, and important questions such as *what* types of estimator to combine and *how* to do it are currently getting considerable attention.

A well-known class of combined models contains methods such as the adaptive mixture of experts (AME) (Jacobs *et al.*, 1991) and hierarchical mixture of experts (HME) (Jordan and Jacobs, 1994), which are both based on the *divide and conquer* approach where a mixture of experts compete to gain responsibility in modelling the output in a portion of the input space. The system's output is obtained as a linear combination of the experts' outputs, where the weights are computed as a parametric function of the inputs by a gating module. The underlying probabilistic model is based on the assumption of mutual exclusivity, i.e. a single expert is responsible for each data point. In mixture models like AME and HME the different experts are usually trained on a single dataset simultaneously by minimising a combined cost function. The final combination of the experts is determined by the gating module, which is constructed during the same training session. When training all the experts on the same dataset with the same data representation, the dependence of errors among experts is high, thus diminishing their collective contribution.

A different aspect of the multiple model concept is introduced by using a committee of classifiers, also called an *ensemble* (Hansen and Salamon, 1990; Perrone and Cooper, 1993). Consider a trained classifier as a realisation of a generic model trained on the given data. Thus, different datasets would yield different realisations, all of which are members of the *post training* distribution of the possible solutions. As the solution space is generally highly degenerate and includes many local minima, it is more robust to use a sample from this solution space (i.e. ensemble of estimators) rather than a single representative (Hansen *et al.*, 1994). The ensemble of classifiers can be averaged to produce an aggregated classifier, or any linear combination of the realisations can be used.

When all the classifiers give similar results, the accuracy of their combined classification depends largely on their bias, because of the bias–variance tradeoff (Geman *et al.*, 1992). Whenever the bias of a generic model is high, the multiple classifications of its ensemble might all be wrong even if the variance is low, in which case no combination operation will help. On the other hand, when the bias is small and the variance is high, one could expect the ensemble to disagree whenever the input signals are ambiguous. In such cases the ensemble's result, i.e. the aggregated classification, will have the same bias but a reduced variance. Hence, combining multiple classifiers can eliminate the need to regularise over-fitted models with high variance (Sollich and Krogh, 1996). Moreover, the classification confidence can be evaluated from the variance which represents the agreement within the ensemble, regarding signals with high classification variance as being treated with suspicion.

Estimation of the optimal combination of the experts should be done in a robust way, i.e. by averaging or cross-validation techniques rather than by parametric estimation based on the same training data (LeBlanc and Tibshirani, 1993). It has been shown that, in order for the combination of experts to be optimal, the experts should be made as independent as possible (Krogh and Vedelsby, 1995; Meir, 1994; Jacobs, 1995; Raviv and Intrator, 1996). Another method which uses multiple classifiers in a sequential training scheme is that of 'Boosting' (Schapire, 1990; Drucker *et al.*, 1994). In this method, which is typically suitable for very large training sets as in optical character recognition (OCR) problems, each classifier is trained on patterns that have been filtered by the previous classifier. Thus, the result is a combination of classifiers that have been trained on statistically different datasets in a sequential process. A more general framework for combining multiple estimators is that of 'stacked generalisation' (Wolpert, 1992), in which each estimator is trained with a different subset of the data and the optimal combination is estimated using cross-validation methods. A formulation of this method for regression estimators was presented by Breiman (1992) and was compared with other methods by LeBlanc and Tibshirani (1993). 'Bagging' is a method which produces an aggregated estimator using bootstrap replicas of the training data (Breiman, 1996). It is reported to be useful whenever the estimator is unstable, i.e. when perturbing the training set can cause significant changes in the constructed classifier. Notice that this condition corresponds to the requirement mentioned earlier of maximum independence among the experts. Several ways have been suggested for making the experts less dependent, for example injecting noise during training, as in the smoothed bootstrap (Raviv and Intrator, 1996).

Since the search for an optimal classifier is tied to the search for an optimal data representation (i.e. an optimal transformation of the input signals with respect to the classification task at hand), it is advisable to examine and possibly use more than one signal representation. In order to maximise both the information extracted from the data and the gain from combining multiple classifiers, we suggest constructing a 'redundant classification environment', which allows for different signal representations to supply a wide coverage of the effective feature space.

3.1   Creating a redundant classification environment

The hierarchical scheme proposed by Shimshoni and Intrator is based on experts which are ensembles of artificial neural networks (ANNs), each of which is associated with a specific data representation and network architecture. The redundancy is pronounced within these ensembles, which are collections of ANN realisations trained on different subsets of the data. Each ensemble is trained using the same data representation, i.e. a unique time–frequency decomposition and smoothing level of the input signals. The network architecture (number of hidden units) is also fixed for all members in an ensemble. The redundancy is even more pronounced, as we use various combinations of time–frequency decompositions, smoothing levels and network architectures to create and train several such ensembles. The different training conditions yield relatively independent classifiers, that might produce different classification results given an ambiguous signal. The *integrated classification machine* (ICM) that is formulated next integrates the different ensembles in this classification environment and produces a final classification which achieves better generalisation performance than the single classifying components; see the results in Section 3.4.

The integrated classification machine is constructed from a hierarchy of classifiers as shown in Figure 6 below. The smallest building block of the ICM is a neural network (feedforward multilayer perceptron with logistic sigmoidal activation functions). All networks are trained to predict the class label of a given signal. As mentioned, we use several representations for the input signals, so that the input layers of the neural networks in the ICM have different dimensionalities (N) according to the input representations used. The hidden layer can contain various numbers (H) of sigmoidal units and the output layer contains two sigmoidal units.

3.1.1   *The network's prediction value*

The desired output of the networks for a given signal can be either $\{1,0\}$ for *natural* events or $\{0,1\}$ for *artificial* events. The sigmoidal (σ) output units of the trained networks will produce continuous values in the range of 0 to 1 according to the network weights:

$$O^l = \sigma \left(\sum_{i=1}^{H} W_{li} \sigma \left(\sum_{j=1}^{N} w_{ij} x_j + w_{i0} \right) + W_{l0} \right) \quad l = 1, 2. \quad (3.1)$$

Let us define the (signed) difference between the two output units, $y = (O^1 - O^2)$, to be the *prediction value* of the network. Hence, $y \in [-1, 1]$ and the predicted class label is given by thresholding y at zero, assigning positive values to the class of *natural* events and negative values to the class of *artificial* events. Each network is trained on T repeated trials, changing only the initial random weights; in our implementation $T = 5$. We define the *prediction value* of the network component in the ICM with respect to a signal x as the average prediction value $y^{\mathrm{NET}}(x)$ from the T training trials: $y^{\mathrm{NET}}(x) = \frac{1}{T} \sum_{t=1}^{T} y_t(x)$.

3.1.2 The ensemble's prediction value

Each ensemble is a collection of B networks, where each network is trained on one of the B replicas, $D_r^b, b = 1, \ldots, B$, of the original dataset D_r for a specific data representation r; we used 30 bootstrap sets, which is fewer than suggested by Efron and Tibshirani (1993) (≈ 200), but was found to be sufficient. Details of the replication of the data into bootstrap sample sets (Efron and Tibshirani, 1993) are given by Shimshoni and Intrator (1998). All the networks in an ensemble share the same data representation and the same network architecture. The *prediction value* of an ensemble with respect to a signal x is defined as the average over all the prediction values of the participating networks, as in the method of 'bagging' (Breiman, 1996):

$$y^{\mathrm{ENS}}(x|D_r) = \frac{1}{B} \sum_{b=1}^{B} y_b^{\mathrm{NET}}(x|D_r^b). \quad (3.2)$$

3.1.3 The integrated prediction value

A collection ($\mathcal{K}$) of ensembles, which use different input representations, i.e. time–frequency decompositions or smoothing levels, and different network architectures, i.e. number of hidden units, form the integrated classification machine (ICM) shown in Figure 6. The *integrated prediction value* of the ICM with respect to a signal x is defined by

$$y^{\star}(x) = \sum_{k \in \mathcal{K}} \alpha_k \, \beta_k(x) \, y_k^{\mathrm{ENS}}(x). \quad (3.3)$$

Here α_k is a prior reliability measure of the kth ensemble, which can be determined from the training data or from prior knowledge, with $\alpha_k = 1/K$ as a default value. The quantity $\beta_k(x)$ is a posterior classification confidence measure that is specific to each signal and is discussed in the next section. Both measures are normalised and together determine the strength of the 'vote' of ensemble k in the 'classification committee' for signal x.

In order to detect signals with ambiguous class membership and to rank the different ensembles in terms of their accuracy of classification, we have constructed a confidence measure for the ensemble's classification. This posterior confidence is based on the variance of the networks' prediction values,

$$\text{CONF}^{\text{ENS}}(x) = \left[\text{Var}\left(y^{\text{NET}}(x) \right) \right]^{-1}, \tag{3.4}$$

where $y^{\text{NET}}(x)$ is the network's prediction value. The CONF score represents the amount of 'agreement' among all the participating networks in the ensemble (Krogh and Vedelsby, 1995; Hansen et al., 1994). It should be remembered that, when the bias of the networks is high, such a measure will not convey confidence (i.e. when all members agree on the same wrong class).

Therefore, one should examine the classification results carefully to check whether or not there is a correlation between the confidence scores and the errors of the combined classification.

3.2   Combining the hierarchy of classifiers

The ICM shown in Figure 6 is a hierarchy of classifiers with two levels at which multiple classifiers are combined. At the first level, for each ensemble, B networks are combined using simple averaging to construct an aggregated ensemble classification

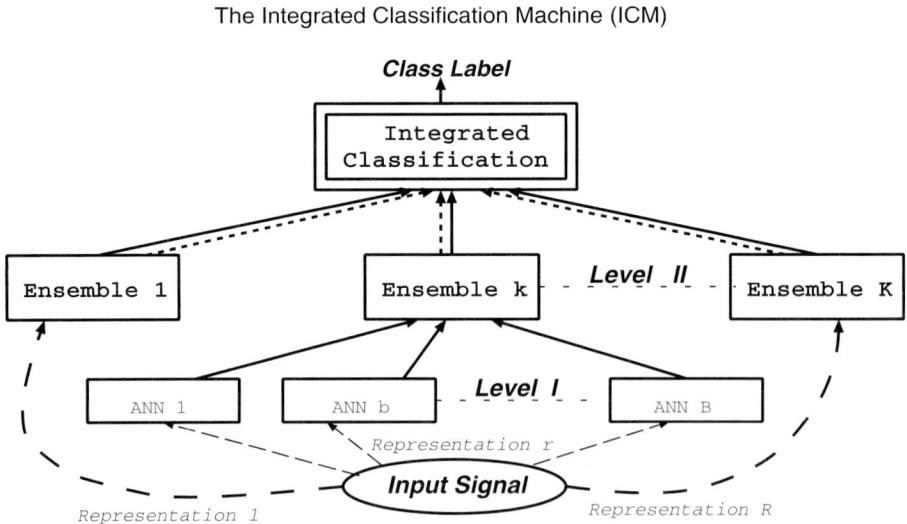

Fig. 6: Several representations of the waveform are fed into different ensembles, then integrated to produce the final classification. (In level II, the regular arrows are the ensembles' prediction values and the dashed arrows are the attached confidence values.)

(equation (3.2)). These networks are in fact multiple realisations of the same classifier trained on different bootstrap samples.

The second level integrates ensembles into the final classification. Integrating ensembles involves an estimation of $\beta_k(x)$, namely the strength of the 'vote' of the kth ensemble (equation (3.3)). This is done by applying a least-squares calculation with nonnegativity constraints (Breiman, 1992). Finally, prior weights can also be used to combine the ensembles.

Unlike the integration of networks into ensembles at the first level, where each network is the same apart from being trained on a different subset of the data, in the second level it is less likely that a fixed weighting will produce better classification results than those of *all* single ensembles. There might be some ensembles which use inferior data representations or less suitable model architectures, which will have disturbing effects on the weighted result. In practice, we have noticed that, for fixed weighting methods like those mentioned above, the integrated classification results were often worse than the best participating single ensemble.

Given a low signal-to-noise ratio and nonstationarity of the signal space, along with a shortage of training data, it is furthermore desirable not to base the choice of averaging weights on the characteristics of the currently available dataset. Therefore, we suggest using the signal x to find the optimal set of weights for its own 'classification committee', namely to apply a nonfixed weighting of the classifiers. One can use the ensemble's prediction value $y^{ENS}(x)$ as the basis for the weighting, or decide how to weight the votes using the classification confidence $CONF^{ENS}(x)$ (Tresp and Taniguchi, 1995). Application of the maximum entropy principle (Jaynes, 1982) suggests that an optimal combination of ensembles should set

$$\beta(x) = \exp\left[-\text{Var}\left(y^{NET}(x)\right)\right]. \tag{3.5}$$

We have used a nonfixed weighting strategy which is a dynamic *winner takes all* procedure. In order to integrate the different ensembles in the ICM for a *specific* signal x, all classifiers (ensembles) are ranked and the optimal *one* is selected. The ranking is governed by the overall reliability of the classifiers and the selection is based on the classification confidence $CONF^{ENS}(x)$ supplied by the ensembles along with their classification (see Figure 6). This approach, which will be elaborated in the next section, is different from the linear nonfixed weighting that was suggested by Tresp and Taniguchi (1995) for combining ANNs, where all classifiers participate in the committee with weights inversely proportional to the variance; this is similar to the use of CONF values here, but the weights are estimated in a different way as they use single ANN classifiers rather than ensembles. The next section discusses a similar approach, but exploits further the variability between members of each ensemble.

3.3   Competing rejection algorithm (CRA)

Generally, a signal is said to be rejected by a classifier if some measure representing the quality of its classification falls below a predefined threshold. Obviously, the higher the threshold, the more signals will be rejected and the smaller will be the misclassification rate over the remaining unrejected signals. Shimshoni and Intrator (1998) have presented an algorithm which performs a sequence of classifications by polling the group of K classifiers with respect to the signal at hand. Each classifier (ensemble) in turn can either classify or reject the signal. A key motivating observation is that some classifiers perform globally better than others. Nevertheless, classifiers can occasionally outperform 'superior' classifiers, and should therefore be given the opportunity to compete and possibly to 'steal' a classification even if the signal is rejected by those 'superior' classifiers.

Implementation of this idea requires a prior reliability ranking of the classifiers and a definition of a rejection criterion. The rejection is done by thresholding the confidence measure $\mathrm{CONF}^{\mathrm{ENS}}(x)$, defined in (3.4), so that a classification is rejected when its confidence value is lower than the 'reject' threshold. Each ensemble k has its own threshold, Θ_k, depending on its accuracy and variability, that fixes the minimum level of confidence 'allowed' for its classifications. Θ_k is calculated as a certain upper percentile of the confidence scores $\mathrm{CONF}^{\mathrm{ENS}}(x)$, of an unlabelled dataset $D^\star$ as follows:

$$\Theta_k = \text{percentile}\{\mathrm{CONF}^k(x), \text{reject-rate}_k\}. \tag{3.6}$$

The 'reject-rate' of a classifier represents its global credibility and can be determined either from its performance on training data or by using subjective information. One simple approach is for all ensembles to have the same credibility, and thus a uniform reject-rate is set, for example based on the upper 20th percentile. When all classifiers reject a signal, it can be either 'globally rejected' or classified by the *ultimate classifier*, which is optionally predefined by the user. The global rejection rate cannot be determined by the user and is usually much smaller than the individual rejection rates.

This selection algorithm, along with the whole ICM structure, is easily scalable, and no retraining is required when new classifiers are added. It is also flexible, in the sense that it is straightforward to incorporate other types of experts, including human ones, as long as they produce suitable prediction values and CONF values.

3.4   Results on seismic data classification

The integrated classification machine was applied to a dataset consisting of 380 seismic events, including all the natural local earthquakes that occurred between January, 1990, and June, 1993, inside an area of $22\,500$ km^2 in the northern part of Israel. The seismic dataset is available from

ftp://www.math.tau.ac.il/\$\sim\$shimsh/pub/seismic-data/.

A similar number of artificial explosion events were randomly sampled from the same spatio-temporal window. All events have magnitude $M_L < 2.7$, while 77% of the events are below 2.0 and the mean magnitude is 1.53. Results from these data (Shimshoni, 1995; Shimshoni and Intrator, 1998) indicate that the ICM could further improve results of the best ensemble by dynamically combining results from inferior ensembles using the algorithms discussed above. Comparison to other ensemble methods and more classical statistical methods further demonstrates the strength of this method as a general mixture of expert machines.

4 Bias control: imposing prior knowledge

Smoothness constraints (Wahba, 1990; Poggio and Girosi, 1990) are often used as variance constraints. Similarly, in training neural networks, weight decay, which attempts to shrink the weights to zero or to a single value (Hinton, 1986; Krogh and Hertz, 1992) is often used. In other statistical contexts, such as regression or generalised linear models, this method is often called shrinkage (James and Stein, 1960). The resulting constraint on the weights effectively reduces the number of free parameters in the model. A brute force method for reducing the number of model parameters is 'optimal brain damage' (Le Cun *et al.*, 1990), in which weights which become smaller than a predefined threshold are set to zero. Other related methods include weight sharing, in which a single weight is shared among many connections in the network (Waibel *et al.*, 1989; Le Cun *et al.*, 1989). An extension of this idea is 'soft weight sharing' which favours irregularities in the weight distribution in the form of multimodality (Nowlan and Hinton, 1992). This approach may improve upon generalisation results obtained by weight elimination. Both these methods make an explicit assumption about the structure of the weight space, independently of the structure of the input space.

We now turn our attention to the second part of prediction error, namely the bias. It may sometimes appear difficult to distinguish between bias constraints and variance constraints. For example, how do we treat smoothness constraints? While it reduces the variance, it clearly enforces a bias towards smooth models. Since we are dealing with additive constraints, as will be clear from the way that parameter estimates are calculated (see equation (4.5)), we can offer a simple distinction. If the update rule leads to no meaningful result when only the additional (bias/variance) constraint is effective, we regard it as a *variance* constraint. When some meaningful result is obtained via this unsupervised constraint only, we regard it as a *bias* constraint. Thus, classical constraints such as smoothness, as well as assumptions about the distribution of the parameters, e.g. favouring small weights via weight decay or favouring particular distributions such as mixtures of Gaussians (Nowlan and Hinton, 1992), are actually variance constraints. The next section discusses a general framework for imposing bias constraints and describes some specific examples of such constraints.

4.1 Projection pursuit

A general framework for additive and semi-linear dimensionality reduction is projection pursuit. It is useful when a low-dimensional representation is embedded in high-dimensional data. The supervised version, called projection pursuit regression (PPR) (Friedman and Stuetzle, 1981), is capable of performing dimensionality reduction by composition, in that it constructs an approximation to the desired response function using a composition of lower-dimensional smooth functions. These functions depend on linear-dimensional projections through the data.

When the dimensionality of the problem is in the thousands, even projection pursuit regression methods are almost always over-parameterised and lead to data over-fitting. At that stage we must impose prior knowledge on the desired low-dimensional representation. The unsupervised version, called exploratory projection pursuit (EPP) (Friedman and Tukey, 1974; Friedman, 1987) becomes relevant. It searches in a high-dimensional space for structure in the form of (semi-)linear projections with constraints characterised by a *projection index*, which measures the goodness of a projection. The projection index may be considered as a universal prior for a large class of problems, or may be tailored to a specific problem based on prior knowledge.

4.2   Brief description of projection pursuit regression

Let (X, Y) be a pair of random variables, with $X \in R^d$ and $Y \in R$. The problem is to approximate the d-dimensional regression surface,

$$f(x) = E[Y|X = x], \tag{4.1}$$

on the basis of n observations $(x_1, y_1), \dots, (x_n, y_n)$.

Projection pursuit regression tries to approximate f by a sum of ridge functions, i.e. functions that are constant along lines:

$$f(x) \simeq \sum_{j=1}^{m} g_j \left(a_j^T x \right). \tag{4.2}$$

The fitting procedure alternates between estimation of directions by $\hat{a}_j$ and estimation of smooth functions by $\hat{g}_j$. At stage j, the sum of squares of the residuals

$$r_{ij}(x_i) = r_{i,j-1}(x_i) - \hat{g}_j \left(\hat{a}_j^T x_i \right) \tag{4.3}$$

is minimised. The process is initialised by setting $r_{i0}(x_i) = y_i$, for each i. Usually, the initial values of $\hat{a}_j$ are taken to be the first few principal components of the data. Estimation of the ridge functions can be achieved by various nonparametric smoothing devices such as locally linear functions (Friedman and Stuetzle, 1981), k-nearest neighbour methods (Hall, 1989), splines or variable degree polynomials. Imposition of a smoothness constraint on $\hat{g}_j$ implies that the actual projection pursuit is achieved by minimising, at iteration j, the sum

$$\sum_{i=1}^{n} r_{ij}^2(x_i) + C(\hat{g}_j), \tag{4.4}$$

for some smoothness measure C.

Since the estimation of the nonparametric ridge functions is not decoupled from the estimation of the projections, over-fitting is very likely to occur in one of the low-order $\hat{g}_j$, thereby invalidating subsequent estimation stages. Obviously, if $\hat{g}_j$ is not well estimated, the search for an optimal projection direction will not yield good results.

Artificial neural network architectures are closely related and offer some properties which improve their variance control in high-dimensional cases. In feedforward neural nets, the family of ridge functions is limited to sigmoidal functions with variable threshold. This avoids the nonparametric or semiparametric estimation of the ridge functions, but may require a large number of projections, as in the example in Section 2.3.1. Secondly, the estimation of several projections is performed concurrently. This allows one to find a low-dimensional representation which cannot be found if the search is done sequentially (Huber, 1985).

4.3 Hybrid attempts at dimensionality reduction

There have been various attempts to combine unsupervised learning with supervised learning (Yamac, 1969; Gutfinger and Sklansky, 1991; Bridle and MacKay, 1992). The formulation discussed below is based on projection pursuit ideas, which generalise many of the classical statistical methods, and, in our case, suggest a well-defined statistical framework that allows formulation and comparison between various methods. Consider the artificial neural network architecture presented in Figure 7. The only

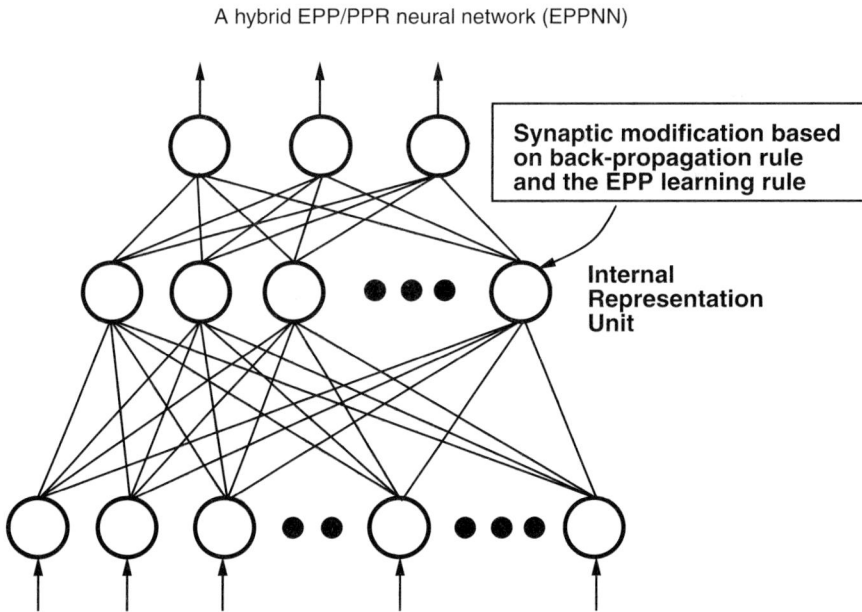

A hybrid EPP/PPR neural network (EPPNN)

Synaptic modification based on back-propagation rule and the EPP learning rule

Internal Representation Unit

Fig. 7: The modification of the hidden units' weights is achieved by backpropagation of the error from the output layer (via the chain rule) and by the gradient of the projection index.

difference from a classical feedforward architecture (Rumelhart *et al.*, 1986) is the additional modification term among the hidden units. We consider the hidden unit representation as a new, reduced-dimensionality representation of the data. As described in the context of projection pursuit regression (Intrator, 1993), we add a penalty term to the energy functional minimised by error backpropagation, for the purpose of measuring directly the goodness of the projections sought by the network. This puts the emphasis on choosing the right prior, as a means to improving the bias–variance tradeoff.

Since our main interest is in reducing over-fitting for high-dimensional problems, our underlying assumption is that the surface function to be estimated can be faithfully represented using a low-dimensional composition of sigmoidal functions, i.e. using a feedforward architecture in which the number of hidden units is *much smaller* than the number of input units. Therefore, the penalty term may be added to the hidden layer only; see Figure 7.

The synaptic modification equations for the hidden units' weights become

$$\frac{\partial w_{ij}}{\partial t} = -\epsilon \left[\frac{\partial \mathcal{E}(w, x)}{\partial w_{ij}} + \frac{\partial \rho(w_1, \dots, w_n)}{\partial w_{ij}} \right.$$

$$\left. + (\text{Contribution of cost/complexity terms}) \right], \qquad (4.5)$$

where $\mathcal{E}$ is the error function and ρ is the explicit measure of goodness of projections (bias constraints), while the contribution of the cost/complexity terms (variance constraints) are also additively imposed.

4.4 Specific bias constraints

Exploratory projection pursuit is based on seeking *interesting* projections of high-dimensional data (Kruskal, 1969; Switzer, 1970; Kruskal, 1972; Friedman and Tukey, 1974; Friedman, 1987; Jones and Sibson, 1987; Huber, 1985). The notion of interesting projections is motivated by an observation that, for most high-dimensional data clouds, most low-dimensional projections are approximately normal (Diaconis and Freedman, 1984). This finding suggests that the important information in the data is conveyed in those directions whose univariate projected distribution is far from Gaussian. Various projection indices differ in the assumptions about the nature of the deviation from normality, and in their computational efficiency. They can be considered as different priors motivated by specific assumptions on the underlying model.

Since the Gaussian distribution maximises entropy, subject to specified mean and variance, it is possible to test deviation from Gaussianity by the difference between the entropy of a Gaussian, with the same variance, and that of the given distribution. This measure is nonnegative and provides a measure of the information content of the distribution relative to the maximal content of a Gaussian with the same variance.

The index is given by

$$J_I(p) = H(p_G) - H(p)$$
$$= \frac{1}{2}\log(2\pi e) + \log(\tau) + \int p(x)\log p(x)dx, \tag{4.6}$$

where p is the probability density function and τ is the standard deviation of the distribution of interest. This index also leads to redundancy reduction and independent component analysis; see below.

As the density $p(x)$ is unknown, it has to be estimated from the data. This is computationally expensive and not very robust. Although it is possible to use the data to estimate the density nonparametrically, for example using the kernel method (Wand, 1994; Viola and Wells, 1995), it is usually preferred to estimate the integral in (4.6) by some other approximation to the density. When the third and fourth cumulants (Kendall and Stuart, 1977),

$$\kappa_3 = \frac{E\left[(x - \bar{x})^3\right]}{\tau^3},$$
$$\kappa_4 = \frac{E\left[(x - \bar{x})^4\right]}{\tau^4} - 3, \tag{4.7}$$

of the distribution are known, it is possible to estimate the integral using some approximation to the probability density function. Comon (1994) proposed using the Edgeworth expansion (Stuart and Ord, 1994), but recently the Gram–Charlier expansion (Stuart and Ord, 1994) has been proposed (Amari *et al.*, 1996) as it explicitly depends on the third and fourth cumulants of the distribution. This Gram–Charlier approximation has the form

$$p(x) \simeq \alpha(x)\left\{1 + \frac{\kappa_3}{3!}H_3(x) + \frac{\kappa_4}{4!}H_4(x)\right\}, \tag{4.8}$$

where

$$\alpha(x) = \frac{1}{\sqrt{2\pi}}\exp(-x^2/2)$$

and $H_k(x)$ are Chebyshev–Hermite polynomials, given in our case by

$$H_3(x) = 4x^3 - 3x,$$
$$H_4(x) = 8x^4 - 8x^2 + 1. \tag{4.9}$$

The exact measure of deviation from a Gaussian distribution is clearly expressed in approximation (4.8) in terms of the skewness and kurtosis of the distribution. If we substitute this approximation into (4.6) we obtain

$$\hat{J}_I(p) = \tau - \frac{(\kappa_3)^2}{2\cdot 3!} - \frac{(\kappa_4)^2}{2\cdot 4!} + \frac{5}{8}(\kappa_3)^2\kappa_4 + \frac{1}{16}(\kappa_4)^3. \tag{4.10}$$

This expansion suggests natural bias constraints on projected distributions. Recently, Intrator (1990) has shown that a BCM neuron can find structure in the input distribution that exhibits deviation from a Gaussian distribution in the form of multimodality in the projected distributions; BCM stands for Bienenstock, Cooper and Munro (Bienenstock *et al.*, 1982). It is a learning rule that was constructed to model early visual cortical plasticity. Current versions of this rule, including mathematical properties, statistical motivation and network extensions, are discussed in Intrator and Cooper (1992). Since clusters cannot be found in the data directly, because of their sparsity, this type of deviation, which is measured by the first three moments of the distribution, is particularly useful for finding clusters in high-dimensional data, and is thus useful for classification or recognition tasks. It thus renders this feature-extraction technique also appropriate for bias constraints. We present some of the constraints here in a form that exhibits their relation to the BCM feature-extraction rule. Connections with independent components analysis and results on natural scene feature extraction are described by Blais *et al.* (1998). Results on classification of faces are described by Intrator *et al.* (1996) and Stainvas *et al.* (1997).

We are interested in finding projections that indicate interesting structure. Furthermore, we are not interested in deviation in the form of asymmetry, but more in deviations in the tails or in manifestations of multimodality. We also want to avoid sensitivity to outliers and therefore we use a rectified activation function (this rectification has little effect on the ability of kurtosis rules to find interesting projections (Blais *et al.*, 1998) denoted by $c = \phi(\mathbf{d} \cdot \mathbf{m})$ and assume that the sigmoid ϕ is a smooth monotone function with a positive output, although a slight negative output is also allowed; ϕ' denotes the derivative of the sigmoid. The rectification is required for all rules that depend on odd moments, because these vanish in a symmetric distribution such as is commonly used in contexts involving natural scenes (Blais *et al.*, 1998).

Skewness 1 This measures deviation from symmetry (Kendall and Stuart, 1977) and is of the form

$$S_1 = E[c^3] / (E[c^2])^{3/2}. \tag{4.11}$$

Maximisation of this measure via gradient ascent uses

$$\nabla S_1 = \frac{1}{\Theta_M^{1.5}} E[c(c - E[c^3]/E[c^2])\phi'\mathbf{d}], \tag{4.12}$$

where Θ_M is defined as $E[c^2]$.

Skewness 2 A similar measure which requires some stabilisation mechanism is given by

$$S_2 = E[c^3] - E^{3/2}[c^2]. \tag{4.13}$$

This measure has a gradient of the form

$$\nabla S_2 = 3E\left[c\left(c - \sqrt{\Theta_M}\right)\phi'\mathbf{d}\right].\tag{4.14}$$

Kurtosis 1 Kurtosis measures deviation from a Gaussian distribution mainly in the tails of the distribution. It has the form

$$K_1 = E[c^4]/E^2[c^2] - 3.\tag{4.15}$$

This measure has a gradient of the form

$$\nabla K_1 = \frac{1}{\Theta_M{}^2}E\left[c\left(c^2 - E[c^4]/E[c^2]\right)\phi'\mathbf{d}\right].\tag{4.16}$$

Kurtosis 2 As before, there is a similar form which requires some stabilisation:

$$K_2 = E[c^4] - 3E^2[c^2].\tag{4.17}$$

This measure has a gradient of the form

$$\nabla K_2 = 4E\left[c(c^2 - 3\Theta_M)]\phi'\mathbf{d}\right].\tag{4.18}$$

With all the above, maximisation of the measure can be used as a goal for projection seeking, so the variable c can be thought of as a (nonlinear) projection of the input distribution on to a certain vector of weights, and the maximisation then defines a learning rule for this vector of weights. Under this framework, it is easy to stabilise the above learning rules by requiring for example that the vector of weights, which we denote by $\mathbf{m}$, has a fixed norm, $\|\mathbf{m}\| = 1$, say. The multiplicative forms of both kurtosis and skewness do not require this type of stabilisation, because of the normalising factor $1/\Theta_M{}^p$ in each rule.

Quadratic BCM The quadratic BCM (QBCM) measure as given by Intrator and Cooper (1992) is of the form

$$\text{QBCM} = \tfrac{1}{3}E[c^3] - \tfrac{1}{4}E^2[c^2].\tag{4.19}$$

Maximising this form using gradient ascent uses the gradient function

$$\nabla\text{QBCM} = E[c(c - \Theta_M)\phi'\mathbf{d}].\tag{4.20}$$

The quadratic BCM rule does not require any additional stabilisation. This turns out to be an important property, since additional information can then be transmitted using the resulting norm of the weight vector $\mathbf{m}$ (Intrator, 1996).

Applicability of such bias constraints has been demonstrated in various real-world applications. Entropy constraints have been used in image compression, with a penalty aimed at minimising the entropy of the projected distributions (Bichsel and Seitz,

1989). BCM constraints have been used in face recognition (Intrator *et al.*, 1996) and more recently for partially occluded and blurred face recognition (Stainvas *et al.*, 1997). These constraints are also useful for acoustic signal classification from wavelet representations (Huynh *et al.*, 1998). Kurtosis constraints have been used in conjunction with reconstruction to find sparse representations (Olhausen and Field, 1996), and recently kurtosis and skewness have been found to be useful for neg-entropy calculations and independent components analysis; see Yang and Amari (1997) for a review.

5 Remarks

Our aim in this chapter has been to suggest several ingredients which may be crucial for multiparameter model estimation. As computational power and database availability increase, real-world problems like these will be encountered increasingly often, and methods such as these can lead to significant improvement in their treatment.

6 References

Amari, S., Cichocki, A. and Yang, H.H. (1996). A new learning algorithm for blind signal separation. In *Advances in Neural Information Processing Systems* **8**, Eds. G. Tesauro, D.S. Touretzky and T. Leen, pp. 757–763. MIT Press, Cambridge, MA.

Baum, E. and Lang, K. (1991). Constructing hidden units using examples and queries. In *Advances in Neural Information Processing Systems* **3**, Eds. R.P. Lippmann, J.E. Moody, and D.S. Touretzky, pp. 904–910. Morgan Kaufmann, San Mateo, CA.

Bichsel, M. and Seitz, P. (1989). Minimum class entropy: A maximum information approach to layered networks. *Neural Networks* **2**, 133–141.

Bienenstock, E.L., Cooper, L.N. and Munro, P.W. (1982). Theory for the development of neuron selectivity: orientation specificity and binocular interaction in visual cortex. *Journal Neuroscience* **2**, 32–48.

Bishop, C.M. (1995). Training with noise is equivalent to Tikhonov regularization. *Neural Computation* **7**, 108–116.

Blais, B.S., Intrator, N., Shouval, H. and Cooper, L.N. (1998). Receptive field formation in natural scene environments: comparison of single cell learning rules. *Neural Computation* **10**, 1797–1813.

Breiman, L. (1992). Stacked regression. Technical Report TR-367, Department of Statistics, University of California, Berkeley.

Breiman, L. (1996). Bagging predictors. *Machine Learning* **24**, 123–140.

Bridle, J.S. and MacKay, D.J.C. (1992). Unsupervised classifiers, mutual information and 'Phantom Targets'. In *Advances in Neural Information Processing Systems* **4**, Eds. J.E. Moody, S. Hanson and R. Lippmann, pp. 1096–1101. Morgan Kaufmann, San Mateo, CA.

Buckheit, J. and Donoho, D.L. (1995). Improved linear discrimination using time-frequency dictionaries. Technical Report. Department of Statistics, Stanford University.

Comon, P. (1994). Independent component analysis: a new concept? *Signal Processing* **36**, 287–314.

Deffuant, G. (1995). An algorithm for building regularized piecewise linear discrimination surfaces: The perceptron membrane. *Neural Computation* **7**, 380–398.

Diaconis, P. and Freedman, D. (1984). Asymptotics of graphical projection pursuit. *Annals of Statistics* **12**, 793–815.

Drucker, H., Cortes, C., Jackel, L., LeCun, Y. and Vapnik, V. (1994). Boosting and other ensemble methods. *Neural Computation* **6**, 1289–1301.

Efron, B. and Tibshirani, R. (1993). *An Introduction to the Bootstrap.* Chapman and Hall, New York.

Elman, J.L. and Zipser, D. (1988). Learning the hidden structure of speech. *Journal of the Acoustical Society of America* **4**, 1615–1626.

Fahlman, S.E. (1988). Faster-learning variations on backpropagation: An empirical study. In *Connectionist Models Summer School.* Eds. T.J. Sejnowski, G.E. Hinton and D.S. Touretzky. Morgan Kaufmann, San Mateo, CA.

Fahlman, S.E. and Lebiere, C. (1990). The cascade-correlation learning architecture. Technical Report CMU-CS-90-100. Carnegie–Mellon University.

Fisher, R.A. (1936). The use of multiple measurements in taxonomic problems. *Annals of Eugenics* **7**, 179–188.

Friedman, J.H. (1987). Exploratory projection pursuit. *J. Amer. Statist. Assoc.* **82**, 249–266.

Friedman, J.H. and Stuetzle, W. (1981). Projection pursuit regression. *J. Amer. Statist. Assoc.* **76**, 817–823.

Friedman, J.H. and Tukey, J.W. (1974). A projection pursuit algorithm for exploratory data analysis. *IEEE Transactions on Computers* C **23**, 881–889.

Geman, S., Bienenstock, E. and Doursat, R. (1992). Neural networks and the bias-variance dilemma. *Neural Computation* **4**, 1–58.

Gutfinger, D. and Sklansky, J. (1991). Robust classifiers by mixed adaptation. *IEEE Transactions on Pattern Analysis and Machine Intelligence* **13**, 552–567.

Hall, P. (1989). On projection pursuit regression. *Annals of Statistics* **17**, 573–588.

Hanley, J.A. and McNeil, B.J. (1982). The meaning and use of the area under a receiver operating characteristic (ROC) curve. *Radiology* **143**, 29–36.

Hansen, L.K., Liisberg, C. and Salamon, P. (1994). The error-reject tradeoff. Can be obtained by FTP from NeuroProse (archive.cis.ohio-state.edu).

Hansen, L.K. and Salamon, P. (1990). Neural networks ensembles. *IEEE Trans. Pattern Anal. Machine Intell.* **12**, 993–1001.

Hastie, T. and Tibshirani, R. (1986). Generalized additive models. *Statistical Science* **1**, 297–318.

Hinton, G.E. (1986). Learning distributed representations of concepts. In *Proceedings of the 8th Annual Conference of the Cognitive Science Society*, pp. 1–12. Erlbaum, Hillsdale, NJ.

Huber, P.J. (1985). Projection pursuit (with discussion). *Annals of Statistics* **13**, 435–475.

Huynh, Q., Cooper, L.N., Intrator, N. and Shouval, H. (1998). Classification of underwater mammals using feature extraction based on time-frequency analysis and bcm theory. *IEEE Trans. Signal Processing* **46**, 1202–1207.

Intrator, N. (1990). A neural network for feature extraction. In *Advances in Neural Information Processing Systems* 2, Eds. D.S. Touretzky and R.P. Lippmann, pp. 719–726. Morgan Kaufmann, San Mateo, CA.

Intrator, N. (1993). Combining exploratory projection pursuit and projection pursuit regression with application to neural networks. *Neural Computation* **5**, 443–455.

Intrator, N. (1996). Neuronal goals: efficient coding and coincidence detection. In *Proceedings of ICONIP Hong Kong. Progress in Neural Information Processing*, Eds. S. Amari, L. Xu, L.W. Chan, I. King and K.S. Leung, **1**, pp. 29–34. Springer-Verlag, Berlin.

Intrator, N. and Cooper, L.N. (1992). Objective function formulation of the BCM theory of visual cortical plasticity: Statistical connections, stability conditions. *Neural Networks* **5**, 3–17.

Intrator, N., Reisfeld, D. and Yeshurun, Y. (1996). Face recognition using a hybrid supervised/unsupervised neural network. *Pattern Recognition Letters* **17**, 67–76.

Jacobs, R.A. (1995). Methods for combining experts' probability assessments. *Neural Computation* **7**, 867–888.

Jacobs, R.A., Jordan, M.I., Nowlan, S.J. and Hinton, G.E. (1991). Adaptive mixtures of local experts. *Neural Computation* **3**, 79–87.

James, W. and Stein, C. (1960). Estimation with quadratic loss. In *Proc. Fourth Berkeley Symp. Math. Stat. Probab.*, volume 1, p. 361–380. University of California Press, CA.

Jaynes, E.T. (1982). On the rationale of maximum entropy methods. *Proc. IEEE* **70**, 939–952.

Jones, M.C. and Sibson, R. (1987). What is projection pursuit? (with discussion). *J. Roy. Statist. Soc.* A **150**, 1–36.

Jordan, M.I. and Jacobs, R.A. (1994). Hierarchical mixtures of experts and the EM algorithm. *Neural Computation* **6**, 181–214.

Kendall, M. and Stuart, A. (1977). *The Advanced Theory of Statistics*, volume 1. MacMillan Publishing, New York.

Krogh, A. and Hertz, J.A. (1992). A simple weight decay can improve generalization. In *Advances in Neural Information Processing Systems* **4**, Eds. J. Moody, S. Hanson and R. Lippmann, pp. 950–957. Morgan Kaufmann, San Mateo, CA.

Krogh, A. and Vedelsby, J. (1995). Neural network ensembles, cross validation, and active learning. In *Advances in Neural Information Processing Systems* **7**, 231–238.

Kruskal, J.B. (1969). Toward a practical method which helps uncover the structure of the set of multivariate observations by finding the linear transformation which optimizes a new 'index of condensation'. In *Statistical Computation*, Eds. R.C. Milton and J.A. Nelder, pp. 427–440. Academic Press, New York.

Kruskal, J.B. (1972). Linear transformation of multivariate data to reveal clustering. In *Multidimensional Scaling: Theory and Application in the Behavioral Sciences, I, Theory*, Eds. R.N. Shepard, A.K. Romney and S.B. Nerlove, pp. 179–191. Seminar Press, New York.

Lang, K.J. and Witbrock, M.J. (1988). Learning to tell two spirals apart. In *Proceedings of the 1988 Connectionists Models*, Eds. D.S. Touretzky, J.L. Ellman, T.J. Sejnowski and G.E. Hinton, pp. 52–59. Cargenie-Mellon University, Pittsburgh.

Le Cun, Y., Boser, B., Denker, J., Henderson, D., Howard, R., Hubbard, W. and Jackel, L. (1989). Backpropagation applied to handwritten zip code recognition. *Neural Computation* **1**, 541–551.

Le Cun, Y., Denker, J. and Solla, S. (1990). Optimal brain damage. In *Advances in Neural Information Processing Systems* **2**, Ed. D.S. Touretzky, pp. 598–605, Morgan Kaufmann, San Mateo, CA.

LeBlanc, M. and Tibshirani, R. (1993). Combining estimates in regression and classification. Can be obtained by FTP fron NeuroProse (`archive.cis.ohio-state.edu`).

Meir, R. (1994). Bias, variance and the combination of estimators: The case of linear least squares.
ftp://archive.cis.ohio-state.edu/pub/neuroprose/meir.bias-variance.ps.Z.

Murphy, P.M. and Aha, D.W. (1992). UCI Repository of machine learning databases. Department of Information and Computer Science. University of California at Irvine. Anonymous ftp from
ics.uci.edu:/usr2/spool/ftp/pub/machine-learning-databases.

Naftaly, U., Intrator, N. and Horn, D. (1997). Optimal ensemble averaging of neural networks. *Network: Computation in Neural Systems* **8**, 283–296.

Nowlan, S.J. and Hinton, G.E. (1992). Simplifying neural networks by soft weight-sharing. *Neural Computation* **4**, 473–493.

Olshausen, B.A. and Field, D.J. (1996). Emergence of simple cell receptive field properties by learning a sparse code for natural images. *Nature* **381**, 607–609.

Perrone, M.P. (1993). *Improving Regression Estimation: Averaging Methods for Variance Reduction with Extensions to General Convex Measure Optimization*. Ph.D. thesis, Institute for Brain and Neural Systems, Brown University.

Perrone, M.P. and Cooper, L.N. (1993). When networks disagree: Ensemble method for neural networks. In *Neural Networks for Speech and Vision*. Ed. R.J. Mammone. Chapman and Hall, London.

Poggio, T. and Girosi, F. (1990). Networks for approximation and learning. *IEEE Proceedings* **78**, 1481–1497.

Raviv, Y. and Intrator, N. (1996). Bootstrapping with noise: An effective regularization technique. *Connection Science, Special issue on Combining Estimators* **8**, 356–372.

Ripley, B.D. (1996). *Pattern Recognition and Neural Networks*. Cambridge University Press.

Rumelhart, D.E., Hinton, G.E. and Williams, R.J. (1986). Learning internal representations by error propagation. In *Parallel Distributed Processing*, Volume 1, Eds. D.E. Rumelhart and J.L. McClelland, pp. 318–362. MIT Press, Cambridge, MA.

Schapire, R. (1990). The strength of weak learnability. *Machine Learning* **5**, 197–227.

Shimshoni, Y. (1995). Classification of seismic signals using ensembles of neural networks. M.Sc. thesis, School of Mathematical Sciences, Tel Aviv University. Anonymous ftp from: ftp.math.tau.ac.il (pub/shimsh/thesis.ps.Z).

Shimshoni, Y. and Intrator, N. (1998). Classifying seismic signals by integrating ensembles of neural networks. *IEEE Trans. Signal Processing* **46**, 1194–1201.
ftp://cns.brown.edu/nin/papers/p.ps.Z.

Sietsma, J. and Dow, R. J.F. (1991). Creating artificial neural networks that generalise. *Neural Networks* **4**, 67–79.

Sollich, P. and Krogh, A. (1996). Learning with ensembles: How over-fitting can be useful. In *Advances in Neural Information Processing Systems* **8**, 190–196.

Stainvas, I., Intrator, N. and Moshaiov, A. (1997). Improving recognition via reconstruction. Preprint (ftp://cns.brown.edu/nin/papers/tradenips.ps.Z).

Stuart, A. and Ord, J.K. (1994). *Kendall's Advanced Theory of Statistics*. Edward Arnold, London.

Switzer, P. (1970). Numerical classification. In *Geostatistics*, Ed. V. Barnett. Plenum Press, New York.

Taniguchi, M. and Tresp, V. (1997). Averaging regularized estimators. *Neural Computation* **9**, 1163–1178.

Tresp, V. and Taniguchi, M. (1995). Combining estimators using non-constant weighting functions. In *Advances in Neural Information Processing Systems* 7, 419–426.

Viola, P. and Wells, W.M. (1995). Alignment by maximisation of mutual information. In *Proceedings of the 5th International Conference on Computer Vision*, pp. 16–23. IEEE, Cambridge, MA.

Wahba, G. (1990). *Splines: Models for Observational Data*. Series in Applied Mathematics, Vol. 59, SIAM, Philadelphia.

Waibel, A., Hanazawa, T., Hinton, G., Shikano, K. and Lang, K. (1989). Phoneme recognition using time-delay neural networks. *IEEE Transactions on ASSP* 37, 328–339.

Wand, M.P. (1994). Fast computation of multivariate kernel estimators. *Journal of Computational and Graphical Statistics* 3, 433–445.

Weigend, A.S., Huberman, B.A. and Rumelhart, D. (1990). Predicting the future: A connectionist approach. *Int. J. Neural Syst.* 1, 193–209.

Wolpert, D.H. (1992). Stacked generalisation. *Neural Networks* 5, 241–259.

Yamac, M. (1969). Can we do better by combining 'supervised' and 'nonsupervised' machine learning for pattern analysis? Ph.D. dissertation, Brown University.

Yang, H.H. and Amari, S. (1997). Adaptive on-line learning algorithms for blind separation—maximum entropy and minimum mutual information. *Neural Computation* 9, 1457–1482.

5

Density Networks

David J.C. MacKay and Mark N. Gibbs

Cavendish Laboratory, University of Cambridge, Madingley Road,
Cambridge, CB3 0HE, United Kingdom
mackay@mrao.cam.ac.uk, mng10@mrao.cam.ac.uk

Abstract

A density network is a neural network that maps from unobserved inputs to observable outputs. The inputs are treated as latent variables so that, for given network parameters, a nontrivial probability density is defined over the output variables. This probabilistic model can be trained by various Monte Carlo methods. The model can discover a description of the observed data in terms of an underlying latent variable space of lower dimensionality. We review results of the application of these models to toy problems with categorical and real-valued observables and to protein data.

1 Density modelling

The most popular supervised neural networks, multilayer perceptrons, are well established as probabilistic models for *regression* and *classification*, both of which are *conditional* modelling tasks: the *input* variables are assumed given, and we *condition* on their values when modelling the distribution over the *output* variables; no model of the density over input variables is constructed. Density modelling (or generative modelling), on the other hand, denotes modelling tasks in which a density over *all* the observable quantities is constructed. Multilayer perceptrons have not conventionally been used to create density models (though belief networks and other neural networks such as the Boltzmann machine do define density models). This chapter discusses how one can use a multilayer perceptron as a density model. This definition of a full probabilistic model with a multilayer perceptron may also prove useful for other interesting problems, for example, the 'missing inputs' problem (Tresp, Ahmad and Neuneier, 1994).

1.1 Traditional density models

A popular class of density models are *mixture models*, which define the probability distribution over observables **t** as a sum of simple densities. These models are 'latent variable' models (Everitt, 1984); each observation $\mathbf{t}^{(n)}$ has an associated *categorical* latent variable $c^{(n)}$ that states which *class* of the mixture that observation comes from. These latent variables $\{c^{(n)}\}$ are not observed, but their values are inferred during the modelling process.

Mixture models might however be viewed as inappropriate models for high-dimensional data spaces such as images or genome sequences. If we imagine modelling a sequence of densities with increasing numbers of independent degrees of freedom, the number of required components in a mixture model has to scale exponentially. Consider, for example, a protein family in which there are two independent correlations: one pair of amino acids in the protein chain are either both hydrophobic, or both hydrophilic, say, and two other amino acids elsewhere in the chain have anti-correlated size. A mixture model would have to use four categories to capture the four valid combinations of these binary attributes, whereas only two independent degrees of freedom are really present. Thus a *combinatorial* representation of underlying variables would seem more appropriate. Luttrell's (1994) partitioned mixture distribution is motivated similarly, but is a different form of quasi-probabilistic model.

These observations motivate the development of density models that have *components* rather than *categories* as their latent variables (Hinton and Zemel, 1994). Let us denote the observables by $\mathbf{t}$. If a density is defined on the latent variables $\mathbf{x}$, and a parameterised mapping is defined from these latent variables to a probability distribution over the observables $P(\mathbf{t}|\mathbf{x}, \mathbf{w})$, then when we integrate over the unknowns $\mathbf{x}$ a nontrivial density over $\mathbf{t}$ is defined, $P(\mathbf{t}|\mathbf{w}) = \int d\mathbf{x}\, P(\mathbf{t}|\mathbf{x}, \mathbf{w}) P(\mathbf{x})$. Simple linear models of this form in the statistics literature are called 'factor analysis' models. In a 'density network' (MacKay, 1995a) $P(\mathbf{t}|\mathbf{x}, \mathbf{w})$ is defined by a more general nonlinear parameterised mapping, and interesting priors on $\mathbf{w}$ may be used. This chapter reviews work on density networks where the observables are categorical (MacKay, 1995a, 1996), and describes preliminary research on the problem of real observables. In Section 2 the model is defined. In Section 3 an implementation of the model using a crude importance sampling method is reviewed, and its application to modelling of protein sequences is described in Section 4. In Section 5, a more sophisticated implementation using the hybrid Monte Carlo method (Neal, 1993) is described and applied to a toy problem with real-valued observables.

2   The density network model

2.1   Description of the model

The latent variables of the model form a vector $\mathbf{x}$ indexed by $h = 1, \ldots, H$; 'h' is mnemonic for 'hidden'. The dimensionality of this hidden space is H but the effective dimensionality of the density in the output space may be smaller, as some of the hidden dimensions may be effectively unused by the model. The relationship between the latent variables and the observables has the form of a mapping from inputs to outputs $\mathbf{y}(\mathbf{x}; \mathbf{w})$, parameterised by $\mathbf{w}$, and a probability of targets given outputs, $P(\mathbf{t}|\mathbf{y})$. The observed data are a set of target vectors $D = \{\mathbf{t}^{(n)}\}_{n=1}^{N}$. To complete the model we assign a prior $P(\mathbf{x})$ to the latent variables (an independent prior for each vector $\mathbf{x}^{(n)}$) and a prior $P(\mathbf{w})$ to the unknown parameters. In the applications that follow the priors over $\mathbf{w}$ and $\mathbf{x}^{(n)}$ are assumed to be spherical Gaussians with widths controlled by hyperparameters α; other distributions could easily be implemented and compared, if desired. In summary, the probability of everything is

$$P(D, \{\mathbf{x}^{(n)}\}, \mathbf{w}|\mathcal{H}) = \prod_n \left[P(\mathbf{t}^{(n)}|\mathbf{x}^{(n)}, \mathbf{w}, \mathcal{H}) P(\mathbf{x}^{(n)}|\mathcal{H}) \right] P(\mathbf{w}|\mathcal{H}), \qquad (2.1)$$

where $\mathcal{H}$ denotes the overall model. It will be convenient to define 'error functions' $G^{(n)}(\mathbf{x}; \mathbf{w})$ as follows:

$$G^{(n)}(\mathbf{x}^{(n)}; \mathbf{w}) \equiv \log P(\mathbf{t}^{(n)}|\mathbf{x}^{(n)}, \mathbf{w}). \qquad (2.2)$$

The function $G^{(n)}$ depends on the nature of the problem. If $\mathbf{t}$ consists of real variables then $G^{(n)}$ might be a summed-squared error between $\mathbf{t}$ and $\mathbf{y}$; in a 'softmax' classifier, where the observations $\mathbf{t}$ are categorical, in general we may have many output groups of different types. The following derivation applies to all cases. As an example, it may be useful to have the following one-layer model in mind. Each observation $\mathbf{t} = t_s \frac{S}{s=1}$ (e.g., a single protein sequence) consists of S categorical attributes that are believed to be correlated; S will be the number of columns in the protein alignment. Each attribute can take one of I discrete values (e.g., $I = 20$), a probability over which is modelled with a softmax logistic distribution,

$$P(\mathbf{t}|\mathbf{x}, \mathbf{w}) = \prod_{s=1}^{S} \left\{ y_{t_s}^{(s)}(\mathbf{x}; \mathbf{w}) \right\}, \qquad (2.3)$$

where

$$y_i^{(s)}(\mathbf{x}; \mathbf{w}) = \exp(a_i^{(s)}(\mathbf{x}; \mathbf{w})) \Big/ \sum_{i'} \exp(a_{i'}^{(s)}(\mathbf{x}; \mathbf{w})). \qquad (2.4)$$

The parameters $\mathbf{w}$ form a matrix of $(H + 1) \times (S \times I)$ weights from the H latent variables $\mathbf{x}$ and one bias unit, which contributes an intercept term to the $S \times I$ outputs:

$$a_i^{(s)}(\mathbf{x}; \mathbf{w}) = w_{i0}^{(s)} + \sum_{h=1}^{H} w_{ih}^{(s)} x_h. \qquad (2.5)$$

The intercept parameter $w_{i0}^{(s)}$ is called the bias of output unit (s, i). The data items $\mathbf{t}$ and their associated latent variables $\mathbf{x}$ are labelled by an index $n = 1, \ldots, N$, not included in the above equations, and the error function $G^{(n)}$ is

$$G^{(n)}(\mathbf{x}^{(n)}; \mathbf{w}) = \sum_s \log y_{t_s^{(n)}}(\mathbf{x}^{(n)}; \mathbf{w}). \qquad (2.6)$$

2.2 Learning in a density network

Having written down the probability of everything in equation (2.1) we can now make any desired inferences by turning the handle of probability theory. Let us aim towards the inference of the parameters $\mathbf{w}$ given the data D, through $P(\mathbf{w}|D, \mathcal{H})$. We can obtain this quantity conveniently by distinguishing two levels of inference.

Level 1: Given $\mathbf{w}$ and $\mathbf{t}^{(n)}$, infer $\mathbf{x}^{(n)}$. The posterior distribution of $\mathbf{x}^{(n)}$ is

$$P(\mathbf{x}^{(n)}|\mathbf{t}^{(n)}, \mathbf{w}, \mathcal{H}) = \frac{P(\mathbf{t}^{(n)}|\mathbf{x}^{(n)}, \mathbf{w}, \mathcal{H}) P(\mathbf{x}^{(n)}|\mathcal{H})}{P(\mathbf{t}^{(n)}|\mathbf{w}, \mathcal{H})}, \tag{2.7}$$

where the normalising constant is

$$P(\mathbf{t}^{(n)}|\mathbf{w}, \mathcal{H}) = \int d^H \mathbf{x}^{(n)} \, P(\mathbf{t}^{(n)}|\mathbf{x}^{(n)}, \mathbf{w}, \mathcal{H}) P(\mathbf{x}^{(n)}|\mathcal{H}). \tag{2.8}$$

This term, which we shall call the 'evidence', is in statistical parlance the marginal likelihood for $\mathbf{w}$ based on the observation $\mathbf{t}^{(n)}$.

Level 2: Given $D = \{\mathbf{t}^{(n)}\}$, infer $\mathbf{w}$.

$$P(\mathbf{w}|D, \mathcal{H}) = \frac{P(D|\mathbf{w}, \mathcal{H}) P(\mathbf{w}|\mathcal{H})}{P(D|\mathcal{H})} \tag{2.9}$$

The data-dependent term here is a product of the normalising constants of the level 1 inferences:

$$P(D|\mathbf{w}, \mathcal{H}) = \prod_{n=1}^{N} P(\mathbf{t}^{(n)}|\mathbf{w}, \mathcal{H}). \tag{2.10}$$

The evaluation of the evidence $P(\mathbf{t}^{(n)}|\mathbf{w}, \mathcal{H})$ for a particular n is a problem similar to the evaluation of the evidence for a supervised neural network (MacKay, 1992). There, the inputs $\mathbf{x}$ are given, and the parameters $\mathbf{w}$ are unknown; we obtain the evidence by integrating over $\mathbf{w}$. In the present problem, on the other hand, the hidden vector $\mathbf{x}^{(n)}$ is unknown, and the parameters $\mathbf{w}$ are conditionally fixed. For each n, we wish to integrate over $\mathbf{x}^{(n)}$ to obtain the evidence.

2.3   The derivative of the evidence with respect to $\mathbf{w}$

The derivative of the log of the evidence (equation 2.8) is

$$\frac{\partial}{\partial \mathbf{w}} \log P(\mathbf{t}^{(n)}|\mathbf{w}, \mathcal{H}) = \frac{1}{P(\mathbf{t}^{(n)}|\mathbf{w}, \mathcal{H})} \int d^H \mathbf{x}^{(n)} \exp(G^{(n)}) P(\mathbf{x}|\mathcal{H}) \frac{\partial}{\partial \mathbf{w}} G^{(n)}$$

$$= \int d^H \mathbf{x}^{(n)} P(\mathbf{x}^{(n)}|\mathbf{t}^{(n)}, \mathbf{w}, \mathcal{H}) \frac{\partial}{\partial \mathbf{w}} G^{(n)}(\mathbf{x}^{(n)}; \mathbf{w}). \tag{2.11}$$

In neural-network terminology, this gradient can thus be written as an expectation of the traditional 'backpropagation' gradient $(\partial/\partial \mathbf{w}) G^{(n)}(\mathbf{x}; \mathbf{w})$, averaging over the posterior distribution of $\mathbf{x}^{(n)}$ found in equation (2.7).

2.4   Higher levels—priors on **w**

We can continue up the hierarchical model, putting a prior on **w** with hyperparameters $\{\alpha_c\}$ which are inferred by integrating over **w**. These priors are important from a practical point of view in order to limit overfitting of the data by the parameters **w**. The complexity of a model is controlled by, among other factors, the number of latent variables that make a significant contribution to the generative model; with a hierarchical prior on **w** making use of hyperparameters $\{\alpha_c\}$, we can automatically infer the appropriate complexity. These priors will also be used to bias the solutions towards ones that are easier for humans to interpret.

3   A simple Monte Carlo implementation

The evidence and its derivatives with respect to **w** both involve integrals over the hidden components **x**. For a hidden vector of sufficiently small dimensionality, a simple Monte Carlo approach to the evaluation of these integrals can be effective. We use importance sampling with the sampler being defined by the prior, $P(\mathbf{x})$.

Let $\{\mathbf{x}^{(r)}\}_{r=1}^{R}$ be random samples from $P(\mathbf{x})$. Then we can approximate the log evidence by

$$\log P(\{\mathbf{t}^{(n)}\}|\mathbf{w}, \mathcal{H}) = \sum_{n} \log \int d^{H}\mathbf{x} \, \exp(G^{n}(\mathbf{x}; \mathbf{w})) P(\mathbf{x}) \tag{3.1}$$

$$\simeq \sum_{n} \log \left[\frac{1}{R} \sum_{r} \exp(G^{n}(\mathbf{x}^{(r)}; \mathbf{w})) \right]. \tag{3.2}$$

Similarly, the derivative can be approximated by

$$\frac{\partial}{\partial \mathbf{w}} \log P(\{\mathbf{t}^{(n)}\}|\mathbf{w}, \mathcal{H}) \simeq \sum_{n} \frac{\sum_{r} \exp(G^{n}(\mathbf{x}^{(r)}; \mathbf{w}))(\partial/\partial \mathbf{w})G^{n}(\mathbf{x}^{(r)}; \mathbf{w})}{\sum_{r} \exp(G^{n}(\mathbf{x}^{(r)}; \mathbf{w}))}. \tag{3.3}$$

This simple Monte Carlo approach loses the advantage that we gained when we rejected mixture models and turned to componential models: this implementation requires a number of samples R that is exponential in the dimension of the hidden space H. More sophisticated methods using stochastic dynamics (Neal, 1993) are described in Section 5.

4   Modelling a protein family

4.1   Introduction

A protein is a sequence of *residues*; each residue is one of the twenty amino acids. A protein family is a set of proteins believed to have the same physical structure but not necessarily having the same sequence of amino acids. In a *multiple sequence alignment*, residues of the individual sequences which occupy structurally analogous positions are aligned into columns. Columns can often be characterised by a predominance of particular amino acids. Lists of marginal frequencies over amino acids in different structural contexts are given by Nakai, Kidera and Kanehisa (1988). Such frequencies correspond to a first-order description of a protein family in which correlations between residues are not modelled.

The development of models for protein families is useful for two reasons. The first is that a good model might be used to identify new members of an existing family, and discover new families, in data produced by genome sequencing projects. The second reason is that a sufficiently complex model might be able to give new insight into the properties of the protein family; for example, properties of the proteins' tertiary structure might be elucidated by a model capable of discovering suspicious inter-residue correlations.

The principal probabilistic model that has been applied to protein families is a hidden Markov model (HMM) (Krogh *et al.*, 1994). Assuming that each state of the HMM corresponds to one column of a multiple sequence alignment, this model is not inherently capable of discovering long-range correlations, as Markov models, by definition, produce no correlations between the observables, given a hidden state sequence. The next-door neighbour of proteins, RNA, has been modelled with a 'covariance model' capable of capturing correlations between base-pairs in antiparallel RNA strands (Eddy and Durbin, 1994). Density networks offer a model capable of discovering general correlations between multiple arbitrary columns in a protein family. Steeg (1997) has developed an efficient statistical test for discovering correlated groups of residues. The density network approach is complementary to Steeg's in that (1) in the density network, a residue may be influenced by more than one latent variable, whereas Steeg's test is specialised for the case where the correlated groups are nonoverlapping, and (2) the density networks developed here define full probabilistic models rather than statistical tests. Here we model the protein families using a density network containing one softmax group for each column; see equations (2.3)–(2.6). The network has only one layer of weights connecting the latent variables $\mathbf{x}$ directly to the softmax groups. We have optimised $\mathbf{w}$ by evaluating the evidence and its gradient and feeding them into a conjugate gradient routine, thereby leading to maximisers of an appropriate likelihood. The random points $\{\mathbf{x}^{(r)}\}$ are kept fixed, so that the objective function and its gradient are deterministic functions during the optimisation. This also has the advantage of allowing one to get away with a smaller number of samples R than might be thought necessary, as the parameters $\mathbf{w}$ can adapt to make the best use of the empirical distribution over $\mathbf{x}$.

4.2   Regularisation schemes

A human prejudice towards comprehensible solutions gives an additional motivation for regularising the model, beyond the usual reasons for having priors. Here we encourage the model to be comprehensible in two ways.

(1) There is a redundancy in the model regarding where it gets its randomness from. Assume that a particular output is actually random and uncorrelated with other outputs. This could be modelled in two ways: its weights from the latent variables could be set to zero, and the biases could be set to the log probabilities; or alternatively the biases could be fixed to arbitrary values, with appropriate connections to unused latent variables being used to create the required probabilities, on marginalisation over the latent variables. In predictive terms, these two models would be identical, but we prefer the first solution, finding it more intelligible. To encourage such solutions we use a prior which only very weakly regularises the biases, so that they are 'cheap' relative to the other parameters.

(2) If the distribution $P(\mathbf{x})$ is rotationally invariant, then the predictive distribution is invariant under corresponding transformations of the parameters $\mathbf{w}$. If a solution can be expressed in terms of parameter vectors aligned with some of the axes (i.e. so that some parameters are zero), then we would prefer that representation. Here we create a nonspherical prior on the parameters by using multiple undetermined regularisation constants $\{\alpha_c\}$, each one associated with a class of weights; cf. the automatic relevance determination model (Neal, 1996; MacKay, 1994, 1995b).

A weight class consists of all the weights from one latent input to one softmax group, so that for a protein with S columns modelled using H latent variables, we introduced SH regularisation constants, each specifying whether or not a particular latent variable has an influence on a particular column. Given α_c, the prior on the parameters in class c is Gaussian with variance $1/\alpha_c$. This prior favours solutions in which one latent input has nonzero connections to all the units in some softmax groups (corresponding to small α_c), and negligible connections to other softmax groups (large α_c). The resulting solutions can easily be interpreted in terms of correlations between columns.

4.3 Method for optimisation of hyperparameters

For given values of $\{\alpha_c\}$, the parameters $\mathbf{w}$ were optimised to maximise the posterior probability. The hyperparameters $\{\alpha_c\}$ were adapted during the optimisation of the parameters $\mathbf{w}$ using a cheap and cheerful method motivated by Gaussian approximations (MacKay, 1992), in which

$$\alpha_c := f \frac{k_c}{\sum_{i \in c} w_i^2}. \tag{4.1}$$

Here k_c is the number of parameters in class c and f is a 'fudge factor' incorporated to imitate the effect of integrating over $\mathbf{w}$ and set to a value between 0.1 and 1.0.

This algorithm could be converted to a correct 'stochastic dynamics' Monte Carlo method (Neal, 1993) by adding an appropriate amount of noise to gradient descent on $\mathbf{w}$ and setting $f = 1$.

4.4   Toy data

A toy dataset was created which imitates a protein family with four columns each containing one of five amino acids (A–E). The 27 data items (Table 1) were constructed to exhibit two correlations between the columns. The first and second columns have a tendency both to be amino acid E together. The third and fourth columns are correlated such that, if one is amino acid A, then the other is likely to be A or B; if one is amino acid B, then the other is likely to be A, B or C; if one is C, then the other is likely to be B, C or D; and if one is D then the other is likely to be C or D. Thus a single underlying dimension runs through the amino acids A, B, C, D; E does not occur in these two columns. The model is given no prior knowledge of the 'spatial relationship' of the columns, or of the ordering of the amino acids. A model that can identify the two correlations in the data is what we are hoping for. Both regularised and unregularised density networks having four latent variables were used to model these data. Unregularised density networks give solutions that successfully predict the two correlations, but the parameters of those models are hard to interpret; see Figure 1(a). There is also evidence of overfitting of the data leading to overconfident predictions by the model. The regularised models, in which all the parameters connecting one input to one softmax group are put in a regularisation class with an unknown hyperparameter α_c, give interpretable solutions that clearly identify the two correlated groups of columns. Figure 1(b) shows the hyperparameters and parameters inferred in a typical solution using a regularised density network. Notice that two of the latent variables are unused in this solution. Of the other two inputs, one has an influence on columns 1 and 2 only, and the other has an influence on columns 3 and 4 only. Thus this model has successfully revealed the underlying 'structure' of the proteins in this family. The parameters are straightforward to interpret also; the first latent variable, when negative, makes E more probable in columns 1 and 2, and makes all other amino acids less probable, with the possible exceptions AExx and EBxx having been memorised; the second latent variable's parameters clearly show the one-dimensional ordering of the amino acids A, B, C, D.

4.5   Results on real data: globins

In MacKay (1995a) a rather ambitious attempt was made to model the joint density of the amino acids in an entire aligned protein sequence using a low-dimensional latent variable space. Data describing 400 proteins in the globin family were received in aligned form by courtesy of Sean Eddy (modelling of unaligned data is possible

Table 1 Toy data for a protein family.

EEAB	EECB	EEBC	EECC	EEAA	EEBA	EEBB	EECD
EEDC	EEDD	AACD	DDDC	CBDD	CCAB	BDCB	ABBC
CBCC	EDAA	ABBA	BCBB	DBAB	AECB	EBBC	BDCC
BCAA	DABA	BCBB					

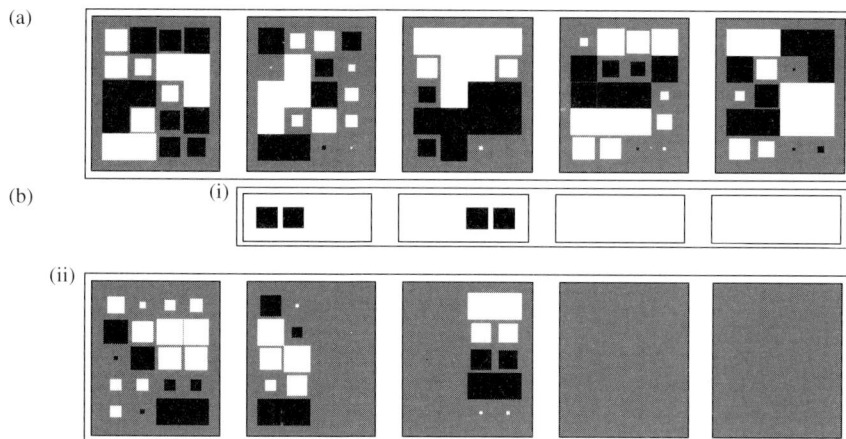

Fig. 1: Parameters and hyperparameters inferred for the toy protein family (a) without regularisation; (b) with adaptive regularisers. (a) Hinton diagram showing parameters **w** of model optimised without adaptive regularisers. Positive parameters are shown by black squares, negative by white. The magnitude of the parameter is proportional to the square's area. This diagram shows, in the five grey rectangles, the projective fields from the bias and the four latent variables to the outputs. In each grey rectangle the influences of one latent variable on the 20 outputs are arranged in a 5×4 grid: in each column the five output units correspond to the five amino acids. It is hard to interpret these optimised parameters. (b) The hyperparameters (i) and parameters (ii) of a hierarchical model with adaptive regularisers. The results are more intelligible and show a model that has discovered the two dimensions that underlie the data. Hyperparameters: each hyperparameter controls all the influences of one latent variable on one column. Square size denotes the value of $\sigma_w^2 = 1/\alpha$ on a log scale from 0.001 to 1.0. The model has discovered that columns 1 and 2 are correlated with each other but not with columns 3 and 4, and vice versa. Parameters: note the sparsity of the connections, making clear the two distinct underlying dimensions of this protein family.

in principle, but harder), with $S = 208$ columns each containing one of $I = 21$ symbols corresponding to 20 amino acids and 'deletion', or else a 'no measurement' symbol. The density network maps from H latent variables to SI outputs, grouped in S softmax groups of I units each. With $H = 20$ latent dimensions this model has about 80 000 parameters. The special case of $H = 0$ latent variables creates a model with independent probabilities over amino acids at each column, which is roughly equivalent to the hidden Markov model. Results for the case of $H = 2$ latent variables are shown in Figure 2. The choice of two dimensions makes the globins easy to visualise. The posterior mean of the latent components, estimated by importance sampling, is displayed for each globin. The globin subfamilies (GLB, HBA, HBB, etc.) are identified by the point style. The subfamilies are seen to be cleanly separated in this representation.

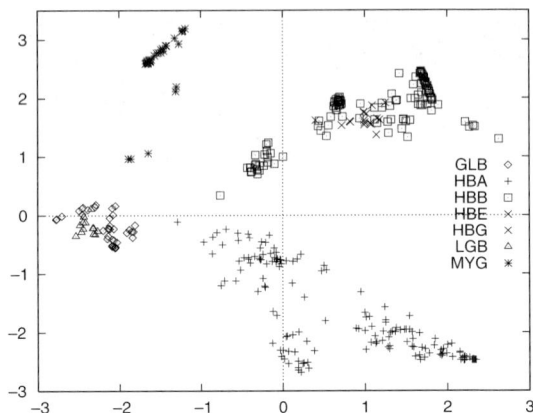

Fig. 2:   Two-dimensional representation of globins. From MacKay (1995a).

4.6   Results on real data: beta sheets

Beta sheets are structures in which two parts of the protein engage in a particular hydrogen-bonding interaction. It would greatly help in the solution of the protein folding problem if we could distinguish correct from incorrect alignments of beta strands.

$N = 1000$ examples were taken from aligned antiparallel beta strand data provided by Tim Hubbard. Density networks with $H = 6$ latent variables were used to model the joint distribution of the 12 residues surrounding a beta hydrogen bond. Our prior expectation is that, if there is any correlation among these residues, it is likely to reflect the spatial arrangement of the residues, with nearby residues being correlated. However, this prior expectation was not included in the model. The hope was that meaningful physical properties such as this would be learned from the data. The parameters of a typical optimised density network are shown in Figure 3. The parameter vectors were compared, column by column, with a large number of published amino acid indices (Nakai *et al.*, 1988) to see if they corresponded to established physical properties of amino acids. Each index was normalised by subtracting the mean from each vector and scaling it to unit length. The similarity of a parameter vector to an index was then measured by the magnitude of their inner product.

Two distinctive patterns reliably emerged in most adapted models, both having a meaningful physical interpretation. First, an alternating pattern can be seen in the influences of the second latent variable (third rectangle from the left). The influences on columns 2, 4, 9 and 11 are similar to each other, and opposite in sign to the influences on columns 3, 5, 10 and 12. This dichotomy between the residues is physically meaningful: residues 2, 4, 9 and 11 are on the opposite side of the beta sheet plane from residues 3, 5, 10 and 12; when these influence vectors were compared with Nakai *et al.*'s (1988) indices, they showed the greatest similarity to indices 57, 17, 7 and 42, which respectively describe the amino acids' polarity, the proportion of residues 100% buried, the transfer free energy to surface, and the consensus normalised hydrophobicity scale.

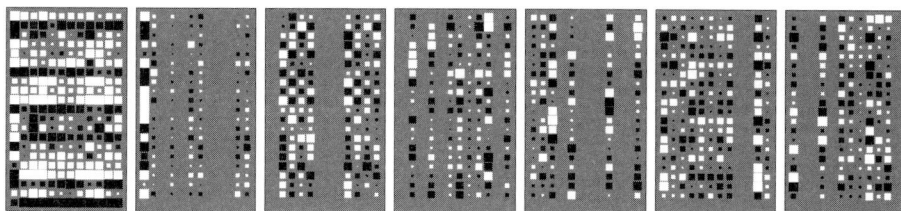

Fig. 3: Parameters **w** of an optimised density network modelling aligned antiparallel beta strands from MacKay (1996). In each grey rectangle the 12 columns represent the 12 residues surrounding a beta hydrogen bond, the first six on one strand and the second six on the other, with the hydrogen bond lying between residues 3, 4, 9 and 10. The 20 rows represent the 20 amino acids, in alphabetical order (A,C,D,...). Each rectangle shows the influences of one latent variable on the 12 × 20 probabilities. The top left rectangle shows the biases of all the output units. There is an additional 21st row in this rectangle for the biases of the output units corresponding to 'no amino acid'. The latent variables were defined to have no influence on these outputs to inhibit the wasting of latent variables on the modelling of dull correlations. The other six rectangles contain the influences of the six latent variables on the output units, of which the second and fifth are discussed in the text.

This latent variable has clearly discovered the inside–outside characteristics of the beta sheet structure: either one face of the sheet is exposed to the solvent (high polarity) or the other face, but not both.

Secondly, a different pattern is apparent in the second rectangle from the right. Here the influences on residues 4, 5, 6, 7, 8 are similar and opposite to the influences on 11, 12, 1, 2. For five of these residues the influence vector shows greatest similarity with index number 21, the normalised frequency of beta turn. What this latent variable has discovered, therefore, is that a beta turn may happen at one end or the other of two antiparallel beta strands, but not both.

Both of these patterns have the character of an 'exclusive-or' problem (Rumelhart and McClelland, 1986). One might imagine that an alternative way to model aligned beta sheets would be to train a *discriminative* model such as a neural network binary classifier to distinguish 'aligned beta sheet' from 'not aligned'. However, such a model would have difficulty learning these exclusive-or patterns. Exclusive-or *can* be learnt by a neural network with one hidden layer and two layers of weights, but it is not a natural function readily produced by such a network. In contrast these patterns are easily captured by the density networks presented here, which have only one layer of weights. It is interesting to note that the two effects discovered above involve competing correlations between large numbers of residues. The inside–outside latent variable produces a positive correlation between columns 4 and 11, for example, while the beta turn latent variable produces a negative correlation between those two columns. These results, although they do not constitute new discoveries, suggest that this technique shows considerable promise.

4.7   Future work

More complex models under development will include additional layers of processing
between the latent variables and the observables. The present model has no way of
knowing that the 21 categories within a softmax group have the same meaning for
all softmax groups. Imagine, for example, that two amino acids are functionally
indistinguishable, and always occur with equal probability. With the present model,
this relationship would have to be learned S times over, once for each softmax group.
If, however, some of the parameters of a second layer were common to all columns of
the protein, the model would be able to generalise amino acid equivalences from one
column to another. This would reduce overfitting and improve predictive performance.
It is hoped that a density network adapted to beta sheet data will eventually be useful
for discriminating correct from incorrect alignments of beta strands. The present work
is not of sufficient numerical accuracy to achieve this.

5   Real density networks and stochastic dynamics

A major weakness of the importance sampling method described in the preceding
sections is that it does not scale well with an increasing number of latent variables.

   In this section we describe a more complex Monte Carlo implementation with
favourable scaling properties. We sample from the joint posterior probability of the
latent variables and the parameters,

$$P(\{\mathbf{x}^{(n)}\}, \mathbf{w} | D, \mathcal{H}) \propto \prod_n \left[P(\mathbf{t}^{(n)} | \mathbf{x}^{(n)}, \mathbf{w}, \mathcal{H}) P(\mathbf{x}^{(n)} | \mathcal{H}) \right] P(\mathbf{w} | \mathcal{H}), \qquad (5.1)$$

using a Gibbs/Metropolis method in which alternately the latent variables $\{\mathbf{x}^{(n)}\}$ are
sampled using a Metropolis method that converges to $P(\{\mathbf{x}^{(n)}\} | \{\mathbf{t}^{(n)}\}, \mathbf{w})$ and then the
parameters are sampled from their distribution given the data and the latent variables,
$P(\mathbf{w} | \{\mathbf{t}^{(n)}\}, \{\mathbf{x}^{(n)}\})$. The sampling is done using the hybrid Monte Carlo method (Neal,
1993) which is reviewed below. The step in which the parameters are sampled is
identical to ordinary Bayesian stochastic training of a neural network with known
inputs.

   One could also implement a simultaneous stochastic sampler which updated the
parameters and the latent variables at the same time as each other. Such an approach,
in which all N latent variable vectors were simultaneously altered, would have the
advantage of reducing random walk behaviour caused by correlations in the joint
posterior distribution, but it would have the disadvantage that, in order to maintain a
reasonable acceptance probability for these more complex proposals, one would have
to reduce the step size of the dynamical simulations.

4.1   A brief review of the hybrid Monte Carlo method

Radford Neal's 'hybrid Monte Carlo' method for neural networks is a sophisticated
Metropolis method, applicable to continuous state spaces, which makes use of gradient

information to reduce random walk behaviour. It will be described here in the context of the general problem of sampling a continuous vector $\mathbf{x}$ from a probability distribution $P(\mathbf{x})$ which can be written in the form

$$P(\mathbf{x}) = \frac{e^{-E(\mathbf{x})}}{Z} \tag{5.2}$$

where not only $E(\mathbf{x})$, but also its gradient with respect to $\mathbf{x}$, can be readily evaluated. It seems wasteful to use a simple random-walk Metropolis method when this gradient is available; the gradient indicates which direction one should go in to find states with higher probability! In the hybrid Monte Carlo method, the state space $\mathbf{x}$ is augmented by *momentum variables* $\mathbf{p}$, and there is an alternation of two types of proposal. The first proposal density randomises the momentum variable, leaving the state $\mathbf{x}$ unchanged. The second proposal density changes both $\mathbf{x}$ and $\mathbf{p}$ using reversible Hamiltonian dynamics as defined by the Hamiltonian

$$H(\mathbf{x}, \mathbf{p}) = E(\mathbf{x}) + K(\mathbf{p}), \tag{5.3}$$

where $K(\mathbf{p})$ is a 'kinetic energy' such as $K(\mathbf{p}) = \mathbf{p}^T \mathbf{p}/2$. Under these dynamics, the momentum variable determines where the state $\mathbf{x}$ goes, and the *gradient* of $E(\mathbf{x})$ determines how the momentum $\mathbf{p}$ changes. The net effect is that, during each of the dynamical proposals, the state of the system moves a distance that goes *linearly* with the computer time, rather than as the square root.

If the simulation of the Hamiltonian dynamics is numerically perfect then the proposals are accepted every time; if the simulation is imperfect, because of finite step sizes for example, then some of the dynamical proposals will be rejected. The rejection rule makes use of the change in $H(\mathbf{x}, \mathbf{p})$, which is zero if the simulation is perfect.

Asymptotically, we obtain samples $(\mathbf{x}^{(t)}, \mathbf{p}^{(t)})$ from the joint density

$$P_H(\mathbf{x}, \mathbf{p}) = \frac{1}{Z_H} \exp[-H(\mathbf{x}, \mathbf{p})] = \frac{1}{Z_H} \exp[-E(\mathbf{x})] \exp[-K(\mathbf{p})]. \tag{5.4}$$

This density is separable, so it is clear that the marginal distribution of $\mathbf{x}$ is the desired distribution $\exp[-E(\mathbf{x})]/Z$. So, simply discarding the momentum variables, we obtain a sequence of samples $\{\mathbf{x}^{(t)}\}$ which asymptotically come from $P(\mathbf{x})$. The source code in Figure 4 describes a hybrid Monte Carlo method which uses the 'leapfrog' algorithm to simulate the dynamics on the function findE(x), whose gradient is found by the function gradE(x).

5.2   Demonstration on a toy real-valued dataset

We created a simple two-dimensional dataset with a nonlinear dependence on a single underlying variable, shown in Figure 5. The points are in fact noisy samples from a semicircle.

We modelled these data with a density network with two latent variables. The parameters were put in just three weight classes (hidden unit biases, input weights

```
g = gradE ( x ) ;                    # set gradient using initial x
E = findE ( x ) ;                    # set objective function too

for l = 1:L                          # loop L times
   p = randn ( size(x) ) ;           # initial momentum is Normal(0,1)
   H = p' * p / 2 + E ;              # evaluate H(x,p)

   xnew = x
   gnew = g ;
   for tau = 1:Tau                   # make Tau 'leapfrog' steps

       p = p - epsilon * gnew / 2 ;  # make half-step in p
       xnew = xnew + epsilon * p ;   # make step in x

       gnew = gradE ( xnew ) ;       # find new gradient
       p = p - epsilon * gnew / 2 ;  # make half-step in p

   endfor

   Enew = findE ( xnew ) ;           # find new objective function
   Hnew = p' * p / 2 + Enew ;        # evaluate new value of H

   dH = Hnew - H ;                   # Decide whether to accept

   if ( dH < 0 )                   accept = 1 ;
   elseif ( rand() < exp(-dH) )    accept = 1 ;
   else                            accept = 0 ;
   endif

   if ( accept )
      g = gnew ;    x = xnew ;    E = Enew ;
   endif
endfor
```

Fig. 4:   Octave source code for the hybrid Monte Carlo method.

and output weights). The output noise level was fixed at $\sigma_x = 0.05$, and the latent variables, parameters and hyperparameters were updated iteratively as follows. First, the latent variables associated with each data point were updated using 200 proposals generated by the hybrid Monte Carlo method. Each proposal used 30 leapfrog steps. Secondly, conditional on the latent variables, the weights were updated using 200 proposals generated by the hybrid Monte Carlo method. Thirdly, the hyperparameters (the mean and variance of the latent variables, and the variance of the three weight classes) were updated; this step might ideally have used Gibbs sampling, but here the hyperparameters were reset to their maximum likelihood values conditional on the parameters. Figure 5 shows the situation after 10 iterations. For typical values of the latent variables, the output of the network is indeed in the semicircular region where the data are located.

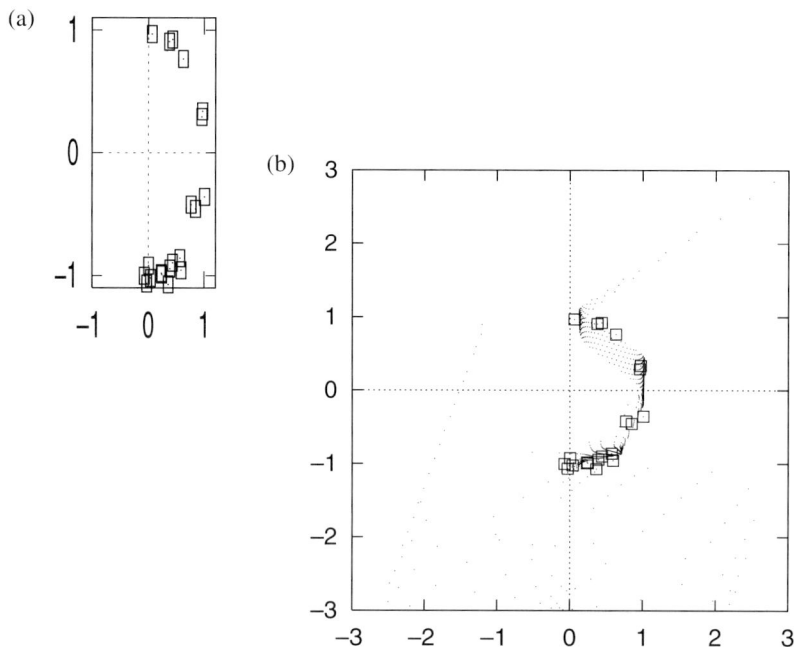

Fig. 5: Density network with real-valued outputs. (a) Data alone; the squares show the 20 data points. (b) State of the network after 10 iterations. The network had two latent variables, 16 hidden units and two outputs. The dots show the output of the network as its latent variables vary in the range ± two standard deviations around their mean.

6 Discussion

Density networks have shown some success as nonlinear latent variable models. It is particularly encouraging that, in applications to beta sheet protein sequences, a natural, separable description of a complex distribution was found, with one latent variable capturing a hydrophobic/hydrophilic effect, and another discovering a beta-turn/no-beta-turn effect. This separation would not have been found if we had not used a hierarchical prior with multiple hyperparameters. The biggest difficulty with density networks is that they require computer-intensive Monte Carlo methods. Importance sampling methods scale very badly with increasing dimensionality; stochastic dynamics methods scale better, but still worse than linearly with the number of variables.

A new latent variable model which can be implemented by local Monte Carlo methods with better scaling properties has recently been introduced by Hinton and Ghahramani (1997); this exciting model is able to discover both categorical and real latent variables. It has been applied to problems with high-dimensional real observables, but not yet to problems with categorical observables.

7 Acknowledgements

We thank Radford Neal, Geoff Hinton, Sean Eddy, Richard Durbin, Tim Hubbard and Graeme Mitchison for invaluable discussions. DJCM gratefully acknowledges the support of this work by the Royal Society Smithson Research Fellowship.

8 References

Eddy, S.R. and Durbin, R. (1994). RNA sequence analysis using covariance models. *Nucleic Acids Research* **22**, 2079–2088.

Everitt, B.S. (1984). *An Introduction to Latent Variable Models.* Chapman and Hall, London.

Hinton, G.E. and Zemel, R.S. (1994). Autoencoders, minimum description length and Helmholtz free energy. In *Advances in Neural Information Processing Systems* **6**, Eds. J.D. Cowan, G. Tesauro and J. Alspector, pp. 3–10. Morgan Kaufmann, San Mateo, California.

Hinton, G.E. and Ghahramani, Z. (1997). Generative models for discovering sparse distributed representations. *Phil. Trans. Royal Soc. B* **352**, 1177–1190.

Krogh, A., Brown, M., Mian, I.S., Sjolander, K. and Haussler, D. (1994). Hidden Markov models in computational biology: Applications to protein modeling, *Journal of Molecular Biology* **235**, 1501–1531.

Luttrell, S.P. (1994). The partitioned mixture distribution: an adaptive Bayesian network for low-level image processing. In *Proc. IEE Vision, Image and Signal Processing* **141**, 251–260.

MacKay, D.J.C. (1992). A practical Bayesian framework for backpropagation networks. *Neural Computation* **4**, 448–472.

MacKay, D.J.C. (1994). Bayesian non-linear modelling for the prediction competition. *ASHRAE Transactions.* **100**, 1053–1062, ASHRAE, Atlanta, Georgia.

MacKay, D.J.C. (1995a). Bayesian neural networks and density networks, *Nuclear Instruments and Methods in Physics Research.* A **354**, 73–80.

MacKay, D.J.C. (1995b). Probable networks and plausible predictions—a review of practical Bayesian methods for supervised neural networks. *Network: Computation in Neural Systems* **6**, 469–505.

MacKay, D.J.C. (1996). Density networks and their application to protein modelling. In *Maximum Entropy and Bayesian Methods, Cambridge 1994*, Eds. J. Skilling and S. Sibisi, pp. 259–268. Kluwer, Dordrecht.

Nakai, K., Kidera, A. and Kanehisa, M. (1988). Cluster analysis of amino acid indices for prediction of protein structure and function. *Prot. Eng.* **2**, 93–100.

Neal, R.M. (1993). Bayesian learning via stochastic dynamics, In *Advances in Neural Information Processing Systems* **5**, Eds. C.L. Giles, S.J. Hanson and J.D. Cowan, pp. 475–482. Morgan Kaufmann, San Mateo, California.

Neal, R.M. (1996). *Bayesian Learning for Neural Networks. Lecture Notes in Statistics 118.* Springer, New York.

Rumelhart, D.E. and McClelland, J.E. (1986). *Parallel Distributed Processing.* MIT Press, Cambridge, Mass.

Steeg, E. (1997). Automated Motif Discovery in Protein Structure Prediction, Ph.D. thesis, Department of Computer Science, University of Toronto.

Tresp, V., Ahmad, S. and Neuneier, R. (1994). Training neural networks with deficient data. In *Advances in Neural Information Processing Systems* **6**, Eds. J.D. Cowan, G. Tesauro and J. Alspector. Morgan Kaufmann, San Mateo, California.

6
Latent Variable Models and Data Visualisation

Christopher M. Bishop and Michael E. Tipping

Microsoft Research
St. George House, 1 Guildhall Street
Cambridge CB2 3NH, U.K.

Abstract

Visualisation is a powerful and widely used technique for data analysis and data mining. For simple datasets a single projection of the data on to a two-dimensional plane, such as that provided by principal component analysis, may prove adequate. In the case of more complex datasets, however, it may be necessary to find multiple plots corresponding to different projection directions and/or different subsets of the data points in order to capture the full complexity of the data. Here we use latent variable models to construct a framework for data visualisation which allows simultaneous soft clustering and projection of the data in a probabilistic setting. We first show how standard principal component analysis can be formulated in terms of maximum likelihood under a latent variable model. Next we extend the formalism to include both mixtures and hierarchical mixtures of principal component models, and derive the corresponding visualisation algorithms. Finally, we illustrate the hierarchical approach to visualisation using datasets obtained from multiphase flows along oil pipelines, and from satellite image data.

1 Introduction

The task of visualising the structure of datasets living in high-dimensional spaces provides an interesting example of a problem which has been addressed by both the statistical and neural computing communities, and indeed a variety of different algorithms has been proposed. Many of these algorithms rely on a projection of the data points *from* the high-dimensional data space *on to* the visualisation space. A familiar example is principal component analysis, or PCA (Jolliffe, 1986). In this chapter we consider an alternative framework for data visualisation based on the construction of a model for the probability distribution of the data. The probability models which will be used express the distribution in the high-dimensional data space in terms of a reduced number of latent, or hidden, variables. For the application to data visualisation we generally consider a two-dimensional latent space. Such models involve a mapping *to* the data variables *from* the latent variables, and are reviewed in Section 2. Once the model has been fitted to the data the transformation can be inverted using Bayes' theorem, leading to a visualisation of the dataset in the latent space. Although standard PCA has some useful characteristics which are important in the context of data visualisation, it does not define a probabilistic model of the data. A key result, derived in Section 3, is that principal component analysis can be reformulated as a specific form of latent variable model. This combines the key properties of standard PCA with the benefits of a probabilistic approach.

One of the limitations of many standard visualisation models, such as PCA, is that they attempt to project the entire dataset on to a single two-dimensional plane. While such algorithms can usefully display the structure of simple datasets, they often prove inadequate in the face of more complex applications. A single two-dimensional projection, even if it were nonlinear, may be insufficient to capture all of the interesting aspects of the dataset. For example, the projection which best separates two clusters may not be the best for revealing internal structure within one of the clusters. This problem can be addressed by exploiting an important advantage of a probabilistic formulation, namely the straightforward extension to *mixture models* (Titterington *et al.*, 1985; McLachlan and Basford, 1988; Tipping and Bishop, 1999a). Compared to the single visualisation projection, a mixture of latent variable models represents a more powerful and flexible density model which integrates clustering with data visualisation. We derive an efficient form of expectation-maximisation (EM) algorithm for training such models in Section 4, and illustrate its application to data visualisation. This approach can be extended further by introducing a *hierarchical* latent variable model. The goal is that the top-level projection should display the entire dataset, perhaps revealing the presence of clusters, while lower-level projections display internal structure within individual clusters, such as the presence of subclusters, which might not be apparent in the higher-level projections. An interesting feature of our algorithm is that it lends itself naturally to an interactive approach which proceeds in a top-down manner guided by the user. This will be illustrated using a simple toy problem. The hierarchical mixture model is derived in Section 5 and is applied in Section 6 to datasets derived from multiphase flows in oil pipelines and from remote-sensing satellite images. Finally, some further implications for models derived in this chapter are discussed in Section 7.

2   Latent variables

We consider the problem of modelling the distribution of data in a d-dimensional space labelled by the vector $\mathbf{t} = (t_1, \dots, t_d)$. The goal of a latent variable model is to express this distribution in terms of a smaller number of latent variables $\mathbf{x} = (x_1, \dots, x_q)$ where $q < d$. This is achieved by first decomposing the joint distribution $p(\mathbf{t}, \mathbf{x})$ into the product of the marginal distribution $p(\mathbf{x})$ of the latent variables and the conditional distribution $p(\mathbf{t}|\mathbf{x})$ of the data variables given the latent variables. We now assume that the conditional distribution factorises over the data variables, so that the joint distribution becomes

$$p(\mathbf{t}, \mathbf{x}) = p(\mathbf{x}) \prod_{i=1}^{d} p(t_i|\mathbf{x}). \qquad (2.1)$$

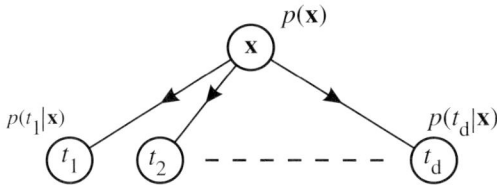

Fig. 1:   Probabilistic graphical representation of the latent variable distribution given by (2.1), in which the data variables $t_1, \ldots, t_d$ are conditionally independent given the latent variables $\mathbf{x}$.

This is the key to the latent variable structure: the data variables $t_1, \ldots, t_d$ are conditionally independent, given the latent variables $\mathbf{x}$, and it is the distribution over the latent variables which induces correlations amongst the data variables. The factorisation property (2.1) can be expressed graphically in terms of a Bayesian network, as shown in Figure 1. Next we express the conditional distribution $p(\mathbf{t}|\mathbf{x})$ in terms of a mapping from latent variables to data variables, so that

$$\mathbf{t} = \mathbf{y}(\mathbf{x}, \mathbf{w}) + \mathbf{u}, \tag{2.2}$$

where $\mathbf{y}(\mathbf{x}, \mathbf{w})$ is a function of the latent variable $\mathbf{x}$ with parameters $\mathbf{w}$, and $\mathbf{u}$ is an $\mathbf{x}$-independent noise process. The distribution $p(\mathbf{u})$, together with (2.2), then defines the conditional distribution $p(\mathbf{t}|\mathbf{x})$. Finally, the latent variable model is completed by specifying the form of the mapping $\mathbf{y}(\mathbf{x}, \mathbf{w})$ and by specifying the marginal distribution $p(\mathbf{x})$. For reasons which will become clear shortly, we consider $p(\mathbf{x})$ to be a *prior* distribution over the latent variables. The desired model for the distribution $p(\mathbf{t})$ of the data is then obtained by marginalising over the latent variables:

$$p(\mathbf{t}) = \int p(\mathbf{t}|\mathbf{x}) p(\mathbf{x}) \, d\mathbf{x}. \tag{2.3}$$

One of the simplest forms of latent variable model is factor analysis (Bartholomew, 1987), which is based on a linear mapping $\mathbf{y}(\mathbf{x}, \mathbf{w})$ of the form

$$\mathbf{t} = \mathbf{W}\mathbf{x} + \boldsymbol{\mu} + \mathbf{u}, \tag{2.4}$$

in which the parameter $\boldsymbol{\mu}$ allows the marginal distribution for $\mathbf{t}$ to have a non-zero mean. The prior distribution $p(\mathbf{x})$ is chosen to be a zero-mean isotropic Gaussian distribution $\mathcal{N}(\mathbf{0}, \mathbf{I})$, while the noise model for $\mathbf{u}$ is also Gaussian with zero mean and a covariance matrix $\boldsymbol{\Psi}$ which is diagonal. The integral in (2.3) now corresponds to the convolution of two Gaussians and hence is easily evaluated to give a distribution $p(\mathbf{t})$ which is also Gaussian, with mean $\boldsymbol{\mu}$ and a covariance matrix given by $\boldsymbol{\Psi} + \mathbf{W}\mathbf{W}^{\mathrm{T}}$. The parameters of the model, comprising $\mathbf{W}$, $\boldsymbol{\Psi}$ and $\boldsymbol{\mu}$, can be determined by maximum likelihood. There is, however, no closed-form analytical solution, and so their values must be determined by iterative procedures. An iterative EM (expectation-maximisation) algorithm for maximising the likelihood function for standard factor analysis was derived by Rubin and Thayer (1982). One of the obvious limitations of the factor analysis model is that the transformation $\mathbf{y}(\mathbf{x}, \mathbf{w})$ from latent space to data space is assumed to be linear. Extensions to nonlinear transformations, however, can lead to more complex training algorithms often requiring Monte Carlo sampling techniques. One approach, which allows flexible nonlinear transformations while leading to a deterministic, computationally efficient learning algorithm, is that of the *generative topographic mapping*, or GTM (Bishop *et al.*, 1998). This can be regarded as a probabilistic formulation of the heuristically derived self-organizing map, or SOM (Kohonen, 1995), which has been widely studied by the neural computing community, providing an interesting example of a fruitful link between neural computation and statistics. While GTM allows general nonlinear transformations, and can be applied to the problem of data visualisation, we shall explore an alternative approach in this chapter based on multiple *linear* models.

3   Probabilistic principal component analysis

Principal component analysis is a well-established technique for dimension reduction, and a chapter on the subject may be found in practically every text on multivariate analysis. Examples of its many applications include data compression, image processing, data visualisation, exploratory data analysis, pattern recognition and time series prediction.

The most common derivation of PCA is in terms of a standardised linear projection which maximises the variance in the projected space (Hotelling, 1933). For a set of observed d-dimensional data vectors $\{\mathbf{t}_n\}$, $n \in \{1 \ldots N\}$, the q *principal axes* $\mathbf{v}_j$, $j \in \{1 \ldots q\}$, are those orthonormal axes on to which the retained variance under projection is maximal. It can be shown that the vectors $\mathbf{v}_j$ are given by the q dominant eigenvectors (i.e. those with the largest associated eigenvalues λ_j) of the sample covariance matrix

$$\mathbf{S} = \frac{1}{N} \sum_{n=1}^{N} (\mathbf{t}_n - \boldsymbol{\mu})(\mathbf{t}_n - \boldsymbol{\mu})^{\mathrm{T}} \tag{3.1}$$

such that $\mathbf{S}\mathbf{v}_j = \lambda_j \mathbf{v}_j$. The q principal components of the observed vector $\mathbf{t}_n$ are given by the vector $\mathbf{u}_n = \mathbf{V}^{\mathrm{T}}(\mathbf{t}_n - \boldsymbol{\mu})$, where $\mathbf{V}^{\mathrm{T}} = (\mathbf{v}_1, \ldots, \mathbf{v}_q)^{\mathrm{T}}$, in which the variables u_j are decorrelated, such that the covariance matrix for $\mathbf{u}$ is diagonal with elements $\{\lambda_j\}$.

A complementary property of PCA, and that most closely related to the original discussions of Pearson (1901), is that, of all orthogonal linear projections $\mathbf{x}_n = \mathbf{V}^{\mathrm{T}}(\mathbf{t}_n - \boldsymbol{\mu})$, the principal component projection minimises the squared reconstruction error $\sum_n \|\mathbf{t}_n - \hat{\mathbf{t}}_n\|^2$, where the optimal linear reconstruction of $\mathbf{t}_n$ is given by $\hat{\mathbf{t}}_n = \mathbf{V}\mathbf{x}_n + \boldsymbol{\mu}$. One limiting disadvantage of both these definitions of PCA is the absence of a probability density model and associated likelihood measure. Deriving PCA from the perspective of density estimation would offer a number of important advantages, many of which will be discussed in Section 7. The key result of this section is to show that principal component analysis may indeed be obtained from a probability model (Tipping and Bishop, 1999b). In particular we show that the maximum-likelihood estimator of $\mathbf{W}$ in (2.4) for a specific form of latent variable model is given by the matrix of scaled and rotated principal axes of the data.

3.1 Relationship to latent variables

Links between principal component analysis and latent variable models have already been noted by a number of authors. For instance Anderson (1963) observed that principal components emerge when the data are assumed to comprise a systematic component plus an independent error term for each variable having common variance σ^2. Empirically, the similarity between the factor loadings and the principal axes has often been observed in situations in which the diagonal elements of $\mathbf{\Psi}$ are approximately equal (Rao, 1955). Basilevsky (1994) further notes that, when the model $\mathbf{WW}^{\mathrm{T}} + \sigma^2\mathbf{I}$ is exact, and therefore equal to $\mathbf{S}$, the factor loadings are identifiable and can be determined analytically through eigendecomposition of $\mathbf{S}$, without resort to iteration. For practical applications we must allow the probability model to be approximate. As well as assuming that the model is exact, such observations do not consider the maximum-likelihood context. By considering a particular case of the factor analysis model in which the noise covariance is isotropic, so that $\mathbf{\Psi} = \sigma^2\mathbf{I}$, we now show that, even when the covariance model is approximate, the maximum-likelihood estimator $\mathbf{W}_{\mathrm{ML}}$ is that matrix whose columns are the scaled and rotated principal eigenvectors of the sample covariance matrix $\mathbf{S}$. An important consequence of this derivation is that PCA may be expressed in terms of a probability density model, which we shall refer to as probabilistic principal component analysis (PPCA).

3.2 The probability model

For the isotropic noise model $\mathbf{u} \sim N(\mathbf{0}, \sigma^2\mathbf{I})$, equation (2.2) implies a probability distribution over $\mathbf{t}$-space for a given $\mathbf{x}$ given by

$$p(\mathbf{t}|\mathbf{x}) = (2\pi\sigma^2)^{-d/2} \exp\left\{-\frac{1}{2\sigma^2}\|\mathbf{t} - \mathbf{Wx} - \boldsymbol{\mu}\|^2\right\}. \tag{3.2}$$

In the case of an isotropic Gaussian prior over the latent variables defined by

$$p(\mathbf{x}) = (2\pi)^{-q/2} \exp\left\{-\frac{1}{2}\mathbf{x}^{\mathrm{T}}\mathbf{x}\right\}, \tag{3.3}$$

we then obtain the marginal distribution of $\mathbf{t}$ in the form

$$p(\mathbf{t}) = \int p(\mathbf{t}|\mathbf{x})p(\mathbf{x})d\mathbf{x} \tag{3.4}$$

$$= (2\pi)^{-d/2}|\mathbf{C}|^{-1/2} \exp\left\{-\frac{1}{2}(\mathbf{t} - \boldsymbol{\mu})^{\mathrm{T}}\mathbf{C}^{-1}(\mathbf{t} - \boldsymbol{\mu})\right\}, \tag{3.5}$$

where the model covariance matrix is

$$\mathbf{C} = \sigma^2 \mathbf{I} + \mathbf{W}\mathbf{W}^\mathrm{T}. \tag{3.6}$$

Using Bayes' rule, the *posterior* distribution of the latent variables $\mathbf{x}$ given the observed $\mathbf{t}$ is given by

$$p(\mathbf{x}|\mathbf{t}) = (2\pi)^{-q/2}|\sigma^{-2}\mathbf{M}|^{1/2}$$
$$\times \exp\left[-\frac{1}{2}\left\{\mathbf{x} - \mathbf{M}^{-1}\mathbf{W}^\mathrm{T}(\mathbf{t} - \mu)\right\}^\mathrm{T}(\sigma^{-2}\mathbf{M})\left\{\mathbf{x} - \mathbf{M}^{-1}\mathbf{W}^\mathrm{T}(\mathbf{t} - \mu)\right\}\right], \tag{3.7}$$

where the posterior covariance matrix is given by

$$\sigma^2\mathbf{M} = \sigma^2(\sigma^2\mathbf{I} + \mathbf{W}^\mathrm{T}\mathbf{W})^{-1}. \tag{3.8}$$

Note that $\mathbf{M}$ is $q \times q$ while $\mathbf{C}$ is $d \times d$.

The log-likelihood associated with the data under this model is

$$\mathcal{L} = \sum_{n=1}^{N} \ln\{p(\mathbf{t}_n)\}$$
$$= -\frac{Nd}{2}\ln(2\pi) - \frac{N}{2}\ln|\mathbf{C}| - \frac{N}{2}\mathrm{Tr}\left(\mathbf{C}^{-1}\mathbf{S}\right), \tag{3.9}$$

where the sample covariance matrix $\mathbf{S}$ of the observed $\{\mathbf{t}_n\}$ is given by (3.1).

In principle, we could determine the parameters for this model by maximising the log-likelihood $\mathcal{L}$ using the EM algorithm of Rubin and Thayer (1982). However, we now show that, for the case of an isotropic noise covariance matrix of the form we are considering, there is an exact analytical solution for the model parameters.

3.3   Properties of the maximum-likelihood solution

Our key result is that the log-likelihood (3.9) is maximised when the columns of $\mathbf{W}$ span the principal subspace of the data. To show this we consider the derivative of (3.9) with respect to $\mathbf{W}$:

$$\frac{\partial \mathcal{L}}{\partial \mathbf{W}} = N(\mathbf{C}^{-1}\mathbf{S}\mathbf{C}^{-1}\mathbf{W} - \mathbf{C}^{-1}\mathbf{W}), \tag{3.10}$$

which may be obtained from standard matrix differentiation results; see Krzanowski and Marriott (1994, p. 133). In Tipping and Bishop (1999b) it is shown that, with $\mathbf{C}$ given by (3.6), the only nonzero stationary points of (3.10) occur for

$$\mathbf{W} = \mathbf{U}_q(\Lambda_q - \sigma^2\mathbf{I})^{1/2}\mathbf{R}, \tag{3.11}$$

where the q column vectors in $\mathbf{U}_q$ are eigenvectors of $\mathbf{S}$, with corresponding eigenvalues in the diagonal matrix Λ_q, and $\mathbf{R}$ is an arbitrary $q \times q$ orthogonal rotation matrix. Furthermore, it is also shown that the stationary point corresponding to the *global maximum* of the likelihood occurs when $\mathbf{U}_q$ comprises the *principal* eigenvectors of $\mathbf{S}$ (i.e. the eigenvectors corresponding to the q largest eigenvalues), and that all other combinations of eigenvectors represent saddle-points of the likelihood surface. Thus, from (3.11), the columns of the maximum-likelihood estimator $\mathbf{W}_{\mathrm{ML}}$ contain the principal eigenvectors of $\mathbf{S}$, with scalings determined by the corresponding eigenvalue and the parameter σ^2, and with arbitrary rotation. It may also be shown that, for $\mathbf{W} = \mathbf{W}_{\mathrm{ML}}$, the maximum-likelihood estimator for σ^2 is given by

$$\sigma_{\mathrm{ML}}^2 = \frac{1}{d-q} \sum_{j=q+1}^{d} \lambda_j, \tag{3.12}$$

which has a clear interpretation as the variance 'lost' in the projection, averaged over the lost dimensions. Note that the columns of $\mathbf{W}_{\mathrm{ML}}$ are not orthogonal since

$$\mathbf{W}_{\mathrm{ML}}^{\mathrm{T}} \mathbf{W}_{\mathrm{ML}} = \mathbf{R}^{\mathrm{T}} (\, \mathbf{\Lambda}_q - \sigma^2 \mathbf{I}) \mathbf{R}, \tag{3.13}$$

which is not diagonal for $\mathbf{R} \neq \mathbf{I}$. In common with factor analysis, there exists a family of solutions related by rotations of the latent variable coordinates. This is also true of many iterative algorithms for determining principal components. However, unlike factor analysis, we can determine $\mathbf{R}$ by noting from (3.13) that the rows of $\mathbf{R}$ correspond to the eigenvectors of $\mathbf{W}_{\mathrm{ML}}^{\mathrm{T}} \mathbf{W}_{\mathrm{ML}}$. That the rotational ambiguity may be resolved is a consequence of the scaling of the eigenvectors by $(\, \mathbf{\Lambda}_q - \sigma^2 \mathbf{I})$.

In summary, we can obtain a probabilistic principal components model by finding the q principal eigenvectors and eigenvalues of the sample covariance matrix. The density model is then given by a Gaussian distribution with mean $\boldsymbol{\mu}$ given by the sample mean, and a covariance matrix $\mathbf{W}\mathbf{W}^{\mathrm{T}} + \sigma^2 \mathbf{I}$ in which $\mathbf{W}$ is given by (3.11) and σ^2 is given by (3.12).

3.4 Application to data visualisation

In standard principal component analysis, the data points are 'visualised' by orthogonal projection on to the principal components plane spanned by the two leading eigenvectors. For our probabilistic PCA model this projection is modified slightly. From (3.7) it may be seen that the *posterior mean* projection of $\mathbf{t}_n$ is given by $\langle \mathbf{x}_n \rangle = \mathbf{M}^{-1} \mathbf{W}^{\mathrm{T}} (\mathbf{t}_n - \boldsymbol{\mu})$. When $\sigma^2 \rightarrow 0$, $\mathbf{M}^{-1} \rightarrow (\mathbf{W}^{\mathrm{T}} \mathbf{W})^{-1}$ and $\mathbf{W}\mathbf{M}^{-1}\mathbf{W}^{\mathrm{T}}$ then becomes an orthogonal projection, so that PCA is recovered, although the density model then becomes singular, and thus undefined. For $\sigma^2 > 0$, the projection on to the manifold becomes skewed towards the origin as a result of the prior over $\mathbf{x}$. As a result of this, $\mathbf{W}\langle \mathbf{x}_n \rangle$ is *not* an orthogonal projection of $\mathbf{t}_n$. We note, however, that information is not lost because of this skewing, since each data point may still be optimally reconstructed from the latent variable by taking this skewing into account. With $\mathbf{W} = \mathbf{W}_{\mathrm{ML}}$ the required reconstruction is given by

$$\hat{\mathbf{t}}_n = \mathbf{W}_{\mathrm{ML}} \{ \mathbf{W}_{\mathrm{ML}}^{\mathrm{T}} \mathbf{W}_{\mathrm{ML}} \}^{-1} \mathbf{M} \langle \mathbf{x}_n \rangle. \tag{3.14}$$

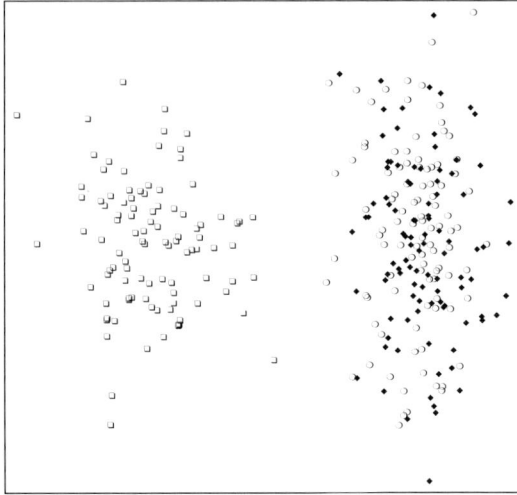

Fig. 2:   Plot of the posterior means of the data points from the toy dataset, obtained from the probabilistic PCA model, indicating the presence of two distinct clusters.

Thus the latent variables convey the information necessary to reconstruct the original data vector optimally, even in the case of $\sigma^2 > 0$. The dataset can therefore be visualised by mapping each data point on to the corresponding posterior mean $\langle \mathbf{x}_n \rangle$ in the two-dimensional latent space. We illustrate the visualisation properties using a toy dataset consisting of 450 data points generated from a mixture of three Gaussians in three-dimensional space. Each Gaussian is relatively flat (has small variance) in one dimension, and two of these clusters lie 'on top' of each other, while the third is well separated from the first two. The structure of this dataset has been chosen in order to demonstrate the benefits of the interactive hierarchical approach developed in Sections 4 and 5. A two-dimensional latent variable model is trained on this dataset, and the result of plotting the posterior means of the data points is shown in Figure 2.

4   Mixtures of principal component analysers

We now extend the latent variable model of Section 3 by considering a *mixture* of probabilistic principal component analysers (Tipping and Bishop, 1999a), so that the probability density takes the form

$$p(\mathbf{t}_n) = \sum_{i=1}^{M_0} \pi_i \, p(\mathbf{t}_n | i), \qquad (4.1)$$

where π_i represent the mixing coefficients, with $\pi_i \geq 0$ and $\sum \pi_i = 1$. For such a model, the log-likelihood associated with the dataset, assuming independent, identically distributed data, is

$$\mathcal{L} = \sum_{n}^{N} \ln \left[p(\mathbf{t}_n) \right] \tag{4.2}$$

$$= \sum_{n}^{N} \ln \left[\sum_{i=1}^{M_0} \pi_i \, p(\mathbf{t}_n | i) \right], \tag{4.3}$$

where $p(\mathbf{t}_n | i)$ is a single PCA model. (Mixtures of standard factor analysis models using the diagonal ($\boldsymbol{\Psi}$) noise covariance matrix have been employed for handwritten digit recognition (Hinton *et al.*, 1997).) Note that a separate mean vector $\boldsymbol{\mu}_i$ is now associated with each component of the mixture, along with the parameters $\mathbf{W}_i$ and σ_i^2.

4.1 EM algorithm

It is straightforward to construct an EM algorithm for estimating the parameters π_i, $\boldsymbol{\mu}_i$, $\mathbf{W}_i$ and σ_i^2. We treat the value of the component label i corresponding to each data point $\mathbf{t}_n$ as 'missing data' (Bishop, 1995). The E-step of the EM algorithm then involves the use of the current parameter estimates to evaluate the posterior probabilities (or 'responsibilities') of the mixture components i for the data points $\mathbf{t}_n$, given by Bayes' theorem:

$$R_{ni} = \frac{p(\mathbf{t}_n | i) \pi_i}{p(\mathbf{t}_n)}. \tag{4.4}$$

In the M-step, the mixing coefficients and component means are re-estimated using

$$\widetilde{\pi}_i = \frac{1}{N} \sum_{n}^{N} R_{ni} \tag{4.5}$$

$$\widetilde{\mu}_i = \frac{\sum_{n}^{N} R_{ni} \mathbf{t}_n}{\sum_{n}^{N} R_{ni}}, \tag{4.6}$$

while the parameters $\mathbf{W}_i$ and σ_i^2 are obtained by first evaluating the weighted covariance matrices, given by

$$\mathbf{S}_i = \frac{\sum_{n=1}^{N} R_{ni} (\mathbf{t}_n - \widetilde{\mu}_i)(\mathbf{t}_n - \widetilde{\mu}_i)^{\mathrm{T}}}{\sum_{n=1}^{N} R_{ni}}, \tag{4.7}$$

and then applying (3.11) and (3.12).

4.2 Interactive data visualisation

We can now apply the mixture formalism to the problem of data visualisation. Once a mixture of probabilistic PCA models has been fitted to the dataset, the procedure for visualising the data points involves plotting each data point $\mathbf{t}_n$ on each of the two-dimensional latent spaces at the corresponding posterior mean position $\langle \mathbf{x}_{ni} \rangle$, given by

$$\langle \mathbf{x}_{ni} \rangle = (\mathbf{W}_i^{\mathrm{T}} \mathbf{W}_i + \sigma_i^2 \mathbf{I})^{-1} \mathbf{W}_i^{\mathrm{T}} (\mathbf{t}_n - \boldsymbol{\mu}_i). \tag{4.8}$$

As a further refinement, the density of 'ink' for each data point is weighted by the corresponding responsibility R_{ni} of model i for that data point, so that the total density of 'ink' is distributed by a partition of unity across the M_0 plots. Thus, each data point is plotted on every component model projection, while, if a particular model takes nearly all of the posterior probability for a particular data point, then that data point will effectively be visible only on the corresponding latent space plot.

We shall regard the single PPCA plot of Section 3 as the top level in a hierarchical visualisation model, in which the mixture model forms the second level. Extensions to further levels of the hierarchy will be developed in Section 5.

The model can be extended to provide an interactive data exploration tool as follows. On the basis of the single top-level plot the user decides on an appropriate number of models to fit at the second level, and selects points $\mathbf{x}^{(i)}$ on the plot, corresponding, for example, to the centres of apparent clusters. The resulting points $\mathbf{y}^{(i)}$ in data space, obtained from $\mathbf{y}^{(i)} = \mathbf{W}\mathbf{x}^{(i)} + \boldsymbol{\mu}$, are then used to initialise the means $\boldsymbol{\mu}_i$ of the respective submodels. To initialise the matrices $\mathbf{W}_i$ we first assign the data points to their nearest mean vector $\boldsymbol{\mu}_i$ and then compute the corresponding sample covariance matrices. This is a hard clustering analogous to k-means and corresponds to approximating the posterior probabilities R_{ni} by replacing the largest posterior probability by 1 and the remainder by 0. For each of these clusters we then find the eigenvalues and eigenvectors of the sample covariance matrix and hence determine the probabilistic PCA density model. Starting from this initialization, the full EM algorithm is then run, as already discussed. The visualisation process can be enhanced further by providing information at the top level on the location and orientation of the latent spaces corresponding to the second level. This can be achieved by considering the orthogonal projection of the latent plane in data space on to the corresponding plane of the parent model, as illustrated in Figure 3.

Consider the application of this procedure to the toy dataset introduced in Section 3. At the top level we observed two apparent clusters, and so we might select a mixture of two models for the second level, with centres initialised somewhere near the centres of the two clusters seen at the top level. This leads to the two-level visualisation plot shown in Figure 4.

5   Hierarchical mixture models

We now extend the mixture representation of Section 4 to give a hierarchical mixture model. Our formulation will be quite general and can be applied to mixtures of any parametric density.

So far we have considered a two-level system consisting of a single latent variable model at the top level and a mixture of M_0 such models at the second level. We can

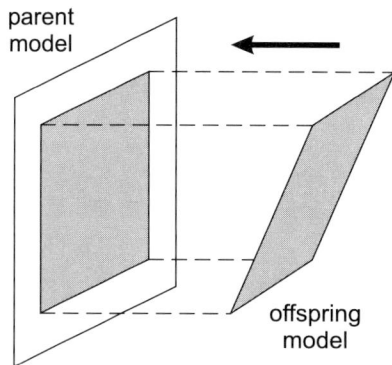

Fig. 3:   Illustration of the projection of one of the latent planes on to its parent plane.

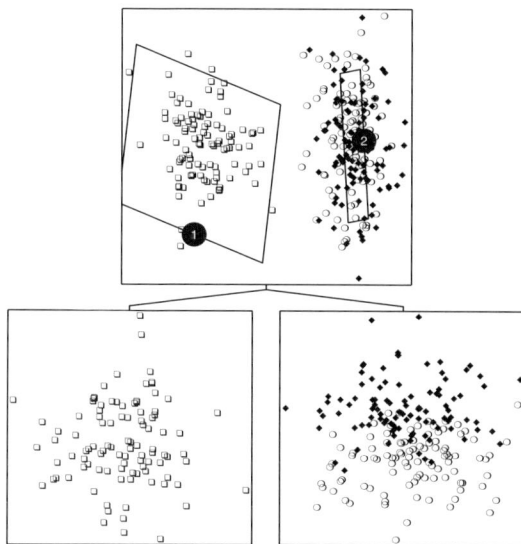

Fig. 4:   The result of applying the two-level visualisation algorithm to the toy dataset. At the second level a mixture of two latent variable models has been fitted and the data plotted on each latent space using the approach described in the text. In addition, the two latent planes have been visualised by projection back on to the top-level model.

now extend the hierarchy to a third level by associating a group $\mathcal{G}_i$ of latent variable models with each model i in the second level. The corresponding probability density can be written in the form

$$p(\mathbf{t}) = \sum_{i=1}^{M_0} \pi_i \sum_{j \in \mathcal{G}_i} \pi_{j|i}\, p(\mathbf{t}|i, j), \qquad (5.1)$$

where the $\{p(\mathbf{t}|i, j)\}$ again represent independent latent variable models, and $\pi_{j|i}$ correspond to sets of mixing coefficients, one set for each i, which satisfy $\sum_j \pi_{j|i} = 1$. Thus each level of the hierarchy corresponds to a generative model, with lower levels giving more refined and detailed representations. This model is illustrated in Figure 5.

The determination of the parameters of the models at the third level can again be viewed as a missing data problem in which the missing information corresponds to labels specifying which model generated each data point. When no information about the labels is provided the log-likelihood for the model (5.1) would take the form

$$\mathcal{L} = \sum_{n=1}^{N} \ln \left\{ \sum_{i=1}^{M_0} \pi_i \sum_{j\in\mathcal{G}_i} \pi_{j|i}\, p(\mathbf{t}|i, j) \right\}. \tag{5.2}$$

If, however, we were given a set of indicator variables z_{ni} specifying which model i at the second level generated each data point $\mathbf{t}_n$ then the log-likelihood would become

$$\mathcal{L} = \sum_{n=1}^{N} \sum_{i=1}^{M_0} z_{ni} \ln \left\{ \pi_i \sum_{j\in\mathcal{G}_i} \pi_{j|i}\, p(\mathbf{t}|i, j) \right\}. \tag{5.3}$$

In fact we only have partial, probabilistic information in the form of the posterior responsibilities R_{ni} for each model i having generated the data points $\mathbf{t}_n$, obtained from the second level of the hierarchy. The corresponding log-likelihood is obtained by taking the expectation of (5.3) with respect to the posterior distribution of the z_{ni} to give

$$\mathcal{L} = \sum_{n=1}^{N} \sum_{i=1}^{M_0} R_{ni} \ln \left\{ \pi_i \sum_{j\in\mathcal{G}_i} \pi_{j|i}\, p(\mathbf{t}|i, j) \right\}, \tag{5.4}$$

in which the R_{ni} are constants. In the particular case in which the R_{ni} are all 0 or 1, corresponding to complete certainty about which model in the second level is

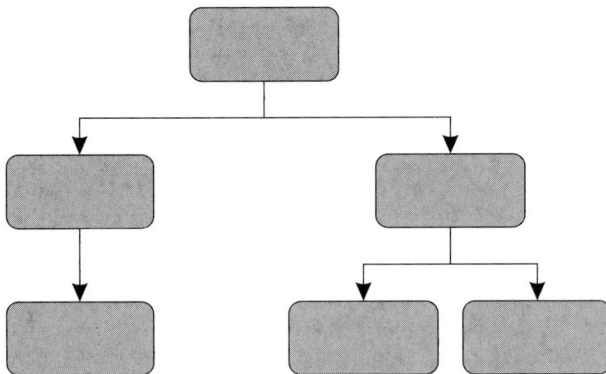

Fig. 5:   The structure of the hierarchical model.

responsible for each data point, the log-likelihood (5.4) reduces to the form (5.3). Maximisation of (5.4) can again be performed using the EM algorithm, as discussed by Bishop and Tipping (1998). This has the same form as the EM algorithm for a simple mixture, discussed in Section 4, except that, in the E-step, the posterior probability that model (i, j) generated data point $\mathbf{t}_n$ is given by

$$R_{ni,j} = R_{ni} R_{nj|i},\qquad(5.5)$$

in which

$$R_{nj|i} = \frac{\pi_{j|i}\, p(\mathbf{t}_n|i, j)}{\sum_{j'} \pi_{j'|i}\, p(\mathbf{t}_n|i, j')}.\qquad(5.6)$$

Note that the R_{ni} are constants determined from the second level of the hierarchy, and the $R_{nj|i}$ are functions of the 'old' parameter values in the EM algorithm. This result automatically satisfies the relation

$$\sum_{j \in \mathcal{G}_i} R_{ni,j} = R_{ni},\qquad(5.7)$$

so that the responsibility of each model at the second level for a given data point n is shared by a partition of unity between the corresponding group of offspring models at the third level. It is straightforward to extend this hierarchical approach to any desired number of levels.

The results of applying this approach to the toy dataset are shown in Figure 6.

6 Applications

We now illustrate the application of the hierarchical visualisation algorithm by considering two real datasets. The first of these arises from a noninvasive monitoring system used to determine the quantity of oil in a multiphase pipeline containing a mixture of oil, water and gas (Bishop and James, 1993). The diagnostic data are collected from a set of three horizontal and three vertical beam-lines along which gamma rays at two different energies are passed. By measuring the degree of attenuation of the gamma rays, the fractional path length through oil and water, and hence gas, can readily be determined, giving 12 diagnostic measurements in total. In practice the aim is to solve the inverse problem of determining the fraction of oil in the pipe. The complexity of the problem arises from the possibility of the multiphase mixture adopting one of a number of different geometrical configurations. Our goal is to visualize the structure of the data in the original 12-dimensional space. A dataset consisting of 1000 points is obtained synthetically by simulating the physical processes in the pipe, including the presence of noise dominated by photon statistics. Locally, the data are expected to have an intrinsic dimensionality of two, corresponding to the two degrees of freedom given by the fraction of oil and the fraction of water, the fraction of gas being redundant. However, the presence of different configurations, as well as the geometrical

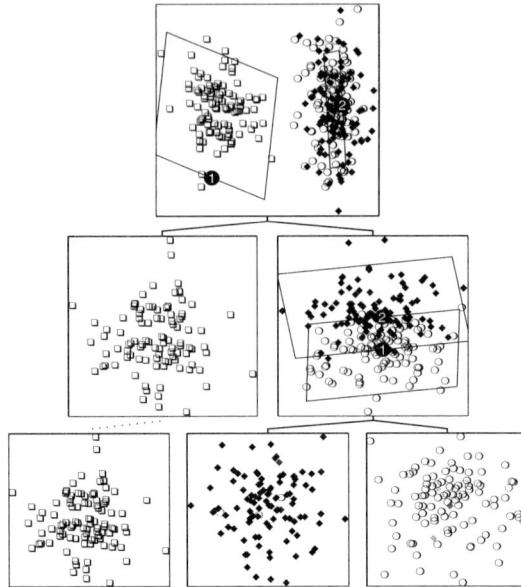

Fig. 6:   Plot of the complete three-level hierarchy for the toy dataset.

interaction between phase boundaries and the beam paths, leads to numerous distinct clusters. It would appear that a hierarchical approach of the kind discussed here should be capable of discovering this structure. Results from fitting the oil flow data using a three-level hierarchical model are shown in Figure 7.

In the case of the toy data, the optimal choice of clusters and subclusters is relatively unambiguous, and a single application of the algorithm is sufficient to reveal all of the interesting structure within the data. For more complex datasets, it is appropriate to adopt an exploratory perspective and investigate alternative hierarchies, through the selection of differing numbers of clusters and their respective locations. The example shown in Figure 7 has clearly been highly successful. Note how the apparently single cluster, number 2, in the top level plot is revealed to be two quite distinct clusters at the second level. Also, data points from the 'homogeneous' configuration have been isolated and can be seen to lie on a two-dimensional triangular structure in the third level. Inspection of the value of σ^2 confirms that this cluster is confined to a nearly planar subspace, as expected from the physics of the diagnostic data for the homogeneous configurations.

As a second example, we consider the visualisation of a dataset obtained from remote-sensing satellite images. Each data point represents a 3×3 pixel region of a satellite land image, and for each pixel there are four measurements of intensity taken at different wavelengths (approximately red and green in the visible spectrum and two in the near infra-red). This gives a total of 36 variables for each data point. There is also a label indicating the type of land represented by the central pixel. This dataset

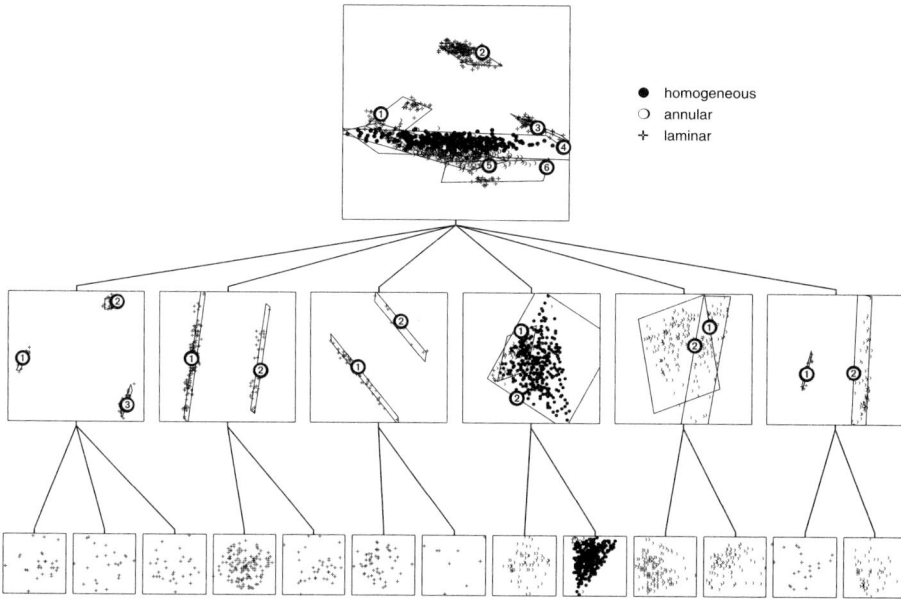

Fig. 7:   Results of fitting the oil data. The symbols denote different multiphase flow configurations corresponding to homogeneous (•), annular (○) and laminar (+).

has previously been the subject of a classification study within the STATLOG project (Michie *et al.*, 1994).

We applied the hierarchical algorithm to 600 data points, with 100 drawn at random from each of six classes in the 4435-point dataset. The result of fitting a three-level hierarchy is shown in Figure 8. Note that the class labels are used only to determine the symbol used to denote the data points and play no role in the maximum likelihood determination of the model parameters. Figure 8 illustrates that the data can be approximately separated into classes, and the 'grey soil' → 'damp grey soil' → 'very damp grey soil' continuum is clearly evident in component 3 at the second level. One particularly interesting additional feature is that there appear to be two distinct and separated clusters of 'cotton crop' pixels, in mixtures 1 and 2 at the second level, which are not evident in the single top-level projection. Study of the original image (Michie *et al.*, 1994, p. 144) indeed indicates that there are *two* separate areas of 'cotton crop', and this is almost certainly the reason for the bimodality. Unfortunately, it is not possible to verify this hypothesis as the supplied data have been deliberately randomised in order that the original image cannot be reconstructed.

7   Conclusions

In this chapter we have considered an approach to density modelling based on latent variables, and in particular we have shown how conventional principal component

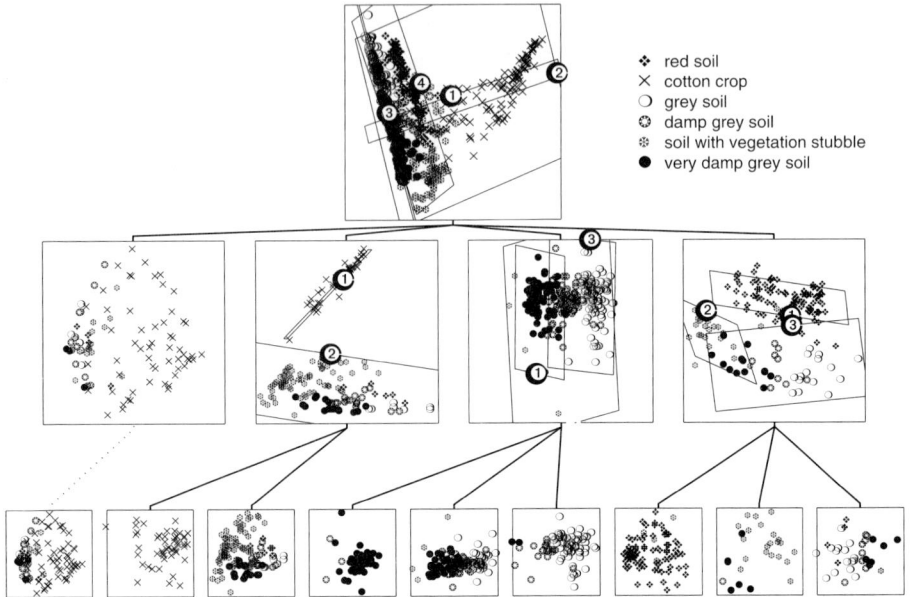

Fig. 8:   Results of fitting the satellite image data.

analysis can be reformulated in terms of a density model. This probabilistic formulation of PCA allows us to consider mixtures, and indeed hierarchical mixtures, of
PCA models. There are, however, many additional benefits of having a probabilistic
formulation of PCA, including the following:

- The definition of a likelihood measure permits comparison with other density-
  estimation techniques and facilitates statistical testing.

- Bayesian inference methods may be applied, e.g. for model comparison, by
  combining the likelihood with a prior.

- If PCA is used to model the class-conditional densities in a classification problem,
  the posterior probabilities of class membership may be computed.

- Missing values can be handled optimally (assuming the data are 'missing at
  random')

- The probability density function gives a measure of the 'novelty' of a new data
  point.

Our main application of probabilistic PCA in this chapter has been to the problem
of data visualisation. This leads to a flexible, interactive framework for visualising
complex datasets in spaces of many dimensions.

The concept of isolating groups of data points for further investigation can be
traced back to Maltson and Dammann (1965), and was further developed by Friedman
and Tukey (1974) for exploratory data analysis in conjunction with projection pursuit. Such subsetting operations are also possible in current dynamic visualisation

software, such as 'XGobi' (Buja *et al.*, 1996). However, in these approaches there are two limitations. First, the partitioning of the data is performed in a *hard* fashion, while the mixture of latent variable models approach permits a *soft* partitioning in which data points can effectively belong to more than one cluster at any given level. Secondly, the mechanism for the partitioning of the data is prone to suboptimality, as the clusters must be fixed by the user based on a single two-dimensional projection. In the hierarchical approach advocated in this chapter, the user selects only a 'first guess' for the cluster centres in the mixture model. The EM algorithm is then utilised to determine the parameters which maximise the likelihood of the model, thus allowing both the centres and the widths of the clusters to adapt to the data in the full multidimensional space. While the user-driven nature of the current algorithm is highly appropriate for the visualisation context, the development of an automated procedure for generating the hierarchy would clearly also be of interest and is currently under investigation.

A software implementation of the algorithm in MATLAB is available from

$$\texttt{http://www.ncrg.aston.ac.uk/PhiVis}$$

8   Acknowledgements

This work was supported by EPSRC grant GR/K51808: *Neural Networks for Visualisation of High-Dimensional Data*. We would like to thank Michael Jordan for useful discussions.

9   References

Anderson, T.W. (1963). Asymptotic theory for principal component analysis. *Ann. Math. Statist.* **34**, 122–148.

Bartholomew, D.J. (1987). *Latent Variable Models and Factor Analysis*. Charles Griffin & Co. Ltd, London.

Basilevsky, A. (1994). *Statistical Factor Analysis and Related Methods*. Wiley, New York.

Bishop, C.M. (1995). *Neural Networks for Pattern Recognition*. Oxford University Press.

Bishop, C.M. and James, G.D. (1993). Analysis of multiphase flows using dual-energy gamma densitometry and neural networks. *Nuclear Instruments and Methods in Physics Research* A **327**, 580–593.

Bishop, C.M., Svensén, M. and Williams, C.K.I. (1998). GTM: the Generative Topographic Mapping. *Neural Computation* **10**, 215–234.

Bishop, C.M. and Tipping, M.E. (1998). A hierarchical latent variable model for data visualisation. *IEEE Trans. Pattern Anal. Machine Intell.* **20**, 281–293.

Buja, A., Cook, D. and Swayne, D.F. (1996). Interactive high-dimensional data visualisation. *Journal of Computational and Graphical Statistics* **5**, 78–99.

Friedman, J.H. and Tukey, J.W. (1974). A projection pursuit algorithm for exploratory data analysis. *IEEE Transactions on Computers* **23**, 881–889.

Hinton, G.E., Dayan, P. and Revow, M. (1997). Modeling the manifolds of images of handwritten digits. *IEEE Transactions on Neural Networks* **8**, 65–74.

Hotelling, H. (1933). Analysis of a complex of statistical variables into principal components. *Journal of Educational Psychology* **24**, 417–441.

Jolliffe, I.T. (1986). *Principal Component Analysis*. Springer-Verlag, New York.

Kohonen, T. (1995). *Self-Organizing Maps*. Springer-Verlag, Berlin.

Krzanowski, W.J. and Marriott, F.H.C. (1994). *Multivariate Analysis Part 2: Classification, Covariance Structures and Repeated Measurements*. Edward Arnold, London.

Maltson, R.L. and Dammann, J.E. (1965). A technique for determining and coding subclasses in pattern recognition problems. *IBM Journal* **9**, 294–302.

McLachlan, G.J. and Basford, K.E. (1988). *Mixture Models: Inference and Applications to Clustering*. Marcel Dekker, New York.

Michie, D., Spiegelhalter, D.J. and Taylor, C.C. (Eds.) (1994). *Machine Learning, Neural and Statistical Classification*. Ellis Horwood, New York.

Pearson, K. (1901). On lines and planes of closest fit to systems of points in space. *The London, Edinburgh and Dublin Philosophical Magazine and Journal of Science, Sixth Series* **2**, 559–572.

Rao, C.R. (1955). Estimation and tests of significance in factor analysis. *Psychometrika* **20**, 93–111.

Rubin, D.B. and Thayer, D.T. (1982). EM algorithms for ML factor analysis. *Psychometrika* **47**, 69–76.

Tipping, M.E. and Bishop, C.M. (1999a). Mixtures of probabilistic principal component analysers. *Neural Computation* **11**, 443–482.

Tipping, M.E. and Bishop, C.M. (1999b). Probabilistic principal component analysis. *J. R. Statist. Soc.* B **61**, 611–622.

Titterington, D.M., Smith, A.F.M. and Makov, U.E. (1985). *Statistical Analysis of Finite Mixture Distributions*. John Wiley, New York.

7

Analysis of Latent Structure Models with Multidimensional Latent Variables

A.P. Dunmur and D.M. Titterington

Department of Statistics, University of Glasgow
Glasgow G12 8QQ, Scotland UK

Abstract

Within the total interface between statistics and neural computing, we shall concentrate in this chapter on the particular area of latent structure models, which has attracted considerable recent interest in the neural-computation literature. We shall review this activity and attempt to relate it to the statistical literature. There are new slants, especially in the form of new models and new computational ideas.

The plan of the chapter is as follows. In Section 1 we outline the general framework of latent structure models, following this in Section 2 with statements of the structures associated with latent trait, latent profile and latent class models, with particular emphasis on the case of multidimensional latent variables. Section 3 discusses maximum likelihood estimation of the parameters in the models, using the EM algorithm, and identifies a problem of computational complexity in the E-step. Section 4 describes how so-called mean-field approximations have been developed in response to this difficulty. Bayesian approaches are described in Section 5, implemented mainly using versions of the Gibbs sampler.

1 Introduction to latent structure models

Bartholomew (1984) identifies the basic formulation of factor analysis models as being given by the relationship

$$f(x) = \int g(x|y)h(y)dy, \tag{1.1}$$

in which $f(x)$ is the joint probability function of p observed variables, $x^T = (x_1, \dots, x_p)$, $h(y)$ is the joint probability density of s unobserved, latent variables, $y^T = (y_1, \dots, y_s)$, and $g(x|y)$ is the obvious conditional probability function. He remarks that $f(x)$ is to be interpreted as a density or a probability mass function, according to the nature of the variables in x. In principle, (1.1) can be generalised to

$$f(x) = \int g(x|y)dH(y), \tag{1.2}$$

where $H(y)$ is a probability measure, and representation (1.2) covers the case of discrete variables, y, as well as continuous.

More recently, Bartholomew (1995) has remarked that this representation covers a variety of models, according to the nature of the variables in x and y. One such taxonomy is given in Table 1, which also appears in Henry (1983).

In fact, (1.1) and (1.2) are representations for a wide class of missing-data problems, in which (x, y) would represent a complete observation within which in practice x is observed and y is not. The modelling of the joint probability function for (x, y) is expressed in terms of the marginal for y and the conditional for x, given y. The other conditional probability function, for y given x, may be of interest in the inference process, and it could be obtained, by Bayes' Theorem, as

$$h(y|x) \propto g(x|y) \, h(y), \tag{1.3}$$

assuming that (1.1) obtains and the factors on the right-hand side of (1.3) are known. The main feature that characterises latent structure models is that the unobserved variables are not real, physical variables that happen to be missing. Instead they represent underlying factors without precise physical definition, but which often, and desirably so, turn out to have meaningful physical interpretation.

A further common feature of latent structure models is an imposition of conditional independence on $g(x|y)$:

$$g(x|y) = \prod_{i=1}^{p} g_i(x_i|y),$$

where x_i is the ith component of x. Bartholomew (1984) comments that this condition (his Condition 1) is fundamental in ensuring that it is the dependence of the x_i on y that accounts fully for the mutual dependence of the x_i. He also imposes a Condition 2, namely that the $g_i(x_i|y)$ are of exponential-family type. This normally ensures amenable parametric analysis, although it is not an issue fundamental to the approach. In proposing a latent structure model, therefore, we have to make three choices:

(i) choose the nature of the latent variable(s), y;
(ii) choose a model for $h(y)$;
(iii) choose a model for $g_i(x_i|y)$.

Table 1 A taxonomy of latent structure models.

Latent	Observed	
	Continuous	Categorical
Continuous	Factor analysis	Latent trait
Categorical	Latent profile	Latent class

In the standard versions of the four cases depicted in Table 1, very natural choices for $h(y)$ and $g_i(x_i|y)$ exist. The latter correspond to regression models, and, in their simplest versions, to generalised linear models with an obvious link function. In factor analysis, with each x_i continuous, it is common for the continuous latent y to be multivariate, and for $g_i(x_i|y)$ to correspond to a Gaussian multiple linear regression model. In standard latent trait analysis, y is univariate, assumed to have a Gaussian distribution, x_i is binary and $g_i(x_i|y)$ corresponds to linear logistic regression. In standard latent class analysis (Lazarsfeld and Henry, 1968), y is a d-category categorical variable, with the corresponding multinomial probabilities making up $h(y)$, x_i is also categorical and $g_i(x_i|y)$ likewise corresponds to a multinomial distribution. As a result, the probability model for x is that of a finite mixture distribution with d components in which the class-conditional densities correspond to independence models and the classes do not have (direct) physical meaning. Finally, in the standard form of latent profile analysis (Gibson, 1959), y again is a single categorical variable and $g_i(x_i|y)$ corresponds to Gaussian linear regression. In the next section, we describe versions of the latent trait, latent profile and latent class models corresponding to multivariate latent variables, referring to relevant statistical publications and also to recent citations from the neural-computation literature.

2   Latent structure models with more than one latent variable

Having more than one latent variable enhances the flexibility of the overall model. We first discuss the case of continuous latent variables and then look in more depth at the categorical case. In this section we assume an independence model for $g(x|y)$ as decreed by Bartholomew (1984) and thus deal with $g_i(x_i|y)$.

2.1   Continuous latent variables—latent trait model

Latent trait analysis corresponds to the case of continuous latent variable(s) and discrete observed variables. (We shall pass over the well-documented case of factor analysis, in which the observed variables are also continuous.) The model is specified in terms of proposals for $h(y)$ and $g_i(x_i|y)$, the second of which is to contribute to an independence model, preferably corresponding to an exponential family. Here it is natural to base the latent trait models on familiar quantal response models and, in particular, on polytomous logistic linear regression.

Example 2.1

Suppose that the ith component of x has b_i possible categories, and that x_i is represented by an indicator vector $x_i = (x_{i1}, \ldots, x_{ib_i})$. Then we can write

$$g_i(x_i|y) = \prod_{k=1}^{b_i} \{P(x_{ik} = 1|y)\}^{x_{ik}},$$

where

$$P(x_{ik} = 1|y) = \frac{\exp\left(w_{0ik} + \sum_{j=1}^{s} w_{jik}y_j\right)}{\sum_{t=1}^{b_i} \exp\left(w_{0it} + \sum_{j=1}^{s} w_{jit}y_j\right)}. \qquad (2.1)$$

Formula (2.1) suffers from lack of identifiability if it is assumed to hold for all $k = 1, \ldots, b_i$. One way of resolving this is to set $w_{ji1} = 0$, for $j = 0, \ldots, s$. This corresponds to the model for polytomous linear logistic regression discussed by Begg and Gray (1984). An alternative suggestion is proposed by Leung (1992), where this multiple latent trait model is discussed in detail. Leung (1992) also proposes that the marginal model for the latent variables be that of a standard, s-dimensional multivariate Gaussian distribution. An EM approach is taken to parameter estimation, and the E-step is dealt with by a Gauss–Hermite quadrature technique that amounts to a discretisation of the latent variables. The fact that the model is unchanged by orthogonal transformation of the latent variables means that the complete set of parameters cannot be updated simultaneously, and Leung (1992) explains how to decompose each full EM cycle into an appropriate set of subiterations. Leung (1992) then applies the methodology to a number of examples, including cases where the fit obtained with two-factor models is much better than that obtained from certain latent-class models. The most basic latent trait model, involving binary observed variables and a single latent variable, is discussed in the context of educational research by Hambleton *et al.* (1991). The extension to more than one latent trait is given by Bartholomew (1987) and De Menezes and Bartholomew (1996). A version based on the probit function, rather than the logistic, is described by Muthen (1978).

From the point of view of this volume, a crucial observation is that the same model underlies what MacKay (1995) calls density networks, although hc takes a Bayesian approach to parameter estimation; see also Chapter 5.

2.2   Categorical latent variables

In this section we define $y = (y_1, \ldots, y_s)$, where each y_j is a categorical variable. In principle, the number of categories, d_j, in the y_j could vary over j, but we shall consider only the case $d_1 = \cdots = d_s = d$. This simplifies the notation and is unlikely to restrict the richness of the available models unduly in practical applications. It is straightforward to extend the discussion to the case of unequal d_j. Each categorical variable y_j is represented by the vector $(y_{j1}, \ldots, y_{jd})$. We introduce the notation $y_u = y_{1u_1} \times \cdots \times y_{su_s}$ where $u = (u_1, \ldots, u_s)$ represents the latent state.

So far as the marginal probability function $h(y)$ is concerned, this structuring of y opens up a wealth of multivariate categorical models, and, in particular, the log-linear class. The overall sample space of y now consists of d^s latent classes, and one might suggest any log-linear model for y, between the extremes of the general d^s-cell multinomial, with $d^s - 1$ free parameters and an independence model, involving

$s(d-1)$ free parameters. In general we may write

$$h(y) = \prod_{\{u\}} h_u^{y_u} \tag{2.2}$$

where $h_u = P(y_u = 1)$. However, the independence model can be further expanded as

$$h(y) = \prod_{j=1}^{s} \prod_{k=1}^{d} h_{jk}^{y_{jk}} \tag{2.3}$$

where $h_{jk} = P(y_{jk} = 1)$. The formalism in (2.2) and (2.3) will prove to be very useful in Section 3.1.

The advantage of this structuring of the latent variables is that interesting qualitative information may result from the nature of the variables. A possible disadvantage is that if $s > 1$ then it would not be possible to fit a distribution with an arbitrary number of latent classes. However, as d and s vary, d^s represents a good variety of positive integers, and the multivariate structure does open up the possibility of parsimoniously parameterised log-linear models, as indicated above. In our detailed discussion of the latent profile and latent class models, we shall tend to restrict attention to the case where the $\{y_j\}$ are independent, ($h(y)$ given by (2.3)) corresponding to most of the existing literature.

For the conditional distribution $g(x|y)$ we can use the notation

$$g_u(x) = g(x|y_u = 1) = g(x|\{y_{ju_j} = 1 : j = 1, \dots, s\}).$$

We may also write $g(x|y) = \prod_{\{u\}} g_u(x)^{y_u}$, and of course $g_u(x)$ typically represents an independence model. The two possible cases are considered next.

2.2.1 *Latent profile model*

In this subsection, we shall formulate, as an example, a simple latent profile model, and we relate it to concepts in the neural-computing literature. Although we shall specialise shortly, we shall initially maintain generality within the model.

Example 2.2 The simplest latent profile model.

As we have hinted, we assume an independence model for $h(y)$. Thus $h(y)$ is given by (2.3). Turning now to $g(x|y)$, which we are also assuming to represent an independence model, arguably the simplest case uses normal linear models, so that

$$E(x_i|y) = \sum_{j=1}^{s} \sum_{k=1}^{d} w_{ik}^{(j)} y_{jk}$$

$$\text{var}(x_i|y) = \sigma_i^2$$

and $x_i|y \sim N, i = 1, \ldots, p$. Thus

$$\log g_i(x_i|y) = \text{const.} - \frac{1}{2}\log(\sigma_i^2) - \frac{1}{2\sigma_i^2}\left(x_i - \sum_{j=1}^{s} w_{ji}^T y_j\right)^2,$$

where w_{ji}^T is the ith row of the matrix W_j which is the $(p \times d)$ matrix with elements $w_{ik}^{(j)}$.

In some cases one might also assume $\sigma_1^2 = \cdots = \sigma_p^2 = \sigma^2$. Example 2.2 corresponds to a version of the standard factor-analysis model in which the latent variables are categorical rather than Gaussian. A structure similar to that in Example 2.2 has been discussed in the neural computing literature, under the terminology of cooperative vector quantisation (CVQ). We acknowledge in particular Ghahramani (1996), who in turns refers back to an unpublished Ph.D. thesis by R. Zemel and to Hinton and Zemel (1994). In its simplest form, vector quantisation associates a data vector x with one of a finite set of categories or classes. As such, the activity is similar to cluster analysis, with each cluster corresponding to a class, and many familiar cluster-analysis algorithms have appeared in the vector-quantisation literature; see, for instance, Cheng and Titterington (1994). In particular, the idea of using a statistical mixture model has been proposed, based on a prescribed number of clusters/mixture components, as well as the special versions that correspond to latent structure analysis. The extra feature of CVQ is the structuring of the latent classes in a multivariate way, as we have set out near the end of Section 2.2. There are assumed to be d^s latent classes, one of which obtains for each x, but they are structured as s sets of d subclasses, and one of each set of subclasses obtains. Ghahramani (1996) effectively assumes the model of Example 2.1, but with $\sigma_1^2 = \cdots = \sigma_p^2 = 1$. The version presented here is more flexible and, therefore, potentially more versatile. Ghahramani (1996) then investigates, as we shall do, the problem of estimating parameters from a set of observed data, and applies the resulting algorithms to a problem in image analysis.

2.2.2   *Latent class model*

In this case, both the observed and latent variables are categorical. The simplest model for $h(y)$ is the independence model set up in (2.3). As in Section 2.1, we represent x_i, the ith component of x, by an indicator vector also to be denoted by x_i. We define

$$g_{ik}^{(u)} = P(x_{ik} = 1|\{y_{ju_j} = 1 : j = i, \ldots, s\}) = P(x_{ik} = 1|y_u = 1),$$

for $i = 1, \ldots, p$, $k = 1, \ldots, b_i$, and for all the d^s possible vectors u. Then

$$g_i(x_i|y) = \prod_{k=1}^{b_i} (g_{ik}^{(u)})^{x_{ik}} = \prod_{\{u\}}\left\{\prod_k \left(g_{ik}^{(u)}\right)^{x_{ik}}\right\}^{y_u},$$

in which the product over $\{u\}$ is over all d^s possible vectors. This form will prove useful in Section 3.1.2.

Altogether, there are $d^s \sum (b_i - 1) + s(d - 1)$ parameters, including those in $h(y)$. It is therefore necessary for identifiability that

$$\left(\prod_i b_i \right) - 1 \geq d^s \sum_i (b_i - 1) + s(d - 1).$$

Latent class models are discussed in detail in Hagenaars (1990), including some discussion of multidimensional latent variables; see also Bartholomew (1987). From the neural-computing literature it is appropriate to highlight Neal (1993).

3   Maximum likelihood estimation: general comments

In this section we discuss maximum likelihood estimation of parameters, particularly in the context of the cases with categorical latent variables, and we highlight computational problems that arise in the implementation of the EM algorithm.

3.1   The EM algorithm with categorical latent variables

As with virtually all incomplete-data problems, explicit maximum likelihood estimates do not exist for the problems considered here, but there are versions of the EM algorithm of Dempster *et al.* (1977) that can be used to maximise the likelihood iteratively. Recall that $g_u(x) = g(x|y_u = 1)$ and, if we use the notation of (2.2), the likelihood of interest, given N independent observations, $\mathbf{x} = (x^{(1)}, \ldots, x^{(N)})$, is

$$L(\theta; \mathbf{x}) = \prod_{n=1}^{N} \sum_{\{u\}} h_u g_u \left(x^{(n)} \right),$$

where θ denotes all parameters within g and h. If we knew the 'true' values of the corresponding latent variables, $\mathbf{y} = (y^{(1)}, \ldots, y^{(N)})$, then we would deal with the *complete-data log-likelihood*

$$\mathcal{L}_c(\theta; \mathbf{x}, \mathbf{y}) = \sum_{n=1}^{N} \sum_{\{u\}} y_u^{(n)} \{\log g_u(x) + \log h_u\}. \tag{3.1}$$

The EM algorithm takes the following form. A sequence of parameter estimates $\{\theta(m)\}$ is generated from an initial $\theta(0)$, using the following double step:

- *E-step*: Given $\theta(m - 1)$, evaluate

$$Q(\theta, \theta(m - 1)) = E\left(\mathcal{L}_c(\theta; \mathbf{x}, \mathbf{y}) | \mathbf{x}, \theta(m - 1) \right).$$

- *M-step*: Choose $\theta = \theta(m)$ to maximise $Q(\theta, \theta(m - 1))$.

In general, the sequence $\{\theta(m)\}$ converges monotonically to a local maximum of the likelihood of interest, L.

In our problem, the E-step amounts to the computation of

$$f_u^{(n)}(m) = E\{y_u^{(n)}|x^{(n)}, \theta(m-1)\} = P(y_{1u_1}^{(n)} = \cdots = y_{su_s}^{(n)} = 1|x^{(n)}, \theta(m-1))$$

$$= g_u(x^{(n)})h_u \Big/ \Big\{\sum_{\{u'\}} g_{u'}(x^{(n)})h_{u'}\Big\}, \tag{3.2}$$

for all u and m, in which all parameter values on the numerator and denominator are based on $\theta(m-1)$. It follows of course that $\sum_u f_u^{(n)}(m) = 1$, for all n, so that the $\{f_u^{(n)}(m)\}$ have the interpretation of weights by which the nth observation is assigned to the latent class represented by u.

In the M-step, we have to maximise

$$Q(\theta, \theta(m-1)) = \sum_n \sum_{\{u\}} f_u^{(n)}(m)\{\log g_u(x^{(n)}) + \log h_u\} \tag{3.3}$$

with respect to the parameters θ within g_u and h_u. This operation will be as easy or difficult as would be the case with complete data, for which each of the $f_u^{(n)}(m)$ would be either 0 or 1. It is natural to try to ensure that this maximisation is explicit. So far as the first term is concerned, this obtains in the linear-Gaussian structure of Example 2.2; see later for details. It would also obtain in many other exponential-family models; cf. Bartholomew (1984). So far as the second term is concerned, recall that it is natural to choose a model for $h(y)$ from the class of log-linear models, and explicit maximisers are available if we use a model for which *direct* estimation is feasible; see for instance §3.4 of Bishop *et al.* (1975). Otherwise, this part of the M-step requires an iterative calculation. One such method is the iterative proportional fitting procedure (IPFP), which has the added characteristic of itself being interpretable as a combination of EM algorithms (Anderson and Titterington, 1995). As a result, even just one complete cycle of the IPFP leads to an increase in the log-likelihood of interest, and could form the basis of a so-called generalised EM (GEM) algorithm (Dempster *et al.*, 1977), which is potentially much more economical than EM and yet should still reach the maximum likelihood estimates.

There is no difficulty when we choose, for h, the independence model (2.3) of Example 2.1, or, at the other extreme, an unstructured, full multinomial (2.2). For the unstructured case, the updated (mth) estimate for the probability associated with the latent class corresponding to u is simply

$$h_u(m) = N^{-1}\sum_n f_u^{(n)}(m). \tag{3.4}$$

For the independence model, the estimates of the parameters in (2.3) are obtained by summing appropriate values in (3.4). For instance, the new estimate of h_{jk} is

$$h_{jk}(m) = \sum_{\{u:u_j=k\}} h_u(m).$$

In the next two subsections we write down the M-steps for parameters within $g(x|y)$ for the latent profile and latent class models, respectively.

3.1.1   *M-step for g in the latent profile model*

It is possible to derive the M-step for the latent profile model in terms of the notation introduced above. However, this over-complicates the situation since we have to calculate the full joint distributions $f_u^{(n)}(m)$. Instead it is simpler to return to the complete-data log-likelihood, the crucial part of which is

$$\sum_{n=1}^{N} \log g(x^{(n)}|y^{(n)}) = \sum_{n=1}^{N} \sum_{i=1}^{p} \log g_i\left(x_i^{(n)}|y^{(n)}\right).$$

Example 3.1   Example 2.2 continued

If we return to the simplest latent profile model, the complete log-likelihood is

$$\sum_{n} \log g(x^{(n)}|y^{(n)}) = \text{const} - \frac{1}{2}N \sum_{i=1}^{p} \log(\sigma_i^2)$$
$$- \frac{1}{2} \sum_{n=1}^{N} \sum_{i=1}^{p} \frac{1}{\sigma_i^2}\left(x_i^{(n)} - \sum_{j=1}^{s} w_{ji}^T y_j^{(n)}\right)^2.$$

This is amenable to explicit maximisation. Here we write down the solution, and refer to Dunmur and Titterington (1998) for the intermediate details. Write $w_i^T = (w_{1i}^T, \ldots, w_{si}^T)$, define a vector $X_i^T = (x_i^{(1)}, \ldots, x_i^{(N)})$ and write $y^{(n)T} = (y_1^{(n)T}, \ldots, y_s^{(n)T})$. Then, were the $\{y^{(n)}\}$ known, w_i would be estimated by the standard least-squares formula

$$\hat{w}_i = (Y^T Y)^{-1} Y^T X_i, \tag{3.5}$$

where $Y^T = (y^{(1)}, \ldots, y^{(N)})$. Consequently, the maximum likelihood estimate of σ_i^2 would be

$$\hat{\sigma}_i^2 = N^{-1}(X_i^T X_i - X_i^T Y \hat{w}_i). \tag{3.6}$$

If a common σ^2 is assumed, then $\hat{\sigma}^2 = p^{-1} \sum_i \hat{\sigma}_i^2$.

In reality, of course, the $\{y^{(n)}\}$ are unknown and, when the M-step is performed, $y^{(n)}$ and $y^{(n)} y^{(n)T}$ are replaced by their expected values, given the data $\{x^{(n)}\}$ and given the current estimates $\theta(m-1)$. Suppose we denote these expectations by $\{\tilde{y}^{(n)}\}$ and $\{\widetilde{y^{(n)} y^{(n)T}}\}$, resulting in the replacement of $Y^T Y$ by $\widetilde{Y^T Y}$ and Y by $\tilde{Y}$, say. Thus, within $\tilde{y}^{(n)}$,

$$\tilde{y}_{jl}^{(n)} = P(y_{jl}^{(n)} = 1|\mathbf{x}, \theta(m-1)).$$

Also, within $\{\widetilde{y^{(n)} y^{(n)T}}\}$ there are elements such as

$$\widetilde{y_{j_1 l_1}^{(n)} y_{j_2 l_2}^{(n)}} = P(y_{j_1 l_1}^{(n)} = y_{j_2 l_2}^{(n)} = 1|\mathbf{x}, \theta(m-1)).$$

The EM iterative stage then amounts to the following modifications of (3.5) and (3.6):

$$\hat{w}_i = (\widetilde{Y^T Y})^{-1} \tilde{Y}^T X_i, \tag{3.7}$$

$$\hat{\sigma}_i^2 = N^{-1}(X_i^T X_i - X_i^T \tilde{Y} \hat{w}_i), \tag{3.8}$$

and these are evaluated for $i = 1, \ldots, p$. Thus $\hat{w}_i$ in (3.7) is simply $w_i(m)$, and so on. The corresponding material in Ghahramani (1996) differs from this, in that (3.7) is replaced by

$$\hat{w}_i = \left\{ \sum_{(n)d=1}^{N} (y^{(n)}\widetilde{y^{(n)T}} y^{(n)}\widetilde{y^{(n)T}}) \right\}^{-1} \left\{ \sum_{n=1}^{N} (y^{(n)}\widetilde{y^{(n)T}}) \tilde{y}^{(n)} x_i^{(n)} \right\}. \tag{3.9}$$

Also, $\sigma_i^2 = 1$ is assumed for all i, so that no analogue of (3.8) appears. Ghahramani and Jordan (1997) extend the methodology to the case where the hidden indicators are not independent but follow a Markov chain. We discuss Bayesian analysis of the model of Ghahramani and Jordan (1997) in Section 5.3.

3.1.2 *M-step for g in the latent class model*

Returning to (3.3) and using the notation of Section 2.2.2, we have to maximise

$$Q(\theta, \theta(m-1)) = \sum_n \sum_{\{u\}} f_u^{(n)}(m) \left\{ \sum_{i=1}^{p} \sum_{k_i=1}^{b_i} x_{ik_i}^{(n)} \log g_{ik_i}^{(u)} + \log h_u \right\}$$

with respect to the parameters of g and h. Estimates of parameters that maximise this are therefore

$$\hat{g}_{ik_i}^{(u)} = \sum_{n=1}^{N} f_u^{(n)}(m) x_{ik_i}^{(n)} \Big/ \sum_{n=1}^{N} f_u^{(n)}(m), \tag{3.10}$$

for all i, k and u. This constitutes the M-step in the EM algorithm, so far as the parameters in $g(x|y)$ are concerned.

3.2   Computational aspects of the E-step

Ghahramani (1996) remarks that, even though the E-step in Example 2.2 requires only univariate and bivariate moments of the missing indicators, conditional on the observed data and the current estimates of the parameters, these moments can be obtained only by first computing the s-variate joint probabilities $\{f_u^{(n)}\}$, as defined in (3.2), and then summing to give the required marginal values. Computation of the complete set of $\{f_u^{(n)}\}$ has complexity of $O(Nd^s)$, which may be impractical in some contexts. The same problem occurs in the context of the corresponding latent class models. As Ghahramani (1996) remarks, these expectations can be approximated by sample averages of realisations each of which is generated by Gibbs sampling. However, this can be very time consuming, and he also develops a deterministic, so-called mean-field approximation, which we describe next.

4   Mean-field approximations in EM algorithms

4.1   General framework

In this section we review the mean-field approximation, in particular in the context of its application in the E-step of the EM algorithm. The original development is due to Zhang, Ghahramani, Jordan and others, as we shall indicate later.

In general, in the E-step, we have to evaluate expectations corresponding to a certain joint probability distribution for the unobserved/latent quantities, conditional on the observables and given current estimates of the parameters $\theta(m-1)$. There are two main ways in which difficulties can arise.

(1) Computation of the expectations may be difficult.
(2) With multidimensional discrete data the number of quantities to be evaluated may be very high, if the quantities involved are multiple indicators.

It is the second problem that confronts us here. Both difficulties can arise in the context of hidden Markov models. For instance, suppose that the $\{y^{(n)}\}$ are true intensities or colours associated with the N pixels making up a pixellated image, that the $\{y^{(n)}\}$ are assumed to follow a Markov random field model, and that only a noisy and possibly otherwise distorted version $\{x^{(n)}\}$ of the image is observed. Suppose also that θ denotes the set of all parameters in the model. Then one can envisage using the EM algorithm to generate estimates of θ, in which case the E-step based on current estimates $\theta(m-1)$ requires one to deal with $P(\{y^{(n)}\}|\{x^{(n)}\}, \theta(m-1))$. In these contexts, this is unmanageable, so far as explicit computations are concerned, because of the intractability of its normalising constant, known as the partition function. The essential feature of mean-field approximations is to replace a complicated joint distribution, $\tilde{p}(y)$, say, by an approximating product of simpler, marginal distributions, and there are a number of rationales by which the approximations can be derived. We suppose that p^* is the proposed approximator of $\tilde{p}$ and we define the product form

$$p^*(y) = \prod_{l=1}^{s} p_l^*(y_l|m_l), \qquad (4.1)$$

where m_l is called the mean-field parameter. Some of the following discussion is based on (4.1) as written. In the simplest version, m_l is chosen to be the expectation of y_l based on the averaging measure p^*, which must therefore be constructed in the light of the original joint distribution $\tilde{p}$. However, for some of the discussion we shall have to specialise (4.1), and we do so as appropriate to the context of the present discussion, in which each y_l is a vector of indicators. Thus we write

$$p^*(y) = \prod_{l=1}^{s} \prod_{k=1}^{d} m_{lk}^{y_{lk}}, \qquad (4.2)$$

within which $\sum_k m_{lk} = 1$, for all l. Equivalently, $\log p^*(y) = \sum_l \sum_k y_{lk} \log m_{lk}$.

It will turn out that $\tilde{p}$ allows us to write, for any particular (l, k),

$$\log \tilde{p}(y) = \sum_k y_{lk} V_{lk}(y_{\partial_l}) + U_l(y),$$

where y_{∂_l} are a small subset of the $\{y_j\}$ that are *neighbours* of y_l, V_{lk} is called a *local field*, and $U_l(y)$ does not involve y_l. We now discuss the rationales for mean-field approximations, and demonstrate their interrelationships.

Rationale 1 (Parisi, 1988)
In terms of distribution $\tilde{p}$, clearly,

$$\tilde{E}_y(y_l) = \tilde{E}_{\{y_j : j \neq l\}} \left(\tilde{E}_{y_l}(y_l | \{y_j : j \neq l\}) \right),$$

where $\tilde{E}_y$ is the expectation with respect to $\tilde{p}(y)$, etc. The approximation is to replace $\{y_j : j \neq l\}$ on the right-hand side of the inner expectation by the corresponding means, thereby nullifying the first expectation, and call all the resulting means m_l. Thus, the parameters being sought satisfy the equations

$$m_l = \tilde{E}_{y_l}(\{y_l | y_j = m_j, j \neq l\}). \tag{4.3}$$

Rationale 2 (Zhang, 1992, 1993)
The rationale now is first to replace the joint distribution $\tilde{p}$ by the product of the associated full conditional distributions, namely,

$$\prod_l \tilde{p}(y_l | \{y_j : j \neq l\}). \tag{4.4}$$

Then, we envisage substituting means on the right-hand side of the conditioning sign, we evaluate the marginal mean of y_l corresponding to the resulting independence model and again denote all means in the subsequent expression by m. The result is again (4.3).

Rationale 3 (Ghahramani, 1996; Ghahramani and Jordan, 1997; Zhang, 1996)
In this approach, the m_l are chosen optimally so as to minimise the Kullback–Leibler directed divergence between p^* and $\tilde{p}$, namely,

$$KL(p^*, \tilde{p}) = \sum_{\{y\}} p^*(y) \log\{p^*(y)/\tilde{p}(y)\}.$$

Substituting $p^*(y)$ from (4.1), we obtain

$$KL(p^*, \tilde{p}) = \sum_{l=1}^{s} \sum_{y_l} p_l^*(y_l) \log p_l^*(y_l) - \sum_{\{y\}} \prod_l \{p_l^*(y_l)\} \log \tilde{p}(y).$$

Although it does not appear to be possible in general to represent the minimiser of this in the form of (4.3), the link can be made in the case where p^* is given by (4.2). Suppose we write

$$KL(p^*, \tilde{p}) = E^* \log p^* - E^* \log \tilde{p}.$$

Then $E^* \log p^* = \sum_l \sum_k m_{lk} \log m_{lk}$ and, for each (l, k),

$$E^* \log \tilde{p} = \sum_r m_{lk} V_{lk}(y_{\partial_l}) + \text{terms not involving } m_l.$$

Next, means denoted by m_{∂_l} are substituted for y_{∂_l}. Minimisation of the resulting $KL(p^*, \tilde{p})$ with respect to m_l, subject to $\sum_k m_{lk} = 1$, gives

$$m_{lk} \propto \exp\{V_{lk}(m_{\partial_l})\}.$$

Rationale 4 (Dayan *et al.*, 1995; Saul *et al.*, 1995)
This approach motivates the mean-field approximation as providing a likelihood bound. Recall that $\tilde{p}(y)$ is, in the context of the E-step, the conditional distribution of the latent variables, y, given the observables, x. Thus, it can be written as

$$\tilde{p}(y) = \tilde{p}_{Y|X}(y|x) = \tilde{p}_{X,Y}(x, y)/\tilde{p}_X(x),$$

showing that the normalising factor is the likelihood function for the observables, evaluated at $\theta(m-1)$. In this discussion it is helpful to write $\tilde{p}(y)$ in Gibbs form. Thus,

$$\tilde{p}(y) = \exp(-\mathcal{E}_y)/\tilde{p}_X(x), \tag{4.5}$$

so that $\tilde{p}_X(x) = \sum_y \exp(-\mathcal{E}_y)$. Also,

$$-\log \tilde{p}_X(x) = \sum_y \mathcal{E}_y \tilde{p}(y) - \left\{ -\sum_y \tilde{p}(y) \log \tilde{p}(y) \right\},$$

a quantity called the *free energy* of $\tilde{p}(y)$ in statistical physics. For any other distribution, which we shall deliberately call $p^*(y)$, it is easy to show that

$$-\log \tilde{p}_X(x) = \left[\sum_y \mathcal{E}_y p^*(y) - \left\{ -\sum_y p^*(y) \log p^*(y) \right\} \right] - KL(p^*, \tilde{p}). \tag{4.6}$$

Formula (4.6) indicates the following points.
  (i) Since $KL(p^*, \tilde{p})$ is nonnegative it demonstrates the fact, well known in statistical physics, that for a given set of $\{\mathcal{E}_y\}$ the free energy is maximised by the Gibbs distribution defined in (4.5).
  (ii) The free energy associated with any other distribution p^* defines a lower bound on the likelihood $\tilde{p}_X(x)$, calculation of which is often not feasible.
  (iii) A sharpest lower bound, for p^* chosen from a particular class of distributions, is achieved by minimising $KL(p^*, \tilde{p})$.

The key consequence of (iii) is the characterisation of the mean-field approximation as providing a sharpest such lower bound on $\tilde{p}_X(x)$. Even the above set of four do not exhaust the rationales for the mean-field approximation; see, for instance, Peterson and Anderson (1987) for a saddle-point approach. Mean-field approximations have been developed and used for various purposes in the statistical-physics literature (Parisi, 1988), in the analysis of certain neural-network models (Amit, 1989; Hertz *et al.*, 1991), and in image restoration (Geiger and Girosi, 1991). Their use in parameter estimation has occurred more recently, for instance in gradient descent algorithms for Boltzmann machines (Peterson and Anderson, 1987); see also Galland (1993). As a way of dealing with the E-step of the EM algorithm, they have been used in the context of hidden Markov chains by Zhang (1992, 1993) and Archer and Titterington (2000), in hidden Markov random fields by Zhang (1992, 1993), and in other latent structure models by Ghahramani (1996) and Ghahramani and Jordan (1997). Saul *et al.* (1995) exploited them in the analysis of a class of belief networks with hidden (missing) variables. In his EM algorithm for data from a hidden Markov random field, Zhang (1992, 1993) uses the mean-field approximation in both the E-step (in order to approximate what we have called $\tilde{p}(y)$), and the M-step, since the complete-data log-likelihood is intractable in this context. Archer and Titterington (2000) compare various methods, including maximum-likelihood methods, in the analysis of hidden Markov chains. They issue a note of caution about the worrying imposition of determinism in substituting means for variables in the mean-field approach, and they give trivial examples that show that the approximation can be disastrous, but they also describe simulations that suggest, in support of the results of Zhang, Ghahramani and Jordan, that the use of mean-field approximations within the EM algorithm can be quite effective. This is an aspect that we revisit in Section 4.4. Finally in this subsection we note that (4.4) is what Besag (1975) calls the pseudo-likelihood, which thereby relates the mean-field approximation to another familiar statistical concept.

4.2   The mean-field approximations for the latent profile example

In this subsection we write down the mean-field approximation for Example 2.2, for which details of the derivation are given by Dunmur and Titterington (1998). Although Example 2.2 was essentially dealt with by Ghahramani (1996), Dunmur and Titterington (1998) found it more transparent to use the approach of Zhang (1992, 1993), described above as *Rationale 2*, in order to derive the approximation.

Example 4.1   Example 2.2 continued

We introduce the 'softmax' vector, defined on positive-valued vectors v by

$$\{\text{softmax}(v)\}_k = e^{v_k} \bigg/ \left(\sum_r e^{v_r} \right),$$

and we define $m_{lk}^{(n)}$ to be the expected value of the binary variable, $y_{lk}^{(n)}$ associated with the mean-field approximating distribution. Then

$$m_l^{(n)} = \text{softmax}\left\{ \tilde{W}_l^T \tilde{\Omega}^{-1}(x^{(n)} - \tilde{x}^{(n)}) + \tilde{W}_l^T \tilde{\Omega}^{-1} \tilde{W}_l m_l^{(n)} - \frac{1}{2}\tilde{\Delta}_l + \log h_l \right\},$$

$$(4.7)$$

for all n and l, where $\tilde{x}^{(n)} = \sum_{j=1}^{s} \tilde{W}_j m_j^{(n)}$, $\tilde{\Omega} = \text{diag}(\tilde{\sigma}_1^2, \ldots, \tilde{\sigma}_p^2)$, and $\tilde{\Delta}_l$ is a vector containing the diagonal elements of $\tilde{W}_l^T \tilde{\Omega}^{-1} \tilde{W}_l$. Formula (4.7) matches corresponding expressions in Ghahramani (1996) and Ghahramani and Jordan (1997), although there is a slight lack of clarity in their notation. The Ns sets of nonlinear equations for the $\{m_l^{(n)}\}$ have to be solved, in principle, in each E-step of the EM algorithm.

4.3 The mean-field approximations for the latent class example

Here also we follow *Rationale 2*, devoting attention to $\tilde{p}(y_l | \{y_j : j \neq l\})$. One way to identify this function which, in more detail, is $P(y_l | \{y_j : j \neq l\}, x, \theta(m - 1))$, is to write down $P(x, y | \theta(m - 1))$, identify the factors within it that involve y_l, and divide them by the appropriate normalising constant. Since the sample space for each individual y_l is small, compared with that of y, computation of the normalising constant is not difficult. Equivalently, one must identify the component of $\log P(x, y | \theta(m - 1))$ that involves y_l. In the case of the latent class model note that, in the notation of Section 2.3, the component of $\log p(x, y | \theta)$ involving y_{lk} is

$$\sum_{i=1}^{p} \sum_{r_i=1}^{b_i} \sum_{\{u:u_l=k\}} y_u x_{ir_i} \log g_{ir_i}^{(u)} + y_{lk} \log h_{lk}.$$

Thus

$$\log p(y_{lk} = 1 | \{y_j : j \neq l\}, x, \theta) = \text{const} + v_{lk},$$

where

$$v_{lk} = \sum_{i=1}^{p} \sum_{r_i=1}^{b_i} \sum_{\{u:u_l=k\}} (y_u/y_{lk}) x_{ir_i} \log g_{ir_i}^{(u)} + \log h_{lk}.$$

From this we obtain the mean-field equations

$$m_l^{(n)} = \text{softmax} \left\{ \left(\sum_{i=1}^{p} \sum_{r_i=1}^{b_i} \sum_{\{u:u_l=k\}} (m_u^{(n)}/m_{lk}^{(n)}) x_{ir_i}^{(n)} \log \tilde{g}_{ir_i}^{(u)} + \log \tilde{h}_{lk} \right) \right\},$$

where, of course, $\sum_k m_{lk}^{(n)} = 1$, for all n and l and $\tilde{g}, \tilde{h}$ refer to the current estimates of the parameters.

4.4 Illustration of the mean-field approximations for the latent profile model

In this section we examine a very simple example with a view to elucidating the comparative behaviour of standard EM and MF, as we shall call the EM algorithm with the mean-field version of the E-step. The two algorithms are different, in general. The convergence properties of EM to the maximum likelihood estimates are in general very satisfactory (Wu, 1983), whereas the approximations and deterministic aspects of MF cast suspicion on its likely properties. On the other hand, the empirical results of Zhang (1992, 1993), Ghahramani (1996) and Ghahramani and Jordan (1997) are very encouraging. The example we use here is deliberately simple, in order to help us pin down the main issues. It is a trivial version of Example 2.2, and in fact concerns Gaussian mixtures.

Example 4.2

We assume that

$$x|y \sim N(W_1^T y_1 + W_2^T y_2, 1),$$

where $y_1^T = (y_{11}, y_{12}) = (1, 0)$ with probability $\frac{1}{2}$ and otherwise is $(0, 1)$. The vector $y_2^T = (y_{21}, y_{22})$ has the same distribution, and is independent of y_1. Thus the marginal distribution of x is that of an equally weighted Gaussian mixture, with means $(w_{11} + w_{21})$, $(w_{11} + w_{22})$, $(w_{12} + w_{21})$ and $(w_{12} + w_{22})$, and with unit variances. We shall assume that the mixing weights and the common variance are known, and we shall create a one-parameter problem by expressing the mean parameters in terms of a single quantity. Specifically, we assume that, for some w, $w_{11} = 3w$, $w_{12} = w$, $w_{21} = 3w$ and $w_{22} = 2w$. Thus the four component means are $6w, 5w, 4w$ and $3w$.

For a given observation, x, let m_1 and m_2 denote the mean-field parameters associated with y_1 and y_2, respectively. Then (4.7) gives, for instance,

$$m_{11} \propto \exp\{3w(x - 3wm_{21} - 2wm_{22}) - \frac{1}{2}(3w)^2\},$$

$$m_{12} \propto \exp\{w(x - 3wm_{21} - 2wm_{22}) - \frac{1}{2}w^2\}$$

and $m_{12} = 1 - m_{11}$. After using $m_{22} = 1 - m_{21}$, we obtain

$$m_{11} = [1 + \exp\{-2w(x - wm_{21}) + 8w^2\}]^{-1}. \tag{4.8}$$

By a similar argument,

$$m_{21} = [1 + \exp\{-w(x - 2wm_{11}) + (7/2)w^2\}]^{-1}. \tag{4.9}$$

Given x and w, equations (4.8) and (4.9) can be solved numerically for m_{11} and m_{21}. Clearly, substitution of (4.9) into (4.8), and (4.8) into (4.9), gives equations in m_{11} alone and m_{21} alone, respectively. In the context of the E-step of the EM algorithm, there will be N such sets of equations, one set for each $x^{(n)}$, and w will be the current estimate $\tilde{w}$. At this point we look at the mechanics of the iterative stage of the EM algorithm, for the time being considering the complete-data log-likelihood for a single observation, x. Since the mixing weights are assumed known, the quantity of interest is just

$$\log p(x|y) = \text{const} - \frac{1}{2}(x - 3wy_{11} - wy_{12} - 3wy_{21} - 2wy_{22})^2$$

$$= \text{const} - \frac{1}{2}aw^2 + xbw,$$

where $a = 9y_{11} + y_{12} + 9y_{21} + 4y_{22} + 18y_{11}y_{21} + 12y_{11}y_{22} + 6y_{12}y_{21} + 4y_{12}y_{22}$ and $b = 3y_{11} + y_{12} + 3y_{21} + 2y_{22}$. In computing a and b we have used the facts that $y_{ij}^2 = y_{ij}$, for all i and j, and $y_{11}y_{12} = y_{21}y_{22} = 0$. In the E-step, each y_{ij} is replaced by some $\tilde{y}_{ij}$ and each $y_{ij}y_{kl}$ by some $\widetilde{y_{ij}y_{kl}}$, giving $-\frac{1}{2}\tilde{a}w^2 + x\tilde{b}w$, say. Furthermore, this is done individually for each $x^{(n)}$, so that the E-step results in the function

$$-\frac{1}{2}\left(\sum_n \tilde{a}^{(n)}\right)w^2 + \left(\sum_n x^{(n)}\tilde{b}^{(n)}\right)w.$$

The M-step then gives the new estimate as $\hat{w} = \sum_n x^{(n)}\tilde{b}^{(n)} / \sum_n \tilde{a}^{(n)}$. So far as the E-step is concerned, the details are as follows.

(i) Standard EM: Here, $\widetilde{y_{ij}y_{kl}} = p(y_{ij} = y_{kl} = 1|x, \tilde{w})$, for all relevant subscripts and, for instance, $\tilde{y}_{11} = \widetilde{y_{11}y_{21}} + \widetilde{y_{11}y_{22}}$. For example, $\widetilde{y_{ij}y_{kl}} \propto \exp\{-\frac{1}{2}(x - 6\tilde{w})^2\}$.

(ii) MF: Here, $\widetilde{y_{ij}y_{kl}} = \tilde{m}_{ij}\tilde{m}_{kl}$. There are a number of possible ways of implementing the MF algorithm: MF_S, where the new estimates of the mean-field parameters are used immediately in subsequent mean-field calculations (sequential); and MF_B, where the new estimates are stored until all the mean-field parameters have been estimated using the old estimates (batch). Thus MF_S corresponds to MF_B in the same way that, in solving linear equations, the Gauss–Seidel method corresponds to the Jacobi method.

A numerical study was carried out based on various sample sizes of $N = 200, 500,$ 1000 averaged over 50 simulations. The true parameter w_{true} and the initial estimate w_{init} were chosen from $0.1, 0.5, 1.0, 2.0, 5, 0$.

Table 2 presents a typical set of results for the estimation of the parameter w from a set of 500 examples averaged over 50 simulations; w_{est} denotes the average of the estimates and RMS denotes the root mean squared error. The effect of the initial estimate was not significant in the EM and MF_S runs. However, in the MF_B runs with larger values of w_{init}, the algorithm had trouble converging to a single solution; it became stuck in limit cycles. This means that the results of the simulation using MF_B are not trustworthy for larger parameter values. The lack of convergence is due to the batch update and the lack of identifiability caused by estimating two degrees of freedom (the mean-field estimates) from a single observation. There is very little difference between the EM and MF_S methods, suggesting that the mean-field approximation is reasonable.

An important point to note is that, as proved by Dunmur and Titterington (1997b), the first correction term to the mean-field terms is given by

$$\frac{1}{2}\langle\Delta\epsilon^2\rangle m(1 - m)(1 - 2m), \tag{4.10}$$

where $\langle\Delta\epsilon^2\rangle$ is the variance of the argument of the softmax function that appears in the mean-field equation (4.7) and m is the mean-field estimate of the individual expectations. Thus, whenever m is close to $0, 0.5$ or 1 the corrections are small. In the case of

Table 2 Table of results $N = 500$, results averaged over 50 simulations, $w_{init} = 0.1$. The figures in brackets give the standard deviation of the estimates, in units according to the final decimal place.

Method	w_{true}	w_{est}	RMS
EM	0.1	0.101(8)	0.0080
MF$_S$	0.1	0.101(8)	0.0079
MF$_B$	0.1	0.101(8)	0.0079
EM	1.0	1.00(1)	0.0143
MF$_S$	1.0	0.99(1)	0.0198
MF$_B$	1.0	0.99(1)	0.0198
EM	2.0	2.00(2)	0.0189
MF$_S$	2.0	2.00(2)	0.0219
MF$_B$	2.0	1.99(2)	0.0164
EM	5.0	5.001(9)	0.0092
MF$_S$	5.0	5.001(9)	0.0094
MF$_B$	5.0	4.51(2)	0.4903

four equally weighted mixtures, the true expectations of the latent variables are all 0.5 and the joint expectations for the different mixtures are 0.25. Hence we would expect the mean-field algorithm to be capable of estimating the individual and therefore the joint expectations reasonably accurately. Further examples and numerical studies are described by Dunmur and Titterington (1998). For instance, they show that, in the case of a mixture of two Gaussian distributions, the EM and MF algorithms are identical. The main computational conclusion to be drawn from the examples concerns the effectiveness of the mean-field approximations within the EM algorithm. Of course, the examples chosen were very simple and were such that there is no problem in applying the standard EM algorithm. The practical advantages of the approximations come to the fore in much more complex problems in which the required latent structure is much less simple, that is, when d^s is large. In those circumstances, the computational demands of solving the equations that determine the mean-field parameters outweigh the difficulties of implementing the E-step in the standard EM algorithm. Dunmur and Titterington (1997b) refine the MF method by incorporating a modification based on the cavity field method of statistical physics (Mezard *et al.*, 1987), and show that, in certain circumstances, this leads to improvements over the basic MF method, but that sometimes it is inferior.

4.5   An example of the application of MF to latent class models

We refrain from carrying out a simulation study here, but instead we illustrate the effect of the model and the mean-field approximation, developed in Section 4.3, on a set of data from a survey on economic and political issues. In this example, all the observables were binary, so that the state space for the observables yields $2^p - 1$ degrees of freedom. We shall assume a model with s binary latent variables, which implies altogether $2^s p + s$ parameters. An unstructured model with 2^s latent classes would involve $2^s(p + 1) - 1$ parameters. Table 3 illustrates the range of models that can be estimated from p-dimensional data. For each (p, s) combination the numbers of parameters in the structured and unstructured models are displayed. For each p, values are given for all s until the *first s* for which one of the models is not identifiable.

The data, presented in Table 4, are taken from Table 3.4 of Hagenaars (1990), page 128. They correspond to counts in a five-way contingency table representing five-dimensional binary data, and the observable variables have the following descriptions.

- (A) Nationality: German or Swiss
- (B) Government should be responsible for equal rights for guest (foreign) workers: yes or no
- (C) Government should be responsible for good medical care: yes or no
- (D) Government should be responsible for good education: yes or no
- (E) Government should be responsible for equal rights for men and women: yes or no

The results of the latent class model for both an unstructured 1×4 latent variable and the structured 2×2 latent variable are presented in Figure 1. The images show the resulting probabilities $g_{ik}^{(u)}$. For the unstructured case, the EM and mean-field algorithms are equivalent. The estimates of h are presented in Table 5 along with the number of iterations needed to reach convergence, the log-likelihood and the Pearson goodness of fit statistic, Q, with the corresponding p-value.

From Figure 1 and Table 5 it can be seen that both the 2×2 structured model and the unstructured model achieve very similar results with the unstructured model having the maximum log likelihood as well as the best fit to the data. This is to be expected since the unstructured model is less constrained than the alternative structured model. The difference between the EM and mean-field algorithms for the

Table 3 Degrees of freedom in the observables and numbers of parameters in (structured, unstructured) models involving p-dimensional binary observables and s binary latent variables.

p	$2^p - 1$	s				
		1	2	3	4	5
2	3	5, 5				
3	7	7, 7	14, 15			
4	15	9, 9	18, 19			
5	31	11, 11	22, 23	43, 47		
6	63	13, 13	26, 27	51, 55	100, 111	
7	127	15, 15	30, 31	59, 63	116, 131	
8	255	17, 17	34, 35	67, 71	132, 143	261, 287

Table 4 Data from Hagenaars (1990), page 128.

E	D	C	B	A	
				1 German	2 Swiss
1 yes	1 yes	1 yes	1 yes	416	119
			2 no	123	28
		2 no	1 yes	92	56
			2 no	26	25
	2 no	1 yes	1 yes	133	68
			2 no	69	26
		2 no	1 yes	159	84
			2 no	85	54
2 no	1 yes	1 yes	1 yes	52	18
			2 no	46	15
		2 no	1 yes	18	18
			2 no	24	9
	2 no	1 yes	1 yes	27	17
			2 no	32	6
		2 no	1 yes	54	32
			2 no	69	48

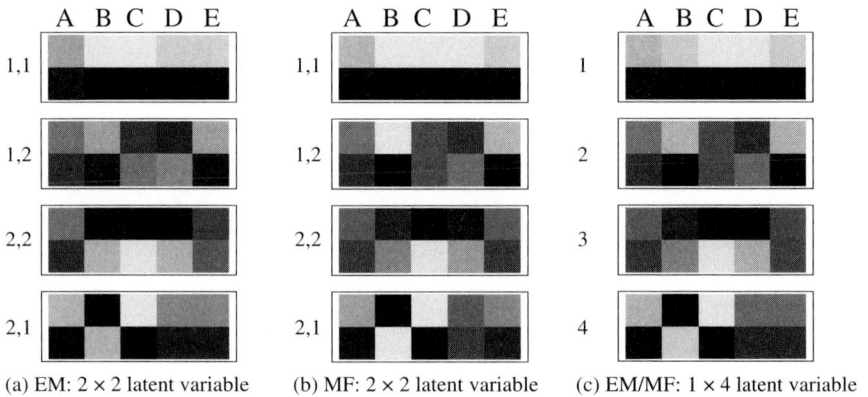

(a) EM: 2 × 2 latent variable   (b) MF: 2 × 2 latent variable   (c) EM/MF: 1 × 4 latent variable

Fig. 1: Results for latent class example showing $g_{ik}^{(u)}$ for two different models and both EM and MF algorithms. Black is 0 and white is 1.

structured model is mainly evident in the probabilities for observable B. Whilst EM gives a larger likelihood than mean-field, the mean-field algorithm actually achieves a better fit to the data. All three models have extracted a latent state in which it is very likely that all five observables will be in state 1. For the unstructured model the probability of being in this state is 0.24, whereas, for the structured model, the probability is $0.29 = 0.41 \times 0.71$. For the mean-field structured model, we can spot

Table 5 Results for latent class example.

Method	EM its	MF its	log(L)	2 × 2 Q	h_{11}	h_{12}	h_{21}	h_{22}
EM	1737	N/A	−6229.79	13.16 ($p = 0.156$)	0.71	0.29	0.41	0.59
MF	795	2.8	−6229.81	12.96 ($p = 0.164$)	0.62	0.38	0.41	0.59

Method	EM its	log(L)	1 × 4 Q	h_1	h_2	h_3	h_4
EM	2532	−6228.67	10.94 ($p = 0.205$)	0.24	0.45	0.17	0.13

a couple of trends in the latent variables. When the first latent variable is in state 1, then observable B is very likely to be 1. When the first latent variable is in state 2, then observable C is in the conjugate state of latent variable 2. When the second latent variable is in state 2 then observable B is in the same state as latent variable 1.

Application of significance tests at the 5% level shows that both the structured and unstructured models provide acceptable fits and also, informally, that the structured model seems adequate when compared against the unstructured model.

5   Bayesian analysis of latent structure models

5.1   General structure and introduction to Gibbs sampling

So far, we have concentrated on maximum likelihood estimation, in which the likelihood of interest is

$$L(h, W; \mathbf{x}) = \prod_n \sum_y h(y) g(x^{(n)} | y, W).$$

Within L, h is used, we hope without danger of confusion, to represent parameters within the probability function h, and W is used to represent the parameters within g: in the discussion of the latent profile illustration, for example, we shall be assuming, for simplicity, that the variance parameters within g are set to unity. In a full Bayesian analysis, a prior is imposed on $\theta = (h, W)$, potentially involving hyperparameters. It is typical that the two sets of parameters, W and h, are distinct, and that they are *a priori* independent. We shall denote the two prior densities by $\pi_1(h)$ and $\pi_2(W)$. From the point of view of the model, the object of prime Bayesian interest is the posterior density $\pi(h, W | \mathbf{x})$, but this is intractable. In principle, especially if Gibbs sampling is to be used in order to create Monte Carlo based inference, it is easier to deal with the joint conditional distribution

$$p(h, W, \mathbf{y} | \mathbf{x}) \propto \prod_n \{g(x^{(n)} | y^{(n)}, W)\} h(\mathbf{y}) \pi_1(h) \pi_2(W). \tag{5.1}$$

Inference about the parameters can be based on the corresponding empirical marginal information. The basic framework of the Gibbs sampling iteration, in which simulated values $h(m-1)$, $W(m-1)$ and $\mathbf{y}(m-1)$ are replaced by values $h(m)$, $W(m)$ and $\mathbf{y}(m)$, is as follows:

    (i) generate $\mathbf{y}(m) \sim p(\mathbf{y}|\mathbf{x}, h(m-1), W(m-1))$;

    (ii) generate $h(m) \sim \pi(h|\mathbf{x}, W(m-1), \mathbf{y}(m))$;

    (iii) generate $W(m) \sim \pi(W|\mathbf{x}, h(m), \mathbf{y}(m))$.

In the case where the different $y^{(n)}$ are independent, stage (i) amounts to generating, for each n,

$$y^{(n)}(m) \sim p\left(y^{(n)}|\mathbf{x}, h(m-1), W(m-1)\right).$$

Furthermore, often $W(m-1)$ is absent from the right-hand side in (ii), and $h(m)$ from (iii).

5.2 Application to the multifactor latent profile model

Here,

$$g\left(x^{(n)}|y^{(n)}, W\right) = N\left(x^{(n)}; \sum_{j=1}^{s} W_j y_j^{(n)}\right), \tag{5.2}$$

where $N(x; \mu)$ denotes the p-variate Gaussian density, with mean vector μ and covariance matrix the identity, evaluated at x. We shall assume independent multinomial distributions for each latent indicator vector as in (2.3), so that

$$h\left(y^{(n)}\right) = \prod_{j=1}^{s} \prod_{k=1}^{d} h_{jk}^{y_{jk}^{(n)}},$$

where, for each j, $\sum_k h_{jk} = 1$. Thus h is represented by the $\{h_{jk}\}$. The Bayesian construction is completed by defining priors for W and the $\{h_{jk}\}$. For W we choose uniform priors, although in general conjugate Gaussian priors could also be used; for the $\{h_{jk}\}$ we use independent Dirichlet priors, such that

$$h_j \sim \text{Dir}(\alpha_{j1}, \dots, \alpha_{jd}),$$

where, for each j and k, $\alpha_{jk} > 0$, and where h_j denotes $\{h_{jk} : k = 1, \dots, d\}$. Thus

$$p(\{h_j\}, W, \mathbf{y}|\mathbf{x}) \propto \prod_n N\left(x^{(n)}; \sum_{j=1}^{s} W_j y_j^{(n)}\right) \prod_{j=1}^{s} \prod_{k=1}^{d} h_{jk}^{\sum_n y_{jk}^{(n)} + \alpha_{jk} - 1}. \tag{5.3}$$

We now outline the workings of the Gibbs sampler in this case. In updating the $\mathbf{y}$ the individual observations are treated separately. For each n, $y^{(n)} = \{y_j^{(n)}\}$. Let e_r

denote the d-vector with 1 in the rth element and 0 elsewhere. Then, for updating $y^{(n)}$ we require the predictive probabilities

$$p(y_j^{(n)} = e_{r_j}, j = 1, \ldots, s | x^{(n)}, W(m-1), \{h_j(m-1)\})$$

$$\propto \left\{ \prod_{j=1}^s h_{jr_j}(m-1) \right\} N\left(x^{(n)}; \sum_{j=1}^s W(m-1)_j e_{r_j}\right). \tag{5.4}$$

The constant of proportionality is adjusted so that the d^s probabilities add up to 1. From these probabilities a new $y^{(n)}$ is simulated. Alternatively, the d $y_j^{(n)}$'s can be generated one by one, using the s probabilities

$$p(y_j^{(n)} = e_{r_j} | x^{(n)}, \{y_k^{(n)}(m-1) : k \neq j\}, W(m-1), \{h_j(m-1)\})$$

$$\propto h_{jr_j}(m-1) \cdot N\left(x^{(n)}; W_j(m-1)e_{r_j} + \sum_{k \neq j} W_k(m-1)y_k^{(n)}(m-1)\right) \tag{5.5}$$

where $y_k^{(n)}(m-1)$ denotes the realisation of $y_k^{(n)}$ within $\mathbf{y}(m-1)$. The parameters W can then be generated from the multivariate Gaussian distribution with density proportional to

$$\exp\left\{ -\frac{1}{2} \sum_{n=1}^N \left(x^{(n)} - \sum_{j=1}^s W_j y_j^{(n)}(m)\right)^T \left(x^{(n)} - \sum_{j=1}^s W_j y_j^{(n)}(m)\right) \right\}.$$

If the W_j are independent parameters, then they can be updated one by one using the relevant conditional distributions, in a similar spirit to what is done in (5.5). Finally, the $h_j(m)$'s can be generated using the independent Dirichlet distributions

$$h_j(m) \sim \text{Dir}\left(\sum_n y_{jr}^{(n)}(m) + \alpha_{jr}; r = 1, \ldots, d\right).$$

Example 5.1

As an illustration we consider a simple case of Example 2.2, in which the weights contain only a single parameter. Specifically, for each j, $W_j = wA_j$, where w is a scalar parameter and the A_j are matrices that are treated as being known. Thus

$$g\left(x^{(n)} | y^{(n)}, w\right) \propto \exp\left\{ -\frac{1}{2}\left(x^{(n)} - B^{(n)}w\right)^T \left(x^{(n)} - B^{(n)}w\right) \right\},$$

where

$$B^{(n)} = \sum_{j=1}^s A_j y_j^{(n)}.$$

In the Gibbs sampling updating of w, therefore, w is generated from the Gaussian distribution with mean μ_w and variance σ_w^2, where

$$\mu_w = \sigma_w^2 \sum_n \sum_j \left(y_j^{(n)}(m) \right)^T A_j^T x^{(n)}$$

and

$$\sigma_w^2 = \left\{ \sum_n \left(\sum_j A_j y_j^{(n)}(m) \right)^T \left(\sum_k A_k y_k^{(n)}(m) \right) \right\}^{-1}.$$

5.3 Factorial hidden Markov models

In this section we consider the models discussed by Ghahramani and Jordan (1997). The model for the observables, given the latent variables, is again given by (5.2), and the latent variables are assumed to be independently distributed, within each observation. However, between observations, each of the s latent variables is assumed to follow a Markov chain model. Thus,

$$p(\mathbf{x}, \mathbf{y}|h, W) = h(\mathbf{y})g(\mathbf{x}|\mathbf{y}, W),$$

where

$$h(\mathbf{y}) = \prod_{j=1}^{s} \prod_{n=2}^{N} y_j^{(n-1)T} H_j y_j^{(n)},$$

in which H_j is a transition matrix, assumed stationary, for the Markov chain followed by the jth latent variable. Thus, for each r, the rth row of H_j contains a set of probabilities $\{h_{j,rt}, t = 1, \ldots, d\}$. In this case, the natural prior for the elements of H_j is a set of sd independent Dirichlet distributions, so that

$$\pi_1(h) \propto \prod_{j=1}^{s} \prod_{r=1}^{d} \left\{ \prod_{t=1}^{d} (h_{j,rt})^{\alpha_{j,rt}-1} \right\},$$

where $\alpha_{j,rt} > 0$ for all j, r, t. For the Gibbs sampler, given $\theta(m-1) = (h(m-1), W(m-1))$, the procedure for generating the parameters W is essentially unchanged. Modifications are, however, required to describe the generation of $y(m)$ and $h(m)$. For $y(m)$, the key probability statement is that, for $n = 2, \ldots, N-1, r_j = 1, \ldots, d$,

$$p\left(y_j^{(n)} = e_{r_j}|\mathbf{x}, \mathbf{y} \setminus y_j^{(n)}, W(m-1), H(m-1) \right)$$

$$= p\left(y_j^{(n)} = e_{r_j}|\mathbf{x}, y_j^{(n-1)}, y_j^{(n+1)}, W(m-1), H(m-1) \right)$$

$$\propto y^{(n-1)T} H_j^{(m-1)} e_{r_j} \cdot e_{r_j}^T H_j^{(m-1)} y_j^{(n+1)}$$

$$\times g\left(x^{(n)}|\{y_k^{(n)}; k \neq j\}, y_j^{(n)} = e_{r_j}, W(m-1) \right),$$

where the constant of proportionality is such that the sum over r_j is unity. The currently updated values for $\{y_k^{(n)}; k \neq j\}$ are used on the right-hand side. For $n = 1$ we choose $y_j^{(1)} = e_1$, for all j, for identifiability reasons, and, for $n = N$, $r_j = 1, \ldots, d$

$$p\left(y_j^{(N)} = e_{r_j}|\mathbf{x}, \mathbf{y} \setminus y_j^{(N)}, W(m-1), H(m-1)\right)$$
$$\propto y^{(N-1)T} H(m-1)_{j} e_{r_j} g\left(x^{(N)}\Big|\left\{y_k^{(N)}; k \neq j\right\}, y_j^{(N)} = e_{r_j}, W(m-1)\right).$$

When generating a $y_j^{(n)}$, current values of all the parameters are used within the above probabilities, as are the most recently generated values of $y_{j'}^{(n')}$, except for $n' = n$ and $j' = j$. The complete set of $\{y_j^{(n)}\}$ should be regenerated in this way, scanning through the possible values of n and, within that, the possible values of j.

Now we turn to the transition probabilities. If $h_{j,r}$ denotes the set of elements in the rth row of the transition matrix H_j, then new values for them can be generated from the independent Dirichlet distributions

$$h_{j,r}(m) \sim \text{Dir}\left(\sum_{n=2}^{N} y_{jt}^{(n-1)}(m) y_{jt}^{(n)}(m) + \alpha_{j,rt}; t = 1, \ldots, d\right).$$

The procedure described in this section is similar to that developed for the nonfactorial hidden Markov chain by Robert *et al.* (1993). The chain $\{\theta(m), \mathbf{y}(m)\}$ created by the Gibbs sampling algorithm is irreducible and aperiodic, because of the positivity of all the full conditional probability functions. For their simpler case, Robert *et al.* (1993) demonstrate that the chain is ergodic, and that the chain corresponding to the (missing) indicators $\{\mathbf{y}(m)\}$ is geometrically ergodic and ϕ-mixing. Although the $\{\theta(m)\}$ do not form a first-order Markov chain, the good convergence characteristics of the $\{\mathbf{y}(m)\}$ chain can be transferred to the $\{\theta(m)\}$ sequence. Straightforward adaptations of their argument exist to cover the factorial case of interest here, as well as the cases covered later in Sections 5.4.1, 5.4.2, 5.4.4 and, with certain reservations (see Rydén and Titterington, 1998), 5.4.5.

5.4 Discussion of other models

The examples considered so far are easy to deal with in that all full conditionals are easily simulated. This is not the case with other models.

5.4.1 *Latent trait models*

Although he uses other terminology for the model, MacKay (1995) proposes a Bayesian analysis of the latent trait model of Example 2.1, assuming spherical Gaussian distributions both for the latent variables, for each n, and for W. MacKay (1995) is particularly interested in making inferences about the latent variables, for a given observation. This involves integration of a joint posterior density over W, and

MacKay adopts Monte Carlo approaches, although not based on Gibbs sampling. In any case, in this example, the Gibbs sampling step for generating W cannot be done neatly, in the same way that Bayesian analysis for the parameters in a linear logistic regression model has no neat solution; see Racine *et al.* (1986) for related work.

Bishop *et al.* (1996) consider a version of this model in which the marginal distribution on the latent space, which we have called $h(y)$, is uniform over a regular grid of finite cardinality, so that there is no unknown parameter in this part of the model; it is assumed that the dimensions of the grid are pre-specified. In fact, they consider the case of continuous observed variables, thereby inventing a form of factor analysis, but assume that the means of the observed variables are linear, not in the latent variables themselves, but in basis functions centred on the grid points in the latent space. The precise form of the basis functions is assumed to be either pre-specified or adjusted 'off-line'. Thus, the only quantities requiring inference are the latent variables and the weights, W, defining the linear relationship. Bishop *et al.* (1996) base their analysis on an EM algorithm, but the Gibbs sampling approach could be used in a straightforward fashion if a Gaussian prior were assumed for W. There would simply be no step involving the sampling of h.

5.4.2   Latent class models

Both components in the model, $h(y)$ and $g(x|y)$, are made up of multinomial models, with the result that the stages in the Gibbs sampling iteration for the associated parameters can be expressed in terms of Dirichlet distributions, provided the priors are chosen from that class. Thus, the computational Bayesian approach is straightforward, provided the underlying model is identifiable; see, for instance, Neal (1993).

5.4.3   Hidden sigmoid belief networks

One reason for the tractability of the factorial hidden Markov model discussed in Section 5.3 is that the index set, $\{1, \ldots, N\}$, is naturally ordered in such a way that the joint probability $h(\mathbf{y})$ can be written in a simple factored form,

$$h\left(y^{(1)}, \ldots, y^{(N)}\right) = \prod_n h(y^{(n)}|pa_n), \tag{5.6}$$

where pa_n denotes a set of *parents* of $y^{(n)}$, all belonging to the set $\{y^{(n')} : n' < n\}$. There are many other processes satisfying (5.6), generally referred to as *belief networks* or *Bayesian networks*. In general the ordering of the superscripts may have nothing to do with 'time'; instead, the ordering is dictated by relationships determining the system of parents. A particular case is the class of sigmoid belief networks described by Saul *et al.* (1996), in which each $y^{(n)}$ is binary and the conditional probabilities on the right-hand side of (5.6) are modelled as linear logistic forms in a set of parameters H. However, in spite of the comparative ease of writing down the joint probability for $\mathbf{y}$, the conditional distributions required in the Gibbs sampling step for generating $h(m)$ are not amenable, rather as in the case of $W(m)$ in Section 5.4.1.

5.4.4 *Hidden Markov mesh random field*

In this and the following subsection, the superscripts identify the locations of N positions on a rectangular lattice, often representing the pixels in a two-dimensional image. The density $h(\mathbf{y})$ provides a model for the true scene and $g(x|y)$ models a noise process. Markov mesh random field models are *causal* or *unilateral*, in that it is possible to express the corresponding $h(\mathbf{y})$ in the factorised form (5.6), where the pixels are ordered as in a raster scan of the image frame, and each $pa(n)$ consists of a small number of neighbouring pixels. As a result, $h(\mathbf{y})$ is easily computed, and allows for easy simulation of realisations of the random field. There are many particular Markov mesh models, see for instance Gray *et al.* (1994), and one in particular, defined by Devijver (1988), involves a single parameter and leads to an $h(\mathbf{y})$ that has the form of a binomial likelihood. Dunmur and Titterington (1997a) show how, by using an obvious beta prior distribution, the Gibbs sampling routine can be easily implemented.

5.4.5 *Hidden Markov random field*

Other Markov random fields do not exhibit the causal properties of the Markov mesh models: these include many models used in statistical image analysis, such as the simple Ising spin-glass model. A major complication in analysing these models is the fact that the partition function in the corresponding formula for $h(\mathbf{y})$ is not computable. As a result, it is impossible exactly to calculate maximum likelihood estimates of parameters or to carry out the Gibbs sampling generation of the corresponding parameters in our general procedure. There are approaches for trying to deal with this *impasse* in the context of maximum likelihood estimation, including the maximum pseudo-likelihood method of Besag (1975) and the use of a Monte Carlo estimate of the partition function by Geyer and Thompson (1992). Rydén and Titterington (1998) investigate these ideas within the type of Bayesian analysis dealt with in this paper; see also Heikkinen and Högmander (1994) and Higdon *et al.* (1997).

6   Other developments in the neural-computation literature

Although, as commented above, the $h(\mathbf{y})$ associated with a belief network is easy to calculate, for a given $\mathbf{y}$ and a given set of weights, W, this is not true if some of the components of $\mathbf{y}$ correspond to hidden units. Unless the hidden variables 'come after' all the visible variables in the ordering, we are in a similar difficulty to that of being faced with an intractable partition function. An alternative to using Monte Carlo methods is to seek analytical approximations, for given x and W. Saul *et al.* (1996) use mean-field approximations to derive a lower bound, in the case of sigmoid belief networks, adapting an approach developed for the so-called Helmholtz machine by Dayan *et al.* (1995) and Hinton *et al.* (1995). Jaakkola and Jordan (1996a) extend the method to the case of noisy-OR networks, and also show how to obtain upper bounds for both types of belief network. In Jaakkola and Jordan (1996b), a way of obtaining upper and lower bounds is described for the Boltzmann machine itself, incorporating a process of pruning the corresponding network until a skeleton is left for which computation of the probability is tractable. For further related work, see Saul *et al.* (1996), Saul and Jordan (1996) and various articles in Jordan (1999). These contributions offer the prospect of approximations to otherwise intractable likelihood functions that might lead to effective parameter estimates, but considerable further development and investigation are required.

7 Acknowledgement

The work in this chapter was carried out with the support of a grant from the UK Engineering and Physical Sciences Research Council.

8 References

Amit, D.J. (1989). *Modelling Brain Function.* Cambridge University Press.

Anderson, N.H. and Titterington, D.M. (1995). Beyond the binary Boltzmann machine. *IEEE Trans. Neural Networks* **6**, 1229–1236.

Archer, G.E.B. and Titterington, D.M. (2000). Parameter estimation for hidden Markov chains. *J. Statist. Plan. Infer*, to appear.

Bartholomew, D.J. (1984). The foundations of factor analysis. *Biometrika* **71**, 221–232.

Bartholomew, D.J. (1987). *Latent Variable Models and Factor Analysis.* Griffin, London.

Bartholomew, D.J. (1995). What is statistics? (with discussion) *J. R. Statist. Soc.* A **158**, 1–20.

Begg, C.B. and Gray, R. (1984). Calculation of polychotomous logistic regression parameters using individualized regressions. *Biometrika* **71**, 11–18.

Besag, J. (1975). Statistical analysis of non-lattice data. *The Statistician* **24**, 179–195.

Bishop, C.M., Svensén, M. and Williams, C.K.I. (1996). EM optimization of latent-variable density models. In *Advances in Neural Information Processing Systems* **8**, Eds. D.S. Touretzky, M.C. Mozer and M.E. Hasselmo. MIT Press, Cambridge, MA.

Bishop, Y.M.M., Fienberg, S.E. and Holland, P.W. (1975). *Discrete Multivariate Analysis.* MIT Press, Cambridge, MA.

Cheng, B. and Titterington, D.M. (1994). Neural networks: a review from a statistical perspective (with discussion). *Statist. Science* **9**, 2–54.

Dayan, P., Hinton, G.E., Neal, R.M. and Zemel, R.S. (1995). The Helmholtz machine. *Neural Computation* **7**, 889–904.

De Menezes, L.M. and Bartholomew, D.J. (1996). New developments in latent structure analysis applied to social attitudes. *J. R. Statist. Soc.* A **159**, 213–224.

Dempster, A.P., Laird, N.M. and Rubin, D.B. (1977). Maximum likelihood estimation from incomplete data via the EM algorithm (with discussion). *J. R. Statist. Soc.* B **39**, 1–38.

Devijver, P.A. (1988). Image segmentation using causal Markov random field models. In *Proc. 4th Int. Conf. Pattern Recognit., Lecture Notes in Computer Science,* Vol. 301, Ed. J. Kittler, pp. 131–143. Springer-Verlag, Berlin.

Dunmur, A.P. and Titterington, D.M. (1997a). Computational Bayesian analysis of hidden Markov Mesh models. *IEEE Trans. Pattern Anal. Machine Intell.* **19**, 1296–1300.

Dunmur, A.P. and Titterington, D.M. (1997b). On a modification to the mean-field EM algorithm in factorial learning. In *Advances in Neural Information Processing Systems* **9**, Eds. M.C. Mozer, M.I. Jordan and T. Petsche, pp. 431–437. MIT Press, Cambridge, MA.

Dunmur, A.P. and Titterington, D.M. (1998). Parameter estimation in latent profile models. *Comp. Statist. Data. Anal.* **27**, 371–388.

Galland, C.C. (1993). The limitations of deterministic Boltzmann machine learning. *Network* **4**, 355–379.

Geiger, D. and Girosi, F. (1991). Parallel and deterministic algorithms from MRF's: surface reconstruction. *IEEE Trans. Pattern Anal. Machine Intell.* **13**, 401–412.

Geyer, C.J. and Thompson, E.A. (1992). Constrained Monte Carlo maximum likelihood for dependent data (with discussion). *J. R. Statist. Soc.* B **54**, 657–699.

Ghahramani, Z. (1996). Factorial learning and the EM algorithm. In *Advances in Neural Information Processing Systems* 7, Eds. G. Tesauro, D.S. Touretzky and T.K. Leen. MIT Press, Cambridge, MA.

Ghahramani, Z. and Jordan, M. I. (1997). Factorial hidden Markov models. *Machine Learning* **29**, 245–273.

Gibson, W. A. (1959). Three multivariate models: factor analysis, latent structure analysis and latent profile analysis. *Psychometrika* **24**, 229–252.

Gray, A.J., Kay, J.W. and Titterington, D.M. (1994). An empirical study of the simulation of various models used for images. *IEEE Trans. Pattern Anal. Machine Intell.* **16**, 507–512.

Hagenaars, J.A. (1990). *Categorical Longitudinal Data.* Sage, London.

Hambleton, R.K., Swaminatham, H. and Rogers, H.J. (1991). *Fundamentals of Item Response Theory.* Sage, London.

Heikkinen, J. and Högmander, H. (1994). Fully Bayesian approach to image restoration with an application in biogeography. *Appl. Statist.* **43**, 569–582.

Henry, N.W. (1983). Latent structure analysis. In *Encyclopedia of Statistical Sciences*, Volume 4, Eds. S. Kotz, N.L. Johnson and C.B. Read, pp. 497–504. Wiley, New York.

Hertz, J., Krogh, A. and Palmer, R.G. (1991). *Introduction to the Theory of Neural Computation.* Addison-Wesley, Redwood City, CA.

Higdon, D.M., Johnson, V.E., Turkington, T.G., Bowsher, J.E., Gilland, D.R. and Jaszczak, R.J. (1997). Fully Bayesian estimation of Gibbs hyperparameters for emission computed tomography data. *IEEE Trans. Medical Imaging* **16**, 516–526.

Hinton, G.E., Dayan, P., Frey, B. and Neal, R.M. (1995). The wake-sleep algorithm for unsupervised neural networks. *Science* **268**, 1158–1161.

Hinton, G.E. and Zemel, R.S. (1994). Autoencoders, minimum description length and Helmholtz free energy. In *Advances in Neural Information Processing Systems* 6, Eds. J. D. Cowan, G. Tesauro and J. Alspector, pp. 3–10. Morgan Kaufmann, San Mateo, CA.

Jaakkola, T.S. and Jordan, M.I. (1996a). Computing upper and lower bounds on likelihoods in intractable networks. MIT AI Memo No. 1571.

Jaakkola, T.S. and Jordan, M.I. (1996b). Recursive algorithms for approximating probabilities in graphical models. MIT Computat. Cognit. Sci. Tech. Rep. 9604.

Jordan, M. I. (Ed.) (1999). *Learning in Graphical Models.* MIT Press, Cambridge, MA.

Lazersfeld, P. and Henry, N.W. (1968). *Latent Structure Analysis.* Houghton Mifflin, Boston.

Leung, S.O. (1992). Estimation and application of latent variable models in categorical data analysis. *Br. J. Math. Statist. Psychol.* **45**, 311–328.

MacKay, D.J.C. (1995). Bayesian neural networks and density networks. *Instr. Meth. in Phys. Res. A* **354**, 73–80.

Mezard, M., Parisi, G. and Virasoro, M.A. (1987). *Spin Glass Theory and Beyond.* Lecture Notes in Physics, 9. World Scientific Press, Singapore.

Muthen, B. (1978). Contributions to factor analysis of dichotomous variables. *Psychometrika* **43**, 551–560.

Neal, R. M. (1993). Probabilistic inference using Markov chain Monte Carlo methods. Tech. Report CRG-TR-93-1, Dept. Comp. Sci., Univ. Toronto.

Parisi, G. (1988). *Statistical Field Theory.* Addison-Wesley, Redwood City, CA.

Peterson, C. and Anderson, J.R. (1987). A Mean Field Theory learning algorithm for neural networks. *Complex Systems* **1**, 995–1019.

Racine, A., Grieve, A.P., Fluhler, H. and Smith, A.F.M. (1986). Bayesian methods in practice: experiences in the pharmaceutical industry (with discussion). *Appl. Statist.* **35**, 93–150.

Robert, C.P., Celeux, G. and Diebolt, J. (1993). Bayesian estimation of hidden Markov chains: a stochastic implementation. *Statist. Prob. Letters* **16**, 77–83.

Rydén, T. and Titterington, D.M. (1998). Computational Bayesian analysis of hidden Markov models. *J. Comp. Graph. Statist.*, **7**, 194–211.

Saul, L.K., Jaakkola, T. and Jordan, M.I. (1996). Mean field theory for sigmoid belief networks. *J. Artific. Intell. Res.* **4**, 61–76.

Saul, L.K. and Jordan, M.I. (1995). Boltzmann chains and hidden Markov models. In *Advances in Neural Information Processing Systems* **7**, Eds. G. Tesauro, D.S. Touretzky and T.K. Leen. MIT Press, Cambridge, MA.

Saul, L.K. and Jordan, M.I. (1996). Exploiting tractable structures in intractable networks. In *Advances in Neural Information Processing Systems* **8**. MIT Press, Cambridge, MA.

Wu, C.F.J. (1983). On the convergence properties of the EM algorithm. *Ann. Statist.* **11**, 95–103.

Zhang, J. (1992). The Mean Field Theory in EM procedures for Markov random fields. *IEEE Trans. Signal Processing* **40**, 2570–2583.

Zhang, J. (1993). The Mean Field Theory in EM procedures for blind Markov random field image restoration. *IEEE Trans. Image Processing* **2**, 27–40.

Zhang, J. (1996). The application of the Gibbs–Bogoliubov–Feynman inequality in mean field calculations for Markov random fields. *IEEE Trans. Image Processing* **5**, 1208–1214.

8

Artificial Neural Networks and Multivariate Statistics

E.B. Martin and A.J. Morris

Centre for Process Analytics and Control Technology,
University of Newcastle, Newcastle-upon-Tyne, NE1 7RU, UK

Abstract

The mechanistic modelling of processes can present major challenges in practice and attention has turned to the sophisticated use of monitored plant data. As a result of this transfer of interest, two major areas of strategic research have emerged since the mid-1980s: multivariate statistics and artificial neural networks. This chapter introduces the key methodologies and describes a number of related issues prior to applying such techniques to complex engineering problems. A number of industrially based examples are presented which contrast multivariate statistical approaches with those of neural network techniques. An additional example is included which focuses upon a situation where linear statistical approaches are not applicable. The first case study compares the techniques of principal components regression, projection to latent structures, orthogonal least squares based regression, feedforward neural networks, radial basis function neural networks and locally weighted regression using, as an example, near-infra-red spectral data. For the second example, some aspects of neural networks are considered and different approaches assessed for the prediction of polymer properties. An industrial application is then presented which is analysed using dynamic artificial neural networks since multivariate statistical techniques are inappropriate for the study. The application of linear projection to latent structures and nonlinear projection to latent structures to the prediction of pH are presented and the results compared. The final case study focuses upon principal components analysis (PCA) and its nonlinear variants. A number of examples are discussed including its application to fault detection and diagnosis.

1   Introduction

Process analysis, monitoring and control rely on the availability of appropriate mathematical models to represent the system of interest. A common but sometimes very demanding approach is to develop a first-principles or mechanistic model of the process based on knowledge of the chemical and physical phenomena underlying the process operation. Empirical data-based modelling is a widely used alternative to mechanistic modelling since it requires less specific knowledge of the process being studied than that needed to develop a first-principles model. Empirical modelling techniques require data (measurements) which are collected on those variables believed to be representative of process behaviour and of the quality or properties of the product. Statistical regression techniques and neural networks are now used routinely in the process industries for building empirical models.

Statistical regression techniques, such as multiple linear regression (MLR), have been used extensively for developing process representations from historical data. It is well known that, when dealing with multivariate problems which involve highly correlated variables, the traditional MLR approach leads to singular solutions or imprecise parameter estimation. Limitations due to measurement noise, correlated variables, unknown variable and noise distribution and dataset dimensionality can be overcome by applying alternative regression techniques such as the multivariate statistical projection based regression tools. In particular, the projection-based techniques of principal components analysis (PCA), principal component regression (PCR) and projection to latent structures or partial least squares (PLS) have found significant application. These techniques surmount some of the problems encountered when using methods such as MLR, including the dimensionality and the collinearity problem, and can also provide a filtering tool for measurement noise. Projection-based techniques can handle highly correlated, noise-corrupted datasets since they are based upon the assumption of dependency (correlation) between the variables and consequently provide the capability to estimate the main underlying structures in terms of a reduced number of *latent variables*. These are linear combinations of the original variables and usually represent the true dimensionality of the system of interest. The linear technique of PCA seeks to explain the variance in a single data matrix, whilst PLS allows models to be developed which relate the process input/output measurements for the prediction of slow, possibly expensive, difficult to measure and infrequent 'quality' variables such as texture, taste, smell, colour, shelf-life, etc. However, it is well known that many industrial processes are highly nonlinear and the development of appropriate mechanistic models can be difficult and time consuming. The problem of identifying, or estimating, a model structure and its associated parameters can be related to the problem of learning a mapping between known input and output spaces. A classical framework for this problem can be found in approximation theory. Almost all approximation, or identification, schemes can be mapped, i.e. expressed as a neural network. For example, the well-known autoregressive moving average with exogenous variables (ARMAX) model can be a single-layer network with inputs comprising lagged system input–output data and prediction errors. In this context a network can be viewed as a function represented by the conjunction of a number of basis functions.

One of the major obstacles to the widespread use of advanced modelling and control techniques is the cost of model development and validation. The utility of neural networks in providing viable process models was demonstrated many years ago where the technique was used to characterise successfully two nonlinear chemical systems as well as to interpret biosensor data. Since that time, studies in neural network theory and applications have flourished, e.g. Bhat *et al.* (1989), IEEE (1988, 1989, 1990) and Hunt *et al.* (1992). A literature search reveals well over 50 different types of network architecture and a number of different neuron-processing functions. This chapter initially reviews one specific topology which has probably been the most prevalent in neural network studies: the feedforward network (often termed backpropagation network) with sigmoidal activation node (hyperbolic tangent) functions. Other nonlinear node activation functions include the monatomic forms such as simple quadratic, etc., and those based on radial basis functions, such as Gaussian distributions or spline functions; see Chapter 3. A one-to-one mapping network, the autoassociative network, is introduced later and used in the development of a nonlinear principal component analysis method for data feature extraction.

2   Multivariate statistical approaches

The effective treatment of noisy and highly correlated process measurements is possible using linear statistical methods such as principal component analysis (Jackson, 1991) to develop a process representation, and projection to latent structures or partial least squares, as it is sometimes known, to develop a model which relates process input–output measurements for the prediction of future 'quality' variables which might be very difficult and/or costly to measure continuously (Geladi and Kowalski, 1986; Höskuldsson, 1988).

2.1   Principal components analysis

The main areas of application of principal components analysis (PCA) in process analysis are data reduction, classification of variables, outlier detection, early warning of potential malfunctions and 'fingerprinting' for fault identification. Let $X = \{x_1, x_2, \ldots, x_m\}$ be an m-dimensional dataset describing either the process variables or the quality information. The first principal component is defined to be the linear combination of the columns of X which accounts for the greatest amount of variability, $t_1 = X p_1$ subject to $|p_1| = 1$, in the original dataset. In m-dimensional space, p_1 defines the direction of greatest variability, and t_1 represents the projection of each object onto p_1. The second principal component is the linear combination defined by $t_2 = X p_2$ which accounts for the next greatest variance subject to $|p_2| = 1$ and subject to the condition that it is orthogonal to the first principal component, t_1:

$$t_2 = E_1 p_2 \quad \text{where} \quad E_1 = (X - t_1 p_1^T).$$

This procedure is essentially repeated until m principal components are calculated. In effect, PCA decomposes the data matrix, X, as

$$X = T P^T = \sum_{i=1}^{m} t_i p_i^T,$$

where the columns of P provide the principal component loadings and those of T the principal component scores. The loadings vector provides information as to which variables contribute the most to individual principal components, i.e. they are the coefficients in the principal components model; information on the clustering of the samples is obtained from the scores, t_i.

The main attribute of PCA which enables the dimensionality of the problem to be reduced is that, if some of the process variables are collinear or strongly correlated, a smaller number of principal components than original variables are required to explain the variability inherent within the data. Consequently it is desirable to exclude some of the components. In addition, the lower-order components often describe the noise in the data. The number of principal components, a, to be retained in the model can be selected according to different heuristic or statistical tools. One possible criterion is to define a cut-off limit for the cumulative percentage of total variation which should be captured by the principal component representation (in many practical situations it might be somewhere between 70% and 90%). The major drawback of this approach is the *a priori* assumption that the pre-specified amount of total variation matches the application requirements, leaving a certain degree of subjectivity in the criterion. An alternative approach based upon statistical procedures is that of cross-validation (Wold, 1978). In cross-validation, the training dataset is split into a number of sub-sets, r say. Initially for a model comprising one latent variable, the first subset of data is omitted and a PCA representation is built on the remaining $r - 1$ subsets of data. The prediction error sum of squares (PRESS) for the omitted subset of data is then computed. The procedure is repeated, replacing the omitted subset until every individual subset has been left out once. The r individual PRESSs are then summed to give the total PRESS. The procedure is repeated for $i = 2, \ldots, a$ principal components and a corresponding total PRESS is calculated. The optimal number of latent variables is chosen to be that which minimises the total PRESS. The final model can be represented as

$$X = TP^T + E = \sum_{i=1}^{a} t_i p_i^T + E$$

where a represents the number of principal components to be retained and E is a matrix of residuals of unfitted variation.

2.2   Projection to latent structures (partial least squares)

Given a reference dataset of N time consistent samples collected on M regressor variables, x_m (the independent variables), and K response variables, y_k (the dependent variables), these can be arranged into an $(N \times M)$ matrix X and an $(N \times K)$ matrix Y, respectively. Typically the X matrix relates to the process variables, or alternatively the wavelengths from a spectrum, whilst the Y matrix corresponds to the quality or reference variables. Standard linear PLS projects the X and Y matrices down on to a subset of latent variables, t and u, which are referred to as the input and output scores, respectively. The objective of the procedure is to fit a linear relationship between the independent and the dependent variables by performing an ordinary least squares regression between each pair of corresponding t and u score vectors:

$$u_j = t_j b_j + e_j, \quad j = 1, 2, \ldots, a,$$

where b_j is the coefficient from the inner linear regression between the jth latent variables t_j and u_j, i.e.

$$b_j = \frac{u_j^T t_j}{u_j^T u_j}.$$

In this way, linear PLS provides a bilinear decomposition of the X and Y matrices into a number of rank-one matrices in a similar manner to principal component analysis (PCA) with a single data matrix. The decomposition can be defined as the product between each pair of input score vectors, t, and predicted output score vectors, $\hat{u}$, and a set of corresponding input and output loading vectors p and q. This reflects the capabilities of the methodology to provide both an approximation model for the regressor variables X and a prediction model for the response variables Y:

$$X = \sum_{j=1}^{a} t_j p_j^T + E$$

$$Y = \sum_{j=1}^{a} \hat{u}_j q_j^T + F,$$

where $\hat{u}_j$ denotes the prediction of the scores u_j in an ordinary least squares (OLS) sense and is given by

$$\hat{u}_j = t_j b_j.$$

E and F are the resulting residual matrices for the regressor and response matrices, X and Y respectively, when a model with a latent variables is used for the approximation of the X matrix and the prediction of the Y matrix. Each pair of latent variables accounts for a certain amount of the variability in the input and output blocks, and usually most of the variance of the X and Y blocks can be accounted for by the first a latent variables, where $a < (M, K)$, while the remaining latent variables typically describe the random noise in the data. Typically, the number of latent variables that provide an adequate description of the data is identified using cross-validation, as described in the preceding section. The 'engine' of the PLS procedure is the non-linear iterative partial least squares (NIPALS) algorithm (Wold, 1966). This algorithm sequentially extracts each pair of corresponding latent variables as a linear combination of the input and output variables, prior to fitting the inner linear regression and then evaluating the linear prediction of the output scores:

$$t_j = \sum_{m=1}^{M} x_m w_{mj} = X w_j$$

$$u_j = \sum_{k=1}^{K} y_k c_{kj} = Y c_j.$$

The input and output loading vectors are computed from the least squares regression of the X and Y matrices on $\hat{u}_j$ and t_j:

$$p_j^T = \frac{t_j^T X}{t_j^T t_j}$$

$$q_j^T = \frac{\hat{u}_j^T Y}{\hat{u}_j^T \hat{u}_j}.$$

The final step in the iterative procedure is to deflate the X and Y matrices:

$$E = X - t_j p_j^T$$
$$F = Y - \hat{u}_j q_j^T.$$

The column vectors w_j and c_j are defined as the weights of the input and output projections and are calculated by performing a least squares regression of $\hat{u}_j$, the predicted output score vectors, and t_j, the input score vector, on the X and Y matrices, respectively:

$$w_j^T = \frac{\hat{u}_j^T X}{\hat{u}_j^T \hat{u}_j}.$$

$$c_j^T = \frac{t_j^T Y}{t_j^T t_j}.$$

Those two equations provide an exchange of information between the two datasets, termed the 'scores exchange' by Geladi and Kowalski (1986). However, it can be shown that the output loadings q regressed on the (deflated) X matrix and on the predicted output scores are the same as the output weights c. Thus the calculation of q can be omitted by setting q equal to c. A detailed description of the algorithm and its statistical properties can be found in Geladi and Kowalski (1986) and Höskuldsson (1988). The advantage of applying PLS in preference to multiple linear regression (MLR) as a linear regression technique arises from the fact that in PLS the input latent variables are linearly independent, i.e. they define a set of orthogonal reference axes and each output latent variable is defined to be orthogonal to the previously extracted input latent variable. Consequently each ordinary least squares regression performed between each pair of input and output score vectors cannot affect the others and cannot be affected by the others. This gives the PLS algorithm the robustness that MLR lacks. When handling ill-conditioned or noisy data, the validated PLS regression model only uses a reduced number of latent variables, thus overcoming the problem of rank deficiency of the input data. A further advantage of omitting the lower-order latent variables, which traditionally explain less of the underlying variability in the data, is that the random noise which is intrinsic to the data is filtered out. In fact these latent variables account for the noise in the variables or the linear dependencies which exist amongst the regressor variables but which are not strongly related to the desired underlying input–output relationship.

3 Neural networks and their topologies

The *IEEE Special Issues on Neural Networks* (1988, 1989, 1990), Hunt *et al.* (1992), Thibault and Grandjean (1991) and Morris *et al.* (1994), provide a good background to neural networks and their applications in system modelling and control. Neural networks have been shown to possess good function approximation capabilities and ease effort in model development (Cybenko, 1989; Girosi and Poggio, 1990; Park and Sandberg, 1991; Wang *et al.*, 1992) and have been applied to process modelling by many researchers (e.g. Bhat and McAvoy, 1990; Chen *et al.*, 1990a,b; Narendra and Parthasarathy, 1990; Willis *et al.*, 1991, 1992a,b; Di Massimo *et al.*, 1992; Morris *et al.*, 1994; Warwick *et al.*, 1992). In developing these neural network models, the knowledge required is usually the process input–output data. Conventional network training results in a so-called 'black box' representation. Potential limitations of conventional neural network models are that they can be difficult to interpret and can lack robustness. These issues can be addressed through stacked regression or statistical bagging methods (Breiman, 1994; Wolpert, 1992), and neuro-fuzzy network approaches (Zhang and Morris, 1996a,b).

3.1 Static feedforward neural networks

The basic feedforward network performs a nonlinear transformation of input data in order to approximate output data. This results in a static network structure. Figure 1 shows a multilayer feedforward network with one hidden layer.

Inputs to a neural network are presented at the input layer. The data from the input neurons is then propagated through the network via the interconnections which are such that every neuron in a layer is connected to every neuron in adjacent layers. It is the structure of the hidden layers which essentially defines the topology of a feedforward network. Each interconnection has associated with it a scalar weight which acts to modify the strength of the signal. The neurons within the hidden layer perform two tasks: they sum the weighted inputs to the neuron and then pass the resulting summation through a nonlinear activation function. In addition to the weighted inputs to

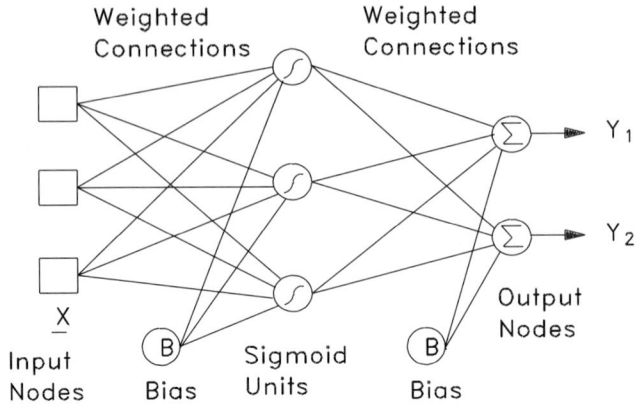

Fig. 1:   A multilayer feedforward neural network.

the neuron, a bias is included in order to shift the space of the nonlinearity. Although several types of activation function can be used, the most commonly used are the sigmoidal (logistic) function and the hyperbolic tangent function. It is also possible to use mixed types of activation function in one network (Gemperline *et al.*, 1991; Zhang *et al.*, 1994; Zhang and Morris, 1995a). The use of network models as predictors may be problematic if care is not taken in specifying the network topology correctly and then training the network properly.

3.2   Radial basis function networks

Radial basis functions (RBF) are widely used within the engineering community as a powerful technique for interpolation in multidimensional space (Micchelli, 1986; Powell, 1985, 1987). A generalised form of the radial basis function has found wide application in areas such as image processing, signal processing, control engineering, etc. A radial basis function is effectively a function which has incorporated within it a distance criterion with respect to a centre. Functions such as this can be used very effectively for interpolation and data smoothing. A recent application of radial basis functions is in the area of neural networks, where they have been used as a replacement for the sigmoidal transfer function. Like most feedforward networks (see Figure 1), RBF networks have three layers, namely an input layer, the hidden layer with the RBF nonlinearity and a linear output layer. The role of the input layer in a RBF network is to distribute all inputs unaltered to each of the hidden layer nodes. The weights on the links between the input layer and the hidden layer are all set to unity and do not change during training. The bias term normally seen in feedforward networks is not required in the RBF network. A radial basis function expansion with n_i inputs, nH hidden layer nodes and a scalar output implements a mapping according to:

$$f(x) = b_0 + \sum_{i=1}^{n_H} w_i g(\|x - c_i\|),$$

where w_i are the parameters or weights, c_i are the RBF centres, n_H is the number of centres and b_0 are the bias parameters. The functional form $g(\cdot)$ is pre-selected, with the centres c_i being some fixed points in n-dimensional space appropriately sampling the input domain. For a given set of centres, the second, or hidden, layer performs a fixed nonlinear transformation which maps the input space on to a new space in a similar manner to that in the sigmoidal feedforward network. Each term $g(\|x - c_i\|)$ forms the activation function in a unit of the hidden layer. The output layer then implements a linear combination on this new space. With the RBF centres regarded as adjustable parameters, a network structure in feedforward form results. The most popular choice for $g(\cdot)$ is the Gaussian density although others such as the thin-plate spline function and B-splines have also been used very effectively.

A number of different methods of placement of the RBF centres have been reported (Hunt and Sbarbaro, 1991). One method is to position the RBF centres evenly spaced out along all dimensions of the input space, which is usually a hypercube with coordinates in the interval $[-1, 1]$. Although this method is simple it has a few drawbacks, especially when the dimension n is large, since the number of RBFs will increase exponentially. An alternative approach clusters the network inputs in the training set and positions the RBF centres at the centres of each of the clusters. For this purpose k-means clustering is used (MacQueen, 1967). This method does not suffer from the same problems as the first approach since it will position RBF centres in those parts of the input space where the data points lie. In contrast to a sigmoidal feedforward network, a RBF network cannot selectively ignore certain inputs. This is a consequence of the way in which the distance measure is incorporated into the network. It is therefore important to take this into account when presenting data to the network and only use those inputs that have a fairly strong, not necessarily linear, correlation with the outputs. Totally uncorrelated data in either the linear or the nonlinear sense should not be presented to the network. For more on RBF networks see Chapter 3.

3.3 Dynamic neural networks

The static approximation characteristics of feedforward networks are of limited use for dynamic process modelling and control. The reason for this is the algebraic representation of the static network compared to the recurrent difference equation representation of the dynamic network. Static networks can only be applied to a window of data or to off-line data, and they cannot generate long-term dynamic predictions of the variables assigned as network inputs. This implies that, if the model of the system being studied involves a history of time-lagged plant inputs and outputs, e.g.

$$x(t - 1), x(t - 2), \ldots, x(t - 10); \quad y(t - 1), y(t - 2), \ldots, y(t - 19),$$

then the network model of the system is likely to be poor irrespective of the network training algorithm. In some situations, appropriate lagged inputs and outputs may be distributed over a very wide dynamic range (stiff systems). Over-specifying the network nodes leads to increasing problems of dimensionality and slow training. A relatively recent development has been the increasing attention being paid to dynamic and recurrent networks. Although these can take several different forms they are all capable of capturing temporal behaviour. Examples are the backpropagation-in-time networks (Werbos, 1990; Su *et al.*, 1992); the Elman network (Elman, 1990), where the hidden neuron outputs at the previous time step are fed back to its inputs through time delay units; locally recurrent network representations (Frasconi *et al.*, 1992; Tsoi and Back, 1994; Ku and Lee, 1995; Zhang and Morris, 1995a,b), where each neuron has one or more delayed feedback loops around itself; globally recurrent networks, where the network output(s) are fed back to the inputs through time delay units; and dynamic, filter-based networks (Willis *et al.*, 1991; Montague *et al.*, 1992), which include filters in the neuron interconnections and which have been shown to provide very good multi-step-ahead predictions. In this way the model of the function of the individual neuron is extended to account for dynamics and dead time. The temporal properties can also be captured by using finite impulse response filters in the network interconnections. The most straightforward way to extend the essentially steady-state mapping of the static network to the dynamic domain is to adopt an approach similar to that taken in linear ARMA (autoregressive moving average) modelling. Here a time series of past process inputs (x) and outputs (y) are used to predict the present process outputs, creating the *time delay network*. Important process characteristics such as system delays can be accommodated by utilising only those process inputs beyond the dead time. Additionally, any uncertainty in delay can be taken account of by using an extended time history of process inputs. Inevitably, a significant number of network inputs result, especially for systems described by sets of 'stiff equations'. It is important to understand, however, that conventional feedforward networks imply that the manipulated process inputs directly affect the plant outputs. This is not true in complex processes where some manipulated inputs affect internal states that go on to affect the system outputs. In addition, if the models are used to predict more than one step ahead in time, and have only been trained for such a task, then the ARMA time series approach is not appropriate. The one-step-ahead ARMA network model does not capture the process dynamics. Essentially, the autoregressive nature of this form of network results in the need to predict $y(t + n)$ from the estimates of $y(t + n - 1)$. Errors in the estimate of y thus accumulate as the prediction horizon increases. The problem of the ARMA network approach can, however, be overcome by minimising the network prediction error over time; that is, the network training minimises the error in the estimate of $y(t + n)$ and all other output predictions up to a specified prediction horizon (Bhat and McAvoy, 1990). This approach is called backpropagation-in-time. Recent studies have shown that the incorporation of dynamics into the network in this manner is highly beneficial in many real process studies. Although perhaps the most concise network representation of a dynamic system is obtained by using network inputs comprising past input and output data, the requirement to model processes over a wide dynamic range (models containing both very large and very small time constants) can result in large network structures leading to network training and convergence problems. In the filter network, the node interconnections are modified to incorporate dynamics inherently within the network. Since a dynamic network model is not autoregressive the prediction problem does not arise. In addition to the sigmoidal processing of nodes, the neurons (or transmission between neurons) can be given dynamic characteristics, for example dynamic processing of the form

$$N(s)\exp(-s\tau_d)/D(s),$$

where $N(s)$ and $D(s)$ are polynomial functions in the Laplace operator s, and the time delay is represented by a Padé approximation. In this way the model of the function of the individual neuron is extended to account for dynamics and dead time. Other workers have modelled the temporal properties using finite impulse response filters. In a dynamic filter network, the filter, usually first-order and linear, is attached after (possibly also before) each hidden neuron. These filters introduce dynamics into the network. The filter parameters can be trained in the same manner as the training of network weights. Willis *et al.* (1991) and Montague *et al.* (1992) have successfully used these networks as software sensors in bioreactor, distillation column and continuous polymerisation reactor predictions and also for model-based predictive control of a distillation tower. Figure 2 shows the structure of a globally recurrent network. The lagged network outputs are fed back to the network input nodes as indicated by the back-shift operator z^{-1}, e.g. $y_{t-1} = z^{-1} y_t$. In this way dynamics are introduced into the network. The network output will depend not only on the network inputs but also on the previous network outputs.

An Elman network is shown in Figure 3. In an Elman network, hidden neuron outputs at the previous time step are fed to all the hidden neurons. Scott and Ray (1993a,b) show good performance of the Elman network in nonlinear systems modelling. In a locally recurrent network the output of a hidden neuron is fed back to its input through one or several units of time delays, as illustrated in Figure 4. Compared with fully recurrent networks, locally recurrent networks have fewer weights and can be trained more quickly (Ku and Lee, 1995). Dynamic networks can be constructed in a number of ways to provide for either a 'global model' or a 'local model'. Global representations are provided by making use of network input signals comprising both lagged process input and lagged output data (a time delay network),

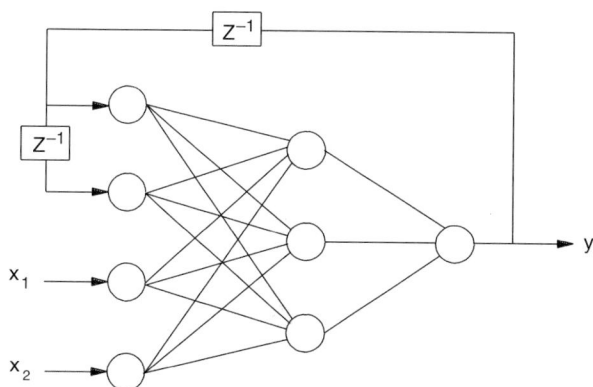

Fig. 2: A globally recurrent network.

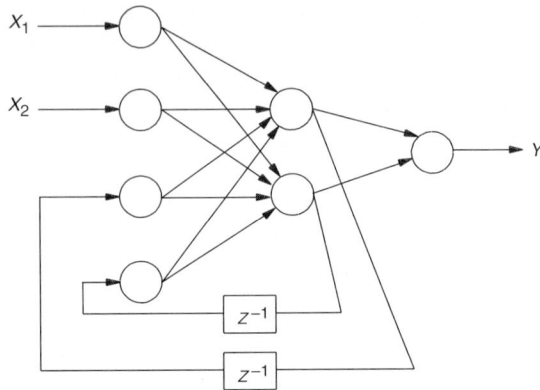

Fig. 3:   An Elman network.

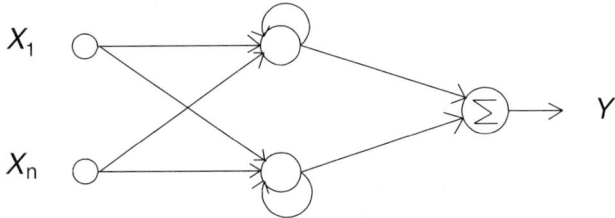

Fig. 4:   A locally recurrent network.

or by the introduction of dynamic behaviour directly into the network itself via node or network interconnection, dynamic relationships (filter network), or by making use of the network 'predicted output' as feedback to the network input (backpropagation-in-time or globally recurrent network).

Local representations are provided by making use of local recurrence around each individual hidden node (locally recurrent network) or by recurrence around each node and between each and all hidden nodes (fully recurrent network). In general, single hidden layer recurrent topologies tend to be used because of the powerful representational capabilities that these networks have. One difficulty with recurrent networks is to be able to determine the best network architecture (i.e. number of hidden units). A method has recently been developed which provides a systematic way of training recurrent networks and uses an orthogonal decomposition approach to minimise the prediction error sequentially (Zhang *et al.*, 1994; Zhang and Morris, 1995a, 1998).

A sequential orthogonal training algorithm for training locally recurrent networks has been developed by Zhang and Morris (1995a,b,c). With this architecture, during network training the first hidden neuron is used to model the relationship between system inputs and outputs, whereas other hidden neurons are added sequentially to model the relationship between inputs and model residuals. When another hidden neuron is added, its output vector can be decomposed into two parts: one is in the space spanned by the output vectors of the previously added hidden neurons and the other part is orthogonal to that space. Its contribution is due only to the orthogonal part. The Gram–Schmidt orthogonalisation algorithm is used at each training step to form a set of orthogonal bases for the space spanned by the hidden neuron outputs. The optimum hidden-layer weights can be obtained through a gradient-based optimisation method while the output-layer weights can be found using least squares regression. Hidden neurons are added sequentially and the training procedure terminates when the model error is lower than a pre-defined level. By this training method, the necessary number of hidden neurons can be found and hence the problem of over-fitting can be avoided. An additional advantage of the sequential network optimisation approach is that the training method allows the development of mixed-order locally recurrent networks. In a mixed-order locally recurrent network, hidden neurons can have different numbers of feedbacks. The objective of incorporating mixed-order hidden neurons is to improve the network representation capabilities. Parsimonious network structures can be generated using approaches developed through canonical decomposition (Wang *et al.*, 1992, 1994) as well as the sequential approach to network hidden layer size determination reviewed above.

3.4   Wavelet networks

Industrial processes can impose a number of problems upon the structures adopted for neural network dynamic modelling caused by varying sampling times, sparse and dense data in different operating regions and the inherent presence of both large and small dynamics. In the case of nonuniformly distributed training data, an efficient way of solving this problem is by learning at multiple resolutions. A higher resolution of input space is used if the data are dense and a lower resolution when sparse. Recently, because of the similarity between the discrete inverse wavelet transform (WT) and a one-hidden-layer neural network, the idea of combining both wavelets and neural networks has been proposed. This has resulted in the wavelet network (Motard and Joseph, 1994; Koulouris *et al.*, 1995; Pati and Krishnaprasad, 1993; Zhang and Benveniste, 1992), a feedforward neural network with one hidden layer of nodes whose basis functions are drawn from a family of orthonormal wavelets. The family of wavelets is derived from the translation and dilations of a single function, the mother wavelet. A schematic of the wavelet network is shown in Figure 5.

The function implemented by the network is given by

$$f(x) = \sum_{i=1}^{N} u_i \Psi[a_i * (x - t_i)],$$

where Ψ is the wavelet transfer function, u represents the output weights and a and t are the dilation and the translation parameters respectively. These are adjusted during network training.

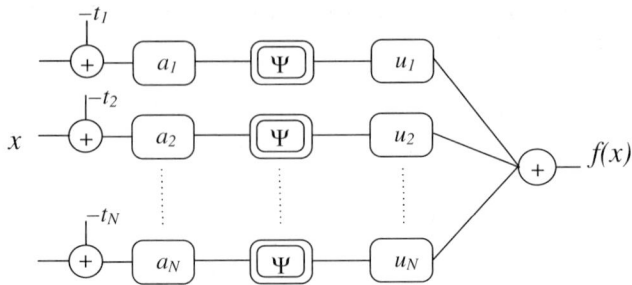

Fig. 5:   A wavelet network.

Wavelets, in addition to forming an orthogonal basis, are capable of explicitly representing the behaviour of a function at various resolutions of input variables. Consequently, a wavelet network is first trained to learn the mapping at the coarsest resolution level. In subsequent stages, the network is trained to incorporate elements of the mapping at higher and higher resolutions. Such hierarchical, multiresolution training has many attractive features for solving engineering problems, such as a more meaningful interpretation of the resulting mapping and more efficient training and adaptation of the network compared to conventional methods. Conventional networks used for learning input–output maps are usually randomly initialised and trained by a backpropagation procedure. The random initialisation makes such learning procedures very inefficient. In contrast, wavelet networks can be initialised by using a regular wavelet lattice consisting of the dilated and translated versions of the mother wavelet. A library of wavelets is constructed according to the available training dataset by selecting a subset from the lattice. Well-established techniques of regressor selection can then be applied for selecting those wavelets most useful for fitting the training data. Since the initialisation is good, training times are significantly reduced. For feedforward networks, the global nature of the activation functions and the fact that these overlap, over a large range of output values, makes adaptation and incremental learning a slow process. Convergence is also not guaranteed, because of the nonlinear nature of the optimisation problem. In addition, these networks provide a value for the output independent of the density of training data. From the point of view of functional analysis, it is well known that functions can be represented as a weighted sum of orthogonal basis functions (for example sinusoids and Legendre polynomials). Unfortunately, most orthogonal functions are global approximators and, like sigmoids, suffer from the disadvantages of approximation using global functions. What is needed is a set of basis functions which are both local and orthogonal. Several orthogonal wavelets with good localisation properties have recently been developed and these offer a good solution to the problem. The resulting wavelet network therefore has all the advantages of true localised learning.

4 Network identification issues—data, network training and model validation

4.1 Plant testing and data analysis

Model fitness-for-purpose and validity are highly dependent upon correctly performed experimental procedures. A number of plant excitation and data analysis prerequisites should be fulfilled for assured neural network identification. An important check is to verify that the inputs have been appropriately selected and are of sufficient excitation. For nonlinear system identification, the input should also cover a wide range of magnitudes. Excitation properties can be investigated either by testing whether the persistent excitation criteria have been met, or by considering the autospectra of the input signals. An input spectrum with a non-zero level over a large spectral range generally ensures suitable experimental conditions offering good properties of identification. These tests provide a satisfactory check of the experimental conditions. For linear systems, a PRBS (pseudo-random binary sequence) signal is usually a good choice for input perturbations. For nonlinear systems, multilevel PRBS input perturbations should be used.

Process and data linearity can be tested visually by observing signal behaviour for different input amplitudes and by considering the symmetry in response to negative and positive input. The coherence spectrum is also a valuable test for linearity. The coherence function expresses the degree of linear correlation in the frequency domain between the system input and output. A coherence function not equal to one indicates the presence of one or more of the following: a disturbance affecting the output, an input not accounted for or a nonlinearity. With carefully designed experimental conditions where disturbances and unconsidered inputs are kept at zero, then a coherence function not equal to one reveals the presence of a nonlinearity.

Finally, as the correct model order is often also not known *a priori*, it makes sense to postulate several different model orders. Based upon these, an error criterion can be computed that indicates which model order to choose. One intuitive approach would be to start with low model orders and gradually increase the model order. F-tests and hypothesis testing on two models of different orders can then be carried out.

4.2 Network training

Neural networks can be trained using a number of training methods, such as the back propagation method (Rumelhart *et al.*, 1986), the conjugate gradient method (private communication from J.A. Leonard and M.A. Kramer), Levenberg–Marquardt optimisation (Marquardt, 1963), or methods based on genetic algorithms (Goldberg, 1989). A problem commonly seen in neural network training, especially with industrial data, is over-fitting. Two strategies are used to overcome this problem. The first uses regularised training where the objective is to penalise excessively large network weights (W) which do not contribute significantly to the reduction of model errors. In this way the trained network has a smooth function surface. In regularised training, the training objective function, J, is modified as follows:

$$J_\lambda = \frac{1}{N} \sum_{t=1}^{N} (\hat{y}(t) - y(t))^2 + \lambda \|W\|^2,$$

where $\hat{y}(t)$ is the network output prediction, $y(t)$ is the actual process output and λ is an adjustable tuning parameter. An intuitive interpretation of this modified objective function is that a weight that does not influence the first term very much will be kept close to zero by the second term. A weight that is important for model fit will, however, not be affected very much by the second term. An appropriate value of λ can be obtained through a cross-validation procedure. For linear models, minimisation of J_λ leads to the well-known ridge regression formula. Regularisation has been widely used in statistical model building and a variety of techniques have been developed, such as ridge regression, principal component regression and partial least squares regression.

The second strategy for preventing over-fitting is an 'early stopping' mechanism in which network training is stopped at a point beyond which over-fitting would occur. This can be explained using Figure 6 which represents a typical neural network learning curve. The vertical axis represents the sum of squared network errors, the horizontal axis represents the network training steps, the solid line is the network error on training data, and the dashed line is the network error on test data. During the initial training stage, both training error and test error decrease quite quickly. As training progresses, the training error will continue to decrease slowly and the testing error will start to increase, sometimes very quickly, after a certain number of training steps. The appropriate point to stop training is the point at which the test error is at its minimum. During network training, both training and test errors are continuously

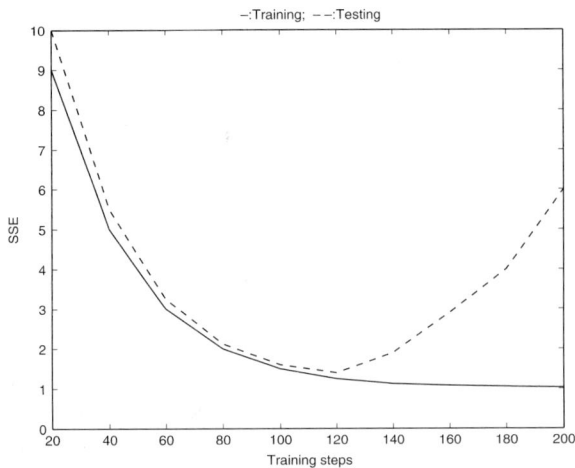

Fig. 6:   Neural network learning curves.

monitored to detect the appropriate stopping point. Early stopping has an implicit effect of regularisation, as shown by Sjoberg *et al.* (1995).

4.3 Assessment of model validity

The purpose of model validation is to verify that the identified model fulfils the modelling requirements according to subjective and objective criteria of good model approximation. A statistical approach for model validation not only involves hypothesis testing with respect to the complexity and model order of an estimated model, but also classification of models with equal orders. It is usually a major objective to obtain a model of least possible complexity within the limits of required model accuracy. A number of criteria have been developed for judging model quality by looking at both model accuracy and model complexity. The residuals of a model represent the misfit between the data and the model. The presence of any information remaining in the residual is an indication that the model might be insufficiently complex or otherwise inappropriate. Residual analysis comprises tests of factors such as:

- independence of residuals;
- normality of the distribution of the residuals;
- zero crossings (changes of sign) of the residual sequence;
- correlation between residuals and inputs.

A number of model validity tests for nonlinear model identification procedures have been developed and applied, such as the statistical chi-squared test (Leonartis and Billings, 1987), the final prediction error criterion (Akaike, 1974), the information theoretic criterion (Akaike, 1974) and the predicted squared error criterion (PSE) (Barron, 1984). The PSE criterion, although originally developed for linear systems, can be applied to feedforward nets providing that they can be approximated by a linear model.

$$\text{Final prediction error (FPE)} = \left(\frac{E}{2N} \right) \frac{N + N_w}{N - N_w}$$

$$\text{Information theoretic criterion (AIC)} = \ln \left(\frac{E}{2N} \right) + 2 \frac{N_w}{N}$$

$$\text{Predicted squared error (PSE)} = \frac{E}{2N} + 2s^2 \frac{N_w}{N},$$

where s^2 is the prior estimate of the true error variance and is independent of the model being considered. It can be seen that these tests make use of functions that strike a balance between the accuracy of model fit (average squared error, E, over N data points, $E\sqrt{2N}$) and the number of adjustable parameters or weights used (N_w). As a result of this, minimisation of these test functions leads to networks (models) that are neither under- nor over-complex. If the model is correct and if the method prerequisites are satisfied, then the residuals should be structureless. In particular, they should be uncorrelated with any other variables, including inputs and outputs.

This is the assumption upon which residual tests are based. A simple check is to plot the residuals versus the fitted values. Such a plot should not reveal any obvious pattern. Another valuable diagram is the histogram of the residual amplitudes, which reveals distributions that differ from the normal distribution.

The validity of a model can be assessed using the high-order correlation tests based on the following properties (Billings and Voon, 1983, 1986; Billings and Zhu, 1995). For linear models, the first two tests are sufficient whilst for nonlinear models all the tests should be used.

$$\Phi_{\epsilon\epsilon}(\tau) = E[\epsilon(t-\tau)\epsilon(t)] = \Delta(\tau) \qquad \forall \tau$$

$$\Phi_{x\epsilon}(\tau) = E[x(t-\tau)\epsilon(t)] = 0 \qquad \forall \tau$$

$$\Phi_{x^2\epsilon}(\tau) = E[x^2(t-\tau) - \overline{x^2(t)}]\epsilon(t) = 0 \qquad \forall \tau$$

$$\Phi_{x^2\epsilon^2}(\tau) = E[x^2(t-\tau) - \overline{x^2(t)}]\epsilon^2(t) = 0 \qquad \forall \tau$$

$$\Phi_{\epsilon(\epsilon x)}(\tau) = E[\epsilon(t)\epsilon(t-1-\tau)x(t-1-\tau)] = 0 \quad \forall \tau,$$

where ϵ is the model residual and $\overline{x^2}$ is the time average of x^2. These tests examine the cross-correlations between the model residuals and the model inputs. Normalisation to give all test statistics a range of $[-1, 1]$ and approximate 95% confidence bands at $1.96/\sqrt{N}$, N being the number of test data, make the tests independent of signal amplitudes and easy to interpret. If the correlation tests are satisfied then the model residuals are a random sequence and are not predictable from the model input. This provides additional evidence of the validity of the identified model.

Finally the number of *'zero crossings of the residuals'* can be checked, where a zero-crossing indicator ζ_k at time k is calculated as

$$\zeta_k = \begin{cases} 1, & \text{if } \epsilon_k\epsilon_{k+1} < 0 \\ 0, & \text{if } \epsilon_k\epsilon_{k+1} > 0. \end{cases}$$

If the model is adequate, then the ζ_k will be independent variables which take on the values 0 and 1 with equal probability. For example, for $N = 1000$, the number of zero crossings should with 95% probability be in the interval $[459, 541]$ to satisfy the above criterion.

4.4   Collinearity of industrial data

Given the well-known industrial data problems of collinearity and noise, the question arises of how to ensure that the neural network is 'optimally trained'. If the resulting neural network is to be used in process control, model validity is questionable since every manipulated variable would have to be changed in such a way that the variable relationships continue to hold. In practice when a feedback loop is introduced, its impact upon data collinearity needs to be assessed. In both model based predictive control (MBPC) and inverse model control (IMC), problems will occur. For example, in the IMC approach, the inverted inputs would have to follow the variable correlations to ensure model validity.

In practice exact collinearity would not generally occur but in the process industries significant collinearity and skewed data are common. The impact this has on multiple linear regression (MLR) is significant due to ill-conditioning; however, neural networks are not as prone to ill-conditioning as MLR. It may therefore be believed that data collinearity is not a problem with neural networks, but this is an incorrect assumption. This can be verified by trying to predict new, unseen data. When the new data are corrupted with measurement noise (as with all industrial data), the derived model often results in a large variance in the prediction; that is, neural networks can enlarge the variance in the presence of collinearity. To overcome this problem, the input variance to the hidden layer should be minimised. Minimising this variance whilst minimising the network training error can be achieved by minimising the performance function

$$E_\lambda = E + \lambda \sum_{i=1}^{n} \sum_{j=1}^{m} \omega_{ij}^2 = E + \lambda \|\omega\|^2,$$

where λ is a penalty scalar on the magnitude of the network weights. Similarly the output variance with respect to the output layer weights should also be minimised. Such an approach is a variation of ridge regression. The problem is to choose the best penalty factors. This is done based on error minimisation, experience and optimisation. Practical process data may not be exactly collinear but can display strong correlations, in which case the penalty factors will introduce *bias* into the prediction whilst reducing the variance. The larger the penalty function, the smaller the variance but the larger the bias. The use of cross-validation is therefore important. The application of principal components analysis (PCA) to the raw data to remove the linear relationships inherent within the data is one way of addressing the problem. One limitation of this approach is that PCA only focuses upon the variability in the inputs and ignores the relationship with the output(s). A lower–order component which only explains a small proportion of the total variability in the inputs may be significant in modelling an output. This can occur with industrial process data where some input variables exhibit high variability but are not correlated with the output(s). In contrast, some variables may be less variable but display significant correlation with the output(s).

5   Case study 1: spectral data calibration

Multivariate statistical methodologies have, for some time, been successfully applied to many types of chemical problem. For example, experimental design techniques have had a major influence on the understanding and improvement of industrial chemical processes. More recently, the field of chemometrics has emerged with the focus upon analysing observational data originating mostly from organic and analytical chemistry, food research and environmental studies. These data tend to be characterised by many measured variables (wavelengths), which are highly correlated, on each of a few observations (spectra). Often the number of wavelengths p greatly exceeds the number of spectra N.

Over the last few years, the use of calibration models, generated from near-infra-red (NIR) spectroscopy, to monitor industrial chemical processes, has received an increasing amount of attention; see for example Brimmer and Hall (1993), Brown (1992), Conlin *et al.* (1998), Hansen and Khettry (1994), Martens and Foulk (1990), Mockel and Thomas (1992), Puebla (1992) and Webster (1994). This approach, com-bined with analytical studies of the process of interest using techniques such as high-pressure liquid chromatography (HPLC) and gas chromatography (GC) can provide calibration models capable of accurately estimating the product quality. These models allow the process to be better monitored and controlled, resulting in a reduction in the variation of the product and hence the amount of off-specification product produced.

A wide range of linear and nonlinear processing methods is available for the analy-sis of spectral data. Linear algorithms include linear principal component regression (LPCR), linear projection to latent structures (LPLS), linear model-based orthogonal least squares decomposition (LORLS) and locally weighted regression (LWR). Non-linear techniques include nonlinear PCR (NPCR), nonlinear PLS (NPLS), nonlinear polynomial models based upon orthogonal least squares decomposition (NPORLS) (Chen *et al.*, 1989), sigmoidal function based feedforward neural networks (SFFN), radial basis function neural networks (RBFN) (Powell, 1987 and Chen *et al.*, 1990a), and wavelet transformation networks (WTN) (Zhang and Benveniste, 1992). The var-ious approaches are investigated and compared on an industrial dataset supplied by an industrial partner within a BRITE-EURAM project, intelligent manufacture of polymers.

5.1 Data description and analysis

In a particular polymer reactor, improvements to the existing on-line monitoring procedures are required for the better control of polymerisation processes. Three-hundred-and-two latex samples were prepared with eight outputs recorded. The reflect-ance of each of the samples was measured in the NIR range $1100–2500\,\mu m$, in $2\,\mu m$ steps, i.e. 700 wavelengths. This raw information was converted to absorbance spectra and then normalised to zero mean and unit standard deviation:

$$\text{Abs} = \log_{10}\left(\frac{100}{\text{reflectance}}\right).$$

Although, for the most part, these curves support the physical property that low absorbancy infers greater reflectance of NIR light by the polymer particles', clear nonlinearities can be observed. The complexity of the structure of the polymer par-ticles, the composition of the polymer products and the large number of input spectra suggest that the application of linear regression techniques may not necessarily be appropriate (Figure 7).

The first step was to allocate the spectra to the training, test and unseen val-idation datasets. One-hundred-and-thirty-seven samples were selected for training, 90 samples for testing and 77 samples for validation. Previous studies have shown that, when samples are measured under comparable operating conditions, the linear

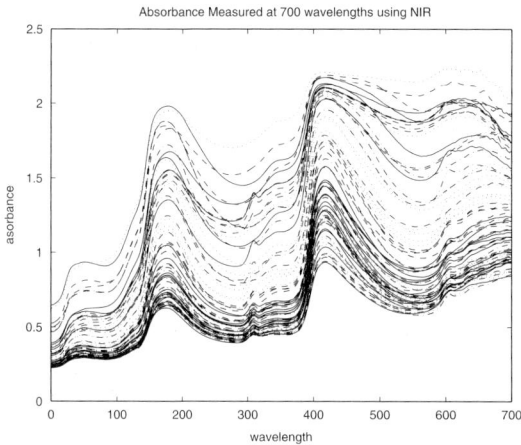

Fig. 7: Absorbance spectra of latexes.

techniques of LPCR, LPLS and ORLS can often provide good results at low computational cost. These results are not necessarily optimal. If the measurement ranges vary to a large extent, advanced statistical modelling methods including nonlinear regression methods and neural network techniques may be needed to obtain a more appropriate representation. Although specific advanced statistical methods and nonlinear regression techniques have been shown to be superior to other approaches for certain problems, the experimental results show that the goodness of the prediction depends strongly upon the data being modelled.

A number of linear and nonlinear regression techniques were used for the calibration of the data, including principal components regression (LPCR), projection to latent structures (LPLS), orthogonal least squares based regression (ORLS), feedforward neural networks (FFN), radial basis function neural networks (RBFN) and locally weighted regression (LWR). The prediction errors resulting from the different processing methods are reported in Table 1 and Table 2 for the linear and nonlinear techniques respectively. The results are presented in terms of the mean sum of squared error (MSSE) and the error range for the training, test and validation datasets as well as the composite dataset comprising 304 samples.

It can be observed that, with the exception of nonlinear PLS, the prediction errors using the nonlinear modelling methods are smaller than those produced from the linear regression techniques. These results imply that the final model should be nonlinear. Both the standard deviations (or sum of squared errors) and the maximum errors are also considered when assessing the performance of a predictor. In some situations, the standard deviation decreases but the maximum error increases as a result of a few data points. Such models cannot be adopted with confidence. Although the polynomial-based nonlinear projection to latent structures (NLPLS) and spline-based NLPLS did not give good results, this does not imply that NLPLS is not suited to the present

Table 1 Prediction errors for solid content using linear empirical techniques.

		ORLS (40 WLs)	LPCR (45 PCs)	LPLS (23 LVs)	LWR
Training (137 samples)	MSSE	9.41	14.06	12.12	7.63
	Error Range	$(-24.6, 52.6)$	$(-37.0, 69.2)$	$(-27.5, 60.7)$	$(-24.6, 44.9)$
Test (90 samples)	MSSE	14.55	13.25	13.69	16.34
	Error Range	$(-41.9, 34.6)$	$(-35.7, 39.3)$	$(-37.2, 39.62)$	$(-54.9, 40.0)$
Validation (77 samples)	MSSE	14.22	13.68	13.30	22.30
	Error Range	$(-37.6, 33.3)$	$(-30.49, 34.4)$	$(-26.6, 41.9)$	$(-30.9, 58.0)$
Total Data Set (304 samples)	MSSE	12.45	13.72	12.89	
	Error Range	$(-41.9, 52.5)$	$(-37.0, 69.2)$	$(-37.2, 60.7)$	

Table 2 Prediction errors for solid content using nonlinear empirical methods.

		NPLS	NPORLS	SFFN	RBFN	WaveNet
Training	MSSE	23.89	6.44	6.65	8.98	7.96
	Error Range	$(-65.5, 97.1)$	$(-19.3, 27.1)$	$(-28.2, 21.9)$	$(-22.4, 35.2)$	$(-22.7, 34.1)$
Test	MSSE	23.96	9.83	11.84	11.23	10.91
	Error Range	$(-59.4, 44.8)$	$(-37.9, 36.8)$	$(-40.4, 38.92)$	$(-39.0, 36.3)$	$(-37.5, 33.6)$
Validation	MSSE	25.43	7.66	11.53	12.12	11.33
	Error Range	$(-61.7, 60.5)$	$(-19.5, 29.1)$	$(-23.4, 42.4)$	$(-25.4, 42.4)$	$(-18.2, 52.6)$
Total Data Set	MSSE		7.89	7.11	10.54	9.90
	Error Range		$(-37.9, 36.8)$	$(-36.4, 40.3)$	$(-39.0, 42.4)$	$(-37.1, 52.9)$

study. There are many alternative factors to be considered, including modification of the initial parameters of the present algorithms or else the use of other versions of NLPLS. The current optimal model was derived using a nonlinear polynomial model based upon orthogonal least squares decomposition (NPORLS). Thirty-three regression variables were included within the model, both linear and nonlinear terms.

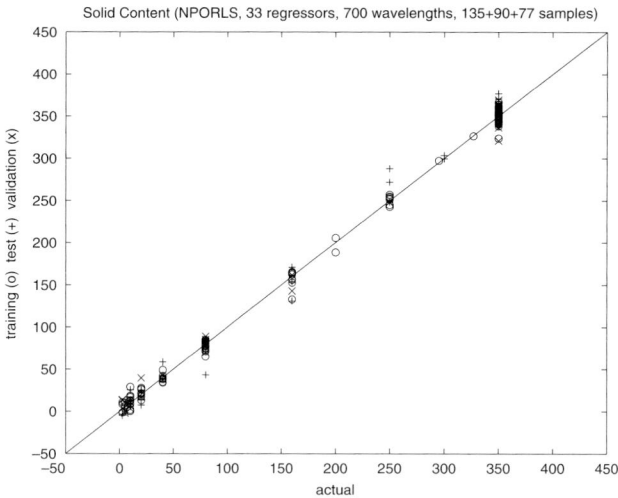

Fig. 8:   Optimal model (NPORLS).

The maximum errors on the training set, test set and validation set were 7.7%, 10.8% and 8.3% of the maximum concentration output respectively. The resulting prediction is shown in Figure 8.

5.2   Bagging regression

Recently, bootstrapping regression has been applied through a technique called bagging predictors (bootstrap aggregated regression) (Breiman, 1996a) to increase the robustness of the estimator. It has been demonstrated that, through the aggregation of bootstrap samples of a predictor, a predictor with better generalisation properties is obtained than when a single predictor is calculated. In practice there is no guarantee that a single predictor is able to encapsulate all the information contained within the data and hence provide the best predictor. Aggregating different predictors can increase the possibility that more information is extracted from the dataset and as a result a more accurate prediction can result. In the context of regression, the aggregation of multiple predictions refers to the averaging of multiple predictors. Since the bootstrap method is easy to implement, bagging predictors provide a simple method for obtaining more robust predictions. The application of bagging predictors to subset regression and some classification procedures is demonstrated by Breiman (1996a). In a similar manner, the use of multiple predictors for prediction has been developed under the name 'stacked modelling or stacking'. The idea of stacking was initiated by Wolpert (1992). The technique has been applied to subset regression (Breiman, 1996b) and backpropagation networks (Sridhar *et al.*, 1996). In stacking, multiple predictors are generated through the use of cross-validation, with the resulting predictor being a linear combination of the multiple predictors. There are two steps in the stacking procedure; the first step is the construction of multiple predictors and the second involves the search for the best linear combination of the multiple predictors.

The method of bagging regression is based upon the generation of new data from the original data by means of random reselection (bootstrapping). For every group of data, one prediction model was developed. These predictors were then aggregated to build a new predictor. The following summarises the steps involved in bagging regression.

(i) A standard bootstrap is initially performed on the original 137 samples, i.e. sampled with replacement.

(ii) A PLS/PCR model was fitted to the new sample of data. Cross-validation was used to select the number of latent variables to include within the analysis.

(iii) Steps 1 and 2 were repeated, 30 times, say.

(iv) The 30 outputs were then averaged, providing the final calibration model.

For illustration of the application of bagging predictors, consider the linear regression model

$$Y = \beta_1 X_1 + \beta_2 X_2 + \beta_3 X_3 + \text{error}.$$

For example, 30 bootstrap models are constructed, resulting in the 30 predictions

$$Y^{*1} = \beta_1^{*1} X_1 + \beta_2^{*1} X_2 + \beta_3^{*1} X_3$$
$$Y^{*2} = \beta_1^{*2} X_1 + \beta_2^{*2} X_2 + \beta_3^{*2} X_3$$

$$\vdots \qquad \vdots$$

$$Y^{*30} = \beta_1^{*30} X_1 + \beta_2^{*30} X_2 + \beta_3^{*30} X_3.$$

Bagged prediction gives

$$Y^* = \frac{1}{3}\sum_{j=1}^{30} Y^{*j} = \frac{1}{3}\sum_{j=1}^{30}\beta_1^{*j}(X_1) + \frac{1}{3}\sum_{j=1}^{30}\beta_2^{*j}(X_2) + \frac{1}{3}\sum_{j=1}^{30}\beta_3^{*j}(X_3).$$

Both 30 predictors and 100 predictors were generated based upon the training set (135 samples). In Table 3 and Table 4 the prediction errors of the bagging regression using the 30 predictors and 100 predictors are compared with those from the original data for the training, test and validation datasets, for PCR and PLS respectively. The prediction errors on the validation set were reduced appreciably.

Figures 9 and 10 illustrate the changes in the standard deviation of the prediction error for bagging LPCR and LPLS, respectively, for an increasing number of predictors. The standard deviation of the error of the jth bagging predictor is given by

$$s_j = \sqrt{\frac{\sum_{i=1}^{n}(\epsilon_{ij} - \overline{\epsilon_j})^2}{n-1}}$$

Table 3 Prediction errors for the single predictor and bagging predictor for PCR.

		LPCR	BAG LPCR (30 Predictors)	BAG LPCR (100 Predictors)
Training	MSSE	14.06	14.63	14.2
	Error Range	$(-37.0, 69.2)$	$(-39.9, 65.4)$	$(-36.6, 75.1)$
Test	MSSE	13.25	12.51	13.3
	Error Range	$(-35.7, 39.3)$	$(-36.7, 35.8)$	$(-44.2, 37.5)$
Validation	MSSE	13.68	12.02	11.7
	Error Range	$(-30.5, 34.4)$	$(-32.0, 37.1)$	$(-28.7, 32.3)$
Whole Data Set	MSSE	13.7	13.4	13.3
	Error Range	$(-37.0, 69.2)$	$(-46.7, 73.2)$	$(-44.2, 75.1)$

Table 4 Prediction errors for the single predictor and bagging predictor for PLS.

		LPLS	BAG LPLS (30 Predictors)	BAG LPLS (100 Predictors)
Training	MSSE	12.12	10.76	10.82
	Error Range	$(-27.5, 60.7)$	$(-30.2, 49.1)$	$(-27.0, 61.9)$
Test	MSSE	13.69	14.27	14.18
	Error Range	$(-37.2, 39.6)$	$(-30.9, 37.3)$	$(-33.9, 41.9)$
Validation	MSSE	13.30	12.32	11.34
	Error Range	$(-26.6, 42.0)$	$(-28.2, 38.7)$	$(-26.4, 36.1)$
Whole Data Set	MSSE	12.88	11.93	12.01
	Error Range	$(-37.2, 60.7)$	$(-36.5, 60.9)$	$(-33.9, 61.9)$

where n is the number of samples, j denotes the jth bagging predictor, the predicted output of which is calculated by averaging the predicted outputs of the j bootstrap predictors. Figures 9 and 10 show that, when the number of aggregated predictors is small, the prediction errors are unstable. Increasing the number of predictors results in a stabilisation of the error standard deviation. To obtain robust calibration models using bagging LPLS requires more aggregated predictors than bagging LPCR. The standard deviation appears to stabilise at around 30 predictors for the training, test and validation datasets, respectively. The study showed that even with just 15 predictors (results not reported), the prediction errors resulting from the bagging regression can be reduced to a stable level. This suggests one approach for finding robust models for calibration using small datasets. As with other approaches, a number of options exist to explore the methodology, e.g. choice of size of datasets, how to aggregate the subpredictors, how to bootstrap the original data, etc. The approach of stacked regression can also be used to provide robust models by aggregating different styles of predictors.

Std. Deviation of Errors V Number of Predictors (BagPCR, 137+90+77 samples)

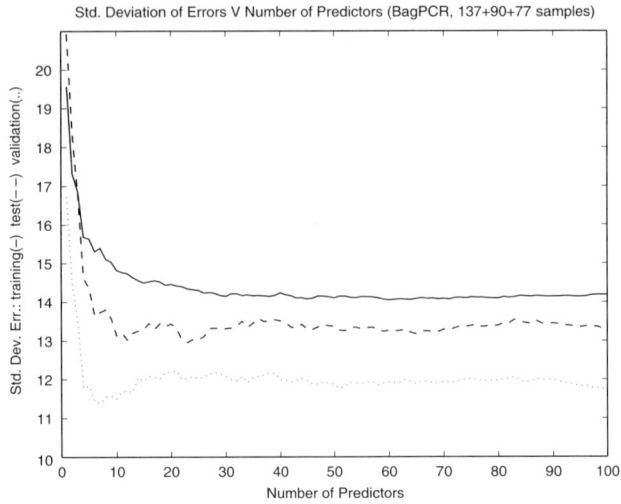

Fig. 9:   Standard deviation versus number of predictors for PCR.

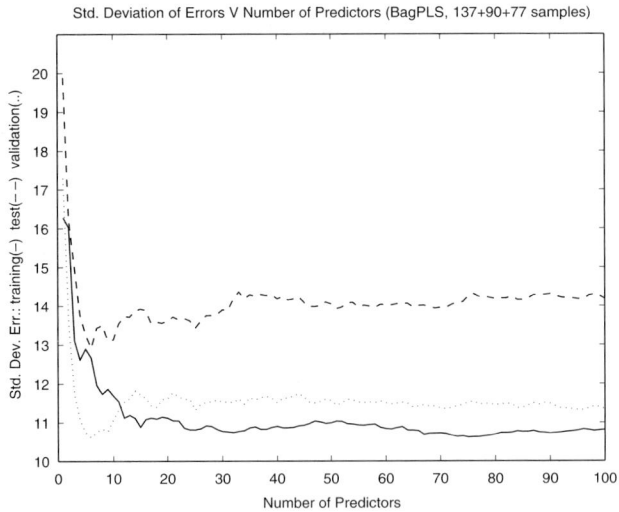

Std. Deviation of Errors V Number of Predictors (BagPLS, 137+90+77 samples)

Fig. 10:   Standard deviation versus number of predictors for PLS.

6   Committees of networks and neural network stacking

For neural networks to become a practical industrial tool in process modelling, not only must they be shown to be able to provide accurate models, but more importantly they must display high levels of robustness. In classical approaches to modelling, it is typically assumed that a certain ideal model is 'true', and based upon certain available evidence a model is accepted or rejected. If a model is rejected because of misfit, better models are sought until the true model is obtained. This approach is appropriate if there is a theory guiding the specification and evaluation of the models. However, such a theory is often lacking for nonlinear systems modelling, especially when building nonlinear models using only a limited amount of system input–output data. When neural network models are built using system input–output data, the developed model is always an approximation of the real system. It is not possible to guarantee that a neural network trained on a limited amount of training data is the true representation (model) of the real system for a variety of reasons. For example, the optimum network structure is difficult to specify; optimisation of network weights results in a local minimum; the minimum found may be highly sensitive; different learning algorithms can lead to different generalisation characteristics; and the different convergence criteria for network training can also lead to different solutions. A promising approach is to combine several models to improve the overall prediction. The emphasis of this approach is on minimising the variance of the predictors and through this, maximising the representational ability of the 'model' to produce future predictions for unseen data. When building neural network models, it is quite possible that different networks perform better in different regions of the input space. Consequently, there is no guarantee that a single predictor is able to represent all the information contained within a dataset and hence provide the best predictor.

Improving prediction accuracy by combining several models has been investigated by a number of researchers including Bates and Granger (1969), Breiman (1992, 1996a), Jacobs *et al.* (1991), Jordan and Jacobs (1994), Raviv and Intrator (1996), Shimshoni and Intrator (1998), Sridhar *et al.* (1996) and Wolpert (1992). Bates and Granger (1969) combined two different models for forecasting a time series and reported improved performance using the combined model. Mixture models such as adaptive mixtures of experts (Jacobs *et al.*, 1991) and hierarchical mixtures of experts (Jordan and Jacobs, 1994) use the divide and conquer approach where a mixture of experts compete to gain responsibility in modelling the output in a given input region. The system's output is obtained as a linear combination of the experts' outputs and the combination weights are computed as a function of the input. The different experts are usually trained on a single dataset simultaneously by minimising a combined cost function, and the final combination of the experts is determined by a gating module which is constructed in the same training session. The gating module comprises a switch which selects a particular single network to pass through the 'gate'. A general framework for combining multiple predictors is stacked generalisation (Breiman, 1992; Wolpert, 1992), where each estimator is trained using a different subset of data. Two approaches for improving prediction accuracy by combining several models have been investigated to enhance the robustness and generalisation properties of single empirical predictors and to increase the possibility that more information is extracted from a dataset:

- *Stacked generalisation (stacked regression)*: Breiman (1992) and Wolpert (1992) proposed a general framework for combining multiple predictors where a number of single predictors are trained with a different subset of the data and which are then combined in an appropriate manner (committees of networks).

- *Bootstrap aggregated regression (bagging)*: Breiman's (1994) concept of bagging is fundamental to the building of robust neural network models from minimal plant data. We assume that, for building stacked neural network models, the number, n, of resamplings should typically be of the order of twenty to thirty. This is realistic in terms of the underlying statistics and the central limit theorem. These findings have been confirmed from previous experience of applying these techniques. For each resampled dataset, a neural network model is developed and these individual neural network models are then combined as shown in Figure 11.

This can be represented by the following relationship:

$$f(X) = \sum_{i=1}^{n} w_i f_i(X),$$

where $f(X)$ is the stacked neural network predictor, $f_i(X)$ is the ith neural network predictor, w_i is the stacking weight for combining the ith neural network, and X is a vector of neural network inputs.

A feature of stacked generalisation is that it attempts to solve simultaneously the problem of model selection and estimation of model combinations to improve model predictions. Stacked generalisation can be viewed as a more general solution to estimating an optimum predictive model while accounting for the bias–variance trade-off. Instead of the selection of a single neural network model, several neural network models are combined to improve model accuracy and robustness. The number of hidden neurons for each individual network in the stacked network is determined by considering a number of neural networks with different numbers of hidden neurons and selecting the one giving the least error on testing data. Each neural network is trained using the Levenberg–Marquardt optimisation algorithm together with an 'early stopping' mechanism to prevent over-fitting. During network training, the algorithm continuously checks the network error on testing data. Training is terminated at the point where the network error on the test data is a minimum, with early stopping used to implement regularisation to improve model robustness.

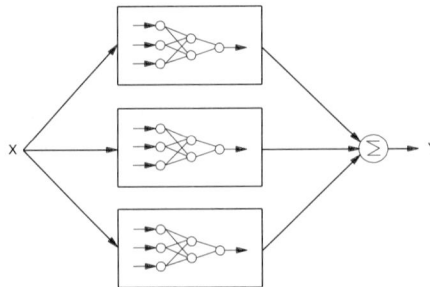

Fig. 11:   A stacked neural network.

The training strategy used has the feature of fast training and not over-fitting corrupting noise. Thirty network representations are typically built. The overall output of the stacked neural network is a weighted combination of the individual neural network outputs. Appropriate determination of the stacking weights is essential for good modelling performance. A simple approach is to take equal weights for the individual networks. This approach was found to give quite good performance. A second way is to obtain the weights through multiple linear regression. However, care needs to be taken because of the highly correlated nature of the individual network predictors since each network is trained to model the same underlying relationship. Breiman (1992) showed that calculation of stacking weights through multiple linear regression does not appear to give good results and he suggested constraining the stacking weights to be nonnegative. Appropriate stacking weights can be obtained through principal component regression (PCR). Determining the weights from PCR can, with care, give improved performance, with the number of principal components to be used being found through cross-validation. Different numbers of principal components can be studied and the resulting model errors on the test data compared. The number of principal components used is then determined based upon the model errors. An additional major industrial problem associated with the building of neural network representations is the availability of training data. Training data are frequently not abundant, but in practice there are numerous industrial situations where it is necessary to build an empirical model using minimal process data. In the approach developed here, and specifically evaluated on a batch polymerisation reactor in the next section, training and test data for each neural network were selected from a small number of batches through bootstrap resampling. The idea of the bootstrap is to assume that a cumulative distribution function (CDF) $\hat{F}_n$, calculated from an observed sample $X_1, X_2, \ldots, X_n$, is sufficiently like the unknown CDF, F, so that a calculation performed using $\hat{F}_n$ can be used as an estimate of the calculation using F: the distribution of the training data obtained through bootstrap resampling is similar to that of the original data distribution.

7 Case study 2: an application of stacked neural networks to polymer property prediction

7.1 Preamble

A major problem in the control of product quality in industrial polymer manufacturing is the lack of suitable on-line product property measurements. Although instruments for measuring product properties might be available, they tend to be off-line in the QA laboratory, thereby introducing substantial measurement delays. Some of these difficult-to-measure variables can however be related to certain more easily measurable variables such as temperature, flows, torques, density, etc. Inferential estimators, or software sensors, of these difficult-to-measure 'quality' variables can then be derived from measurements of the more easily measured process variables. The key step in inferential estimation is to establish a relationship between the difficult-to-measure quantities and the more easily measured process variables. One popular approach is through the use of a first principles mechanistic model of the process and state estimation techniques such as the extended Kalman filter (EKF). This approach, however, requires a detailed understanding of the process and consequently model development is usually very demanding and time consuming. To overcome this difficulty, especially in industrial manufacturing, neural network representations based on monitored plant data can be developed. As a result of the learning capability of a neural network, the relationship between polymer quality variables and the on-line measured variables in the reactor can be identified from the reactor operational data.

An important requirement of inferential estimators is that they should not only provide satisfactory estimation accuracy but also be robust to new plant data. The accuracy and robustness of a neural network model is strongly influenced by the availability of training data. When the amount of training data is limited, the network tends to over-fit and hence exhibit significant generalisation errors. To build an accurate and robust neural network model, ideally a large amount of training data should be made available. With today's on-line process data monitoring facilities, it is generally assumed that good process data are readily available, but in practice in many industrial plants the collection of sufficient, appropriate 'good quality data' is still a real problem. In addition, product 'quality' information is usually limited by analyser constraints, laboratory costs and long time delays before the measurement is made available. Limited process data is a serious problem in the development of accurate and robust network representations.

7.2   Inferential estimation of a quality variable

The robust modelling approach for minimal process data is demonstrated by application to a batch polymerisation process. Here, the on-line measured process variables are temperatures and coolant flow rates (Zhang *et al.*, 1997). Material property variables, number average molecular weight, M_n, and weight average molecular weight, M_w, are in practice only available from the QA laboratory and not measured through the batch. Significant improvements in production performance could however be achieved by estimating these properties from the on-line process measurements. The evolution of the material properties during the course of a production run are mainly determined by the initial batch recipe: T_0, the reactor jacket outlet temperature and I_0, the initial initiator weight. Different batch recipes will lead to different material properties and different heat generation profiles. Correlation analysis of the reactor operational data indicates that there is a strong relationship between the material properties and the temperatures, the coolant flow and the processing time. These are used to estimate the material quality variables. The nominal batch time for this process is 180 time units. In this study, data from nine batches were used to develop neural network based inferential estimators. In designed experiment tests for the nine batches, off-line material property measurements are taken every 20 time units to provide nine sets of property analyses. Two additional batches, with different batch recipes from the nine batches, were used to validate the neural network based inferential estimator. Batch recipes for the 11 batches are given in Table 5. Stacked neural network estimators were developed for estimating the material quality parameters M_n and M_w. Each estimator comprised 30 neural network representations. Training data for each neural network were selected from batches 1, 2, 3, 5 and 7 through the standard bootstrap, i.e. resampling with replacement (Efron, 1982). Data from batches 4, 6, 8 and 9 were then used as the test dataset. Batches 10 and 11 formed the unseen validation dataset. The number of hidden neurons for each individual network in the stacked network was determined by considering several neural networks where the number of hidden neurons ranged from 5 to 25. The network giving the least error on the test data was selected. Each neural network was trained using the Levenberg–Marquardt optimisation algorithm together with a cross-validation-based 'early stopping' mechanism to prevent over-fitting. The training strategy implemented has the advantages of speed and not over-fitting the noise in the data. The weights for combining the individual neural networks were determined through PCR. Two principal components were retained in the PCR models.

Estimates of M_n and M_w from the stacked neural network models on the validation data are plotted in Figure 12. Here the solid lines represent the measured values and the dotted lines represent the estimated values. The estimates from the stacked neural networks provide acceptable predictions of the actual material property profiles. The resulting predictions are more robust than those achievable from a single network predictor as seen by comparing the histograms in Figures 13 and 14. These show the MSE (mean squared error) for the individual neural network models for the estimation of M_n on the training, test and validation data. It can be observed that these models give variable performances. Indeed, it can be seen that the eighth and ninth neural network models give similar performance for the estimation of M_n on both the training and test data. However, their performance on the validation data differs drastically. The 'best' individual network on training and test data is the twelfth network, but its performance

Table 5 Batch recipes.

Batch No.	T_0	I_0
1	343	2.5
2	348	3.0
3	338	2.0
4	343	2.8
5	346	2.0
6	350	1.8
7	332	3.5
8	340	2.6
9	345	2.6
10	342	2.2
11	335	2.4

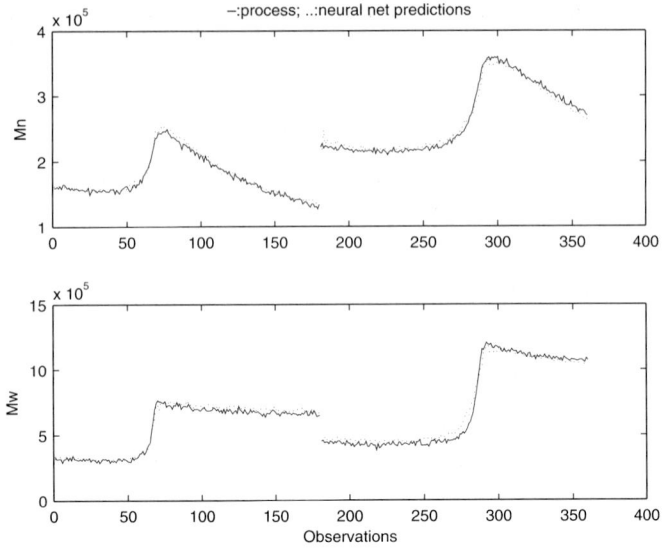

Fig. 12:   Stacked estimation of M_n and M_w (Batch 10: 0 to 180; Batch 11: 181 to 361).

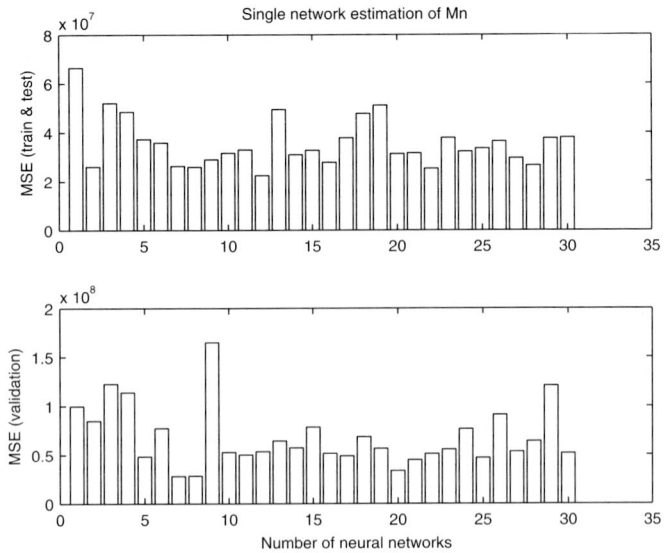

Fig. 13:   Single network MSE errors for the training, test and validation datasets for the estimation M_n.

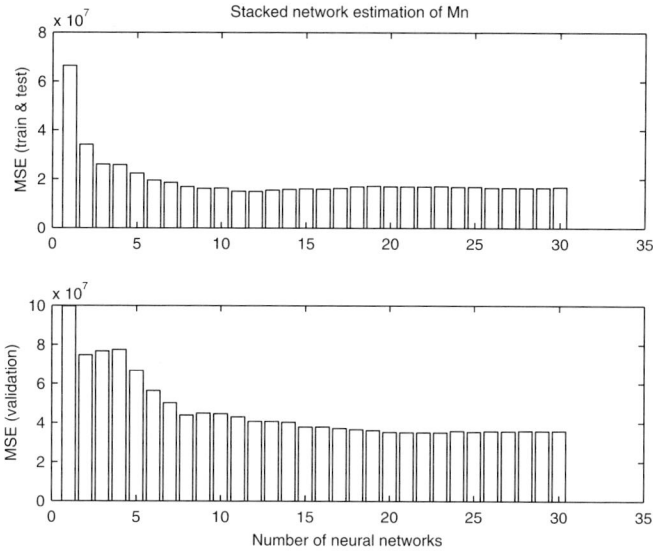

Fig. 14: Stacked network MSE errors for the training, test and validation datasets for the estimation M_n.

on the validation data is not the 'best' amongst all the networks. These studies, together with many others, indicate the nonrobust nature of single neural network models even when they might be selected as being theoretically 'the best'. Figure 14 shows the MSE of the stacked neural network models for estimating M_n on the training, testing and validation data. The x-axis in each plot represents the number of neural networks in the stacked model. It can be observed that the errors generally decrease with the number of individual neural network representations and gradually approach stable levels. The model errors of stacked neural network models on training, test and validation data become consistent. This is in sharp contrast to the single neural network models shown in Figure 13. This clearly demonstrates that stacked neural network models are more robust than single neural network models. By comparing Figures 13 and 14, it can be observed that the estimation accuracy of the stacked network on the unseen validation data is much higher than for most of the individual networks. From Figure 14 it can be concluded that the model errors stabilise after employing 20 networks. From experience on a wide range of case studies, it can be concluded that stacking 20–30 networks is usually sufficient to stabilise the model errors. In practice it is recommended that 30 networks are developed. A major problem in the industrial application of neural network models is the current lack of confidence bounds. This is particularly the case for nonlinear systems, where nonparametric confidence bounds are required (Martin and Morris, 1996a; Shao *et al.*, 1997a,b). Bootstrap techniques can be used to estimate the standard error of model predictions (Efron and Tibshirani, 1993; Tibshirani, 1996). Based on the estimated standard error, confidence bounds for neural network predictions can be calculated. The 95% confidence bounds for the stacked neural network estimation on the two validation batches (10 and 11) are shown in Figure 15. For the sake of clarity, the estimates and confidence bounds are only shown at 10 minute intervals. The narrower the confidence bounds, the higher is the confidence in the estimation. Although predictions for both validation batches are acceptable, Figure 15 clearly shows that the predictions for batch 10 have tighter confidence bounds than those for batch 11. By examining the batch recipes in Table 5, it can be seen that the recipe for batch 10 is closer to the recipes of the five training batches than for batch 11. The scaled distance between the recipe for batch 10 and those for the training batches is 0.2913, whilst for batch 11 it is 0.7315. This explains why the confidence bounds are tighter for batch 10 compared with batch 11. A significant benefit from using the bootstrap resampled datasets is that confidence bounds for the model predictions can be calculated automatically. The results indicate that bagging can lead to improved expected performance, with a substantial improvement for small training datasets. For large datasets, the benefits of bagging are less clear cut. It is conceivable that for large datasets bagging may degrade performance. For small training datasets, the combined model residuals are 'whitened', giving rise to an overall better performance. Indeed, the empirical evidence suggests that the benefits of aggregating models outweighs the decline in performance of the individually trained representations. When sufficient training data are available, a model performance close to the theoretical limit (the inherent variability of the data) can be achieved, and aggregation of an ensemble of models does not provide any significant improvement in overall performance.

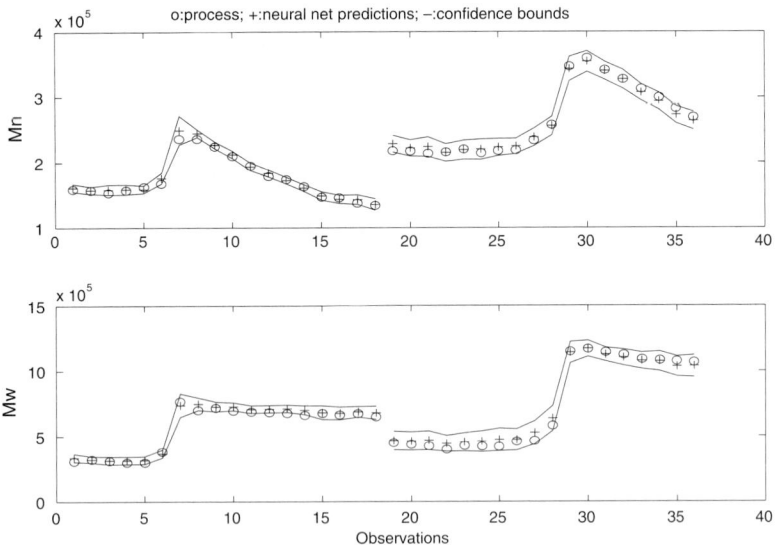

Fig. 15:   Stacked neural network estimations with confidence bounds (Batch 10: 1 to 18; Batch 11: 19 to 36).

8   Applications of dynamic neural networks

8.1   Neural networks in state estimation

A problem of present and increasing industrial interest where artificial neural networks may be of benefit is that of improving the quality of on-line information available to plant operators for plant control and optimisation. Major problems exist in the chemical and biochemical industries (paralleled in the food processing industries) concerned with the on-line estimation of parameters and variables that quantify process behaviour. The fundamental problem is that the key 'quality' variables cannot be measured at a rate which enables their effective regulation. This can be because of limited analyser cycle times or a reliance on off-line laboratory assays. An obvious solution to such problems could be realised by the use of a model along with secondary process measurements, to infer product quality variables (at the rate at which the secondary variables are available) that are either very costly or impossible to measure on-line. Hence, if the relationship between quality measurements and on-line process variables can be captured then the resulting model can be utilised within a control scheme to enhance process regulation. The concept is known as inferential estimation or soft-sensing (Tham *et al.*, 1991). Historically, with varying degrees of success, linear models, adaptive models and process specific mechanistic models have been used to perform this task. Many industrial processes exhibit directionality in their dynamics, and such responses are often found in polymerisation reactors. In many artificial neural network modelling studies such complex dynamics are not modelled accurately, with the network 'averaging' the different directional responses. The dynamic network architectures used in our work are capable of implicitly taking account of this type of response. Figure 16 shows data from a dynamic neural network model development where the directional dynamics of the process can be easily observed (Turner *et al.*, 1996).

Since continuous processes tend to operate at a nominal steady state, it could be argued that the problem of estimation or inferential measurement is roughly linear. In this case a fixed model or adaptive linear time series representation for state estimation would be quite adequate. The same cannot be said of processes from which are demanded regular product grade changes, and processes which involve batch or fed-batch reactions. Here the sometimes quite severe nonlinearities associated with the different phases of reactor operation cause the performance of adaptive mechanisms to degrade significantly. This is where the fast development times of neural network based process representations (models) provide a method of capturing the nonlinear dynamics in a relatively straightforward manner. Both simulation and industrial applications of neural network modelling have demonstrated their significant potential in process modelling and control.

A potential problem with using neural network model based estimators is that network models might have been identified using data collected from the plant which may have some of its control loops still closed. The resulting model will then have been identified with correlated data and will not be representative of the underlying

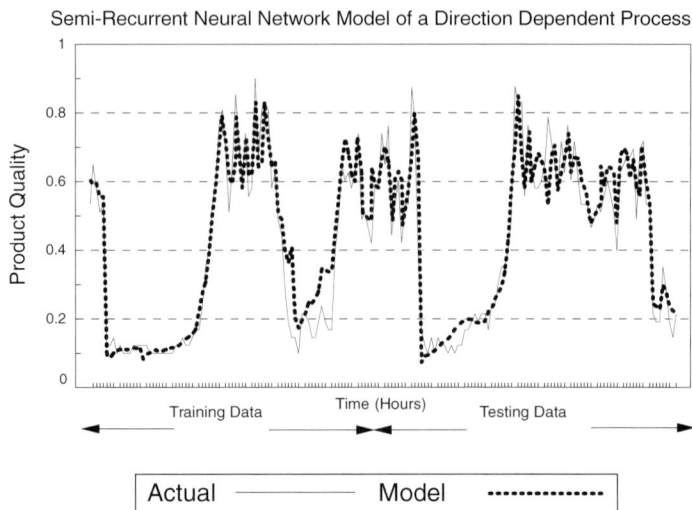

Semi-Recurrent Neural Network Model of a Direction Dependent Process

Fig. 16:   Prediction of direction dependent dynamics.

process behaviour. When such a model is used within a feedback control loop (manual or automatic) it will be subject to new process data that are further correlated, and the model predictions will degrade. In this case it is important to identify a new network using the loop data now available, and when predictions from this new network are deemed better than those of the previous model it should replace the old model. Whilst inferential estimation schemes, operating in 'open-loop', can be used to assist process operators with the availability of fast and accurate product quality estimates, the possibility of closed loop inferential control becomes very appealing. Such a control strategy could be implemented manually or with conventional regulators. Here the inferred estimates of the controlled output are used directly for feedback control. The effective elimination of a time delay caused by the use of an on-line analyser or the need to perform off-line analysis affords the opportunity of tight product control, even through the use of standard industrial controllers. Consequently, reductions in product variability caused by process disturbances, and corresponding reductions in off-specification product, can be achieved.

9   Nonlinear projection to latent structures

9.1   Preamble

When linear projection to latent structures is applied to a nonlinear problem, the minor latent variables cannot necessarily be discarded since they may not only describe the noise in the data but also significant amounts of the variance–covariance structure. Information relating to the nonlinear structure of the problem may be accounted for by a combination of latent variables. These may be both the higher-order and lower-order latent variables calculated from linear PLS. This scenario may help address the problem of fitting a nonlinear structure using linear PLS. However, in practice, identifying the relevant latent variables to incorporate into the model requires additional analysis and, more importantly, the final model might contain too many components to be useful.

Several attempts have been made aimed at producing a nonlinear variant of the linear PLS algorithm by integrating nonlinear features within the linear framework. A simple approach for developing a nonlinear PLS regression model is to extend the input matrix by including nonlinear combinations of the original variables (such as logarithms, square values, cross-products, etc.) and then performing a linear PLS on the extended input matrix and the output matrix (Wold *et al.*, 1989). This approach can be generalised by applying nonlinear transformations to the input and the output variables (Mejdell *et al.*, 1991). Here, process knowledge can help in selecting the most appropriate and viable nonlinear transformations, in an engineering sense. However, if there is no *a priori* knowledge about the underlying nonlinear relationships that exist between some of the variables, then there is no limitation to the number (and kind) of transformations that might be applied. Thus, if datasets are pre-treated in this way, the number of nonlinear terms can increase excessively, resulting in large input and output matrices, and the results become difficult to interpret. Furthermore, there are a number of issues which must be considered when performing individual transformations on the regressor and predicted variables which restrict the use of this approach (Wold, 1992). The main issues relate to the fact that, although nonlinear models are highly flexible, models generated from transformations of individual variables are only reliable if the number of observations is much greater than the original number of variables. In addition, when nonlinear transformations are applied, the underlying assumption of independence between variables, which is necessary to ensure model robustness, is rarely true in practice.

An alternative and more structured approach to the development of a nonlinear PLS model is to modify the NIPALS algorithm by introducing a nonlinear function which relates the output scores u to the input scores t, without modifying the input and output variables, i.e. retaining the input and output matrices X and Y in their original form. This procedure has been investigated and described in detail by Wold *et al.* (1989) and is reviewed in the next section. Several different algorithms have been published including those described by Wold *et al.* (1989), Frank (1990), Wold (1992), Qin and McAvoy (1992) and Saunders (1992).

9.2   Polynomial projection to latent structures

Building upon the approach of linear projection of latent structures as described previously, Wold *et al.* (1989) proposed a nonlinear (polynomial) PLS regression algorithm which retains the framework of the linear PLS algorithm, including the orthogonality of the predictor latent variables t, but which for each dimension (j) modifies the linear inner relation between the predictor (t) and the response (u) latent variables to a nonlinear relationship:

$$u_j = f_j(t_j) + e_j.$$

Furthermore, they suggest that any nonlinear function $u = f(Xw)$ that is continuous and differentiable with respect to the weights w can be used to fit the inner model. In particular, they propose a quadratic polynomial relation for the inner mapping:

$$u_j = c_{0,j} + c_{1,j}t_j + c_{2,j}t_j^2 + e_j.$$

This approach can lead to further modifications of the NIPALS algorithm. In fact, since the projection coefficients of the input block, the weights w, are derived from the correlation of the u scores with the input X matrix, using a nonlinear function to relate each pair of latent variables affects the calculations of the inner mapping as well as those of the outer mapping. To take this into account, Wold proposed updating the weights of the input outer relationship (w) by means of a Newton–Raphson linearisation of the inner relation, i.e. a first-order Taylor series expansion of the quadratic inner relationship, and then solving it with respect to the weight increments Δw. The input weights updating procedure proposed by Wold *et al.* (1989) can be summarised as follows. Write the nonlinear mapping between t and u as

$$u = f(t) + e = f(X, w, c) + e,$$

where $f(\cdot)$ is a continuous function differentiable with respect to w (and c), and c are the parameters of the function $f(\cdot)$. The above relationship can be approximated by means of a Taylor linearisation

$$u = f_{00} + \left.\frac{\partial f}{\partial c}\right|_{00} \Delta c + \left.\frac{\partial f}{\partial w}\right|_{00} \Delta w.$$

Starting with the Taylor linearisation of the nonlinear inner mapping, it is possible to derive three different procedures for updating the input weights w which are more related to the Taylor series expansion of the nonlinear inner mapping than the procedure originally proposed by Wold *et al.* (1989). This is a consequence of writing the Taylor series expansion in matrix form and solving it by OLS regression of the correction factors Δw on the matrix Z and the vector u

$$u = Z\Delta w.$$

These three developments of the weights updating procedure differ from each other in the way in which the matrix Z, containing the known terms of the Taylor series expansion of the function $f(\cdot)$, is constructed and, consequently, the way in which the correction factors Δw are regressed (Baffi *et al.*, 1998). In this work, the quadratic error based algorithm is compared with the quadratic PLS algorithm proposed by Wold *et al.* (1989) and the traditional linear PLS algorithm. These algorithms are compared on the basis of their performances when applied to an industrial-based pH simulation model.

10 Case study 3: simulation of an industrial pH problem

pH-neutralisation systems are known to be highly nonlinear and can exhibit severe time-varying behaviour. For this reason they have been used as a benchmark for testing control algorithms (Henson and Seborg, 1994; Johansen and Foss, 1997; Kim *et al.*, 1997; Kavšek-Biasizzo *et al.*, 1997). In this work, a dynamic model for the pH process described by Henson *et al.* (1994) is used. The process consists of a tank where a strong acid (HNO_3) is neutralised by a strong base ($NaOH$) in the presence of a buffer stream ($NaHCO_3$). For the purposes of this work the composition of the inlet streams (acid, base and buffer) have been kept fixed whilst the flow rates have been changed randomly and then kept constant until the process reached steady state, i.e. a constant value of the pH for the outlet stream. The flow rate of the outlet stream was also changed according to the inlet flow rates in order to maintain a constant level in the tank. A dataset of 999 points was generated by randomly changing the inlet flow rates (Q1, Q2, Q3) and recording the value of the pH on the outlet stream at steady state. The composition of the inlet streams was kept constant, and no noise was added to the pH measurements. The flow rates Q1, Q2, Q3 and Q4 were used as the predictor variables, and the pH measurements as the response variable. The dataset was split into two subsets, a training dataset comprising 799 observations and a testing dataset of 200 values.

The results are presented in Tables 6, 7 and 8 for the percentage of variability explained for the X- and Y-blocks for the three algorithms. Overall, the total percentage of variability explained for each of the three methods is similar, although for the error-based quadratic algorithm an additional 3% of the variability is explained within the Y-block. The major difference between the algorithms is that the majority of the variability is explained by one latent variable for the error-based approach, whilst for the other two approaches two latent variables are required to explain in excess of 90% of the total variance. Even with four latent variables, which may result in over-fitting, neither the linear nor the original quadratic approaches achieve the same level of variability explained as that achieved by one latent variable for the error-based quadratic approach. In contrast, the level of variability explained for the X-block is seen to increase more rapidly for the linear and original quadratic PLS algorithms. For one latent variable, 60.6%, 28.3% and 10.1% of the total variability is explained

Table 6 Percentage of variability explained—linear PLS algorithm.

LV	*X*-block		*Y*-block	
	% Variance captured	Cumulative % variance captured	Variance captured	Cumulative % variance captured
1	60.5851	60.5851	42.6297	42.6297
2	31.9094	92.4945	49.6840	92.3138
3	7.5055	100.0000	3.4998	95.8136
4	0.0000	100.0000	0.0018	95.8154

Table 7 Percentage of variability explained—original quadratic PLS algorithm.

LV	X-block		Y-block	
	% Variance captured	Cumulative % variance captured	Variance captured	Cumulative % variance captured
1	28.3636	28.3636	80.0813	80.0813
2	62.0486	90.4121	10.6662	90.7475
3	9.5879	100.0000	5.2643	96.0118
4	0.0000	100.0000	0.0022	96.0140

Table 8 Percentage of variability explained—error-based quadratic PLS algorithm.

LV	X-block		Y-block	
	% Variance captured	Cumulative % variance captured	Variance captured	Cumulative % variance captured
1	10.1552	10.1552	98.7295	98.7295
2	80.8207	90.9760	0.1324	98.8619
3	8.5568	99.5328	0.1784	99.0404
4	0.4672	100.0000	0.0007	99.0410

by the linear, original quadratic and error-based approach, respectively. These results are also reflected in the mean squared errors and the mean squared prediction errors for the Y-block.

Figures 17, 18 and 19 illustrate the final predictions for the test dataset for the three approaches for one latent variable. The predictions clearly illustrate the improvements that are achievable through the application of the error-based approach. A one latent variable model using the error-based approach provides a sufficiently robust final model for prediction. This has clear advantages if the final objective of the model is for routine process monitoring purposes at operator level. Another advantage is that the error-based technique appears to out-perform the existing algorithms in terms of explaining the overall variability in the data (Baffi *et al.*, 1998).

11   Neural network projection to latent structures

A number of attempts have been made to realise nonlinear PLS algorithms which use sigmoidal feedforward neural networks or radial basis function networks to fit the inner mapping between the input and the output latent variables. However, some of these neural network PLS algorithms are only qualitatively comparable with PLS as a projection-based regression technique in contrast to those algorithms which attempt

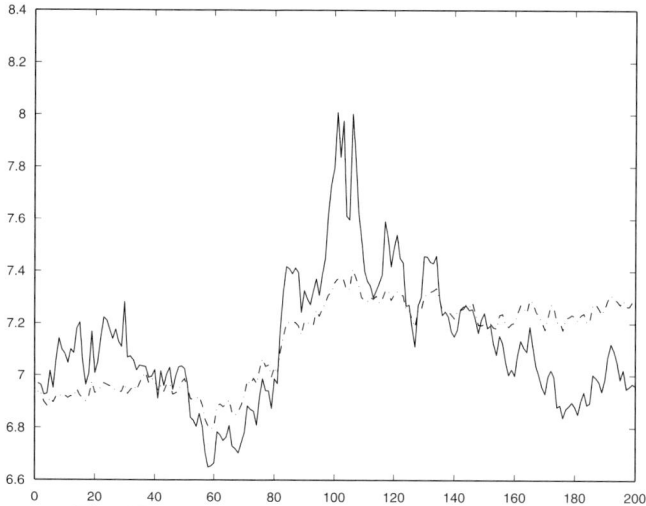

Fig. 17:   Results for the Wold linear PLS approach.

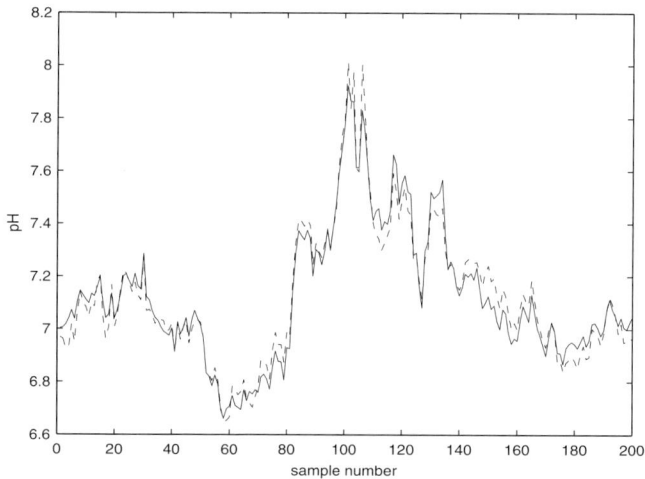

Fig. 18:   Results for the Wold quadratic PLS approach.

to follow the PLS calculation algorithm and seek to tackle the problem of providing a general regression tool for the approximation of the nonlinear mapping between input and output latent variables in PLS. However, these algorithms lack an updating procedure for the weights of the outer mapping. This is acceptable as far as the relationship between each pair of latent variables is slightly nonlinear. If the inner mapping is affected by a strong nonlinearity the approximation given by this approach is no longer reliable since the use of a nonlinear function to relate each pair of input–output scores affects the calculations of the inner mapping as well as those of the

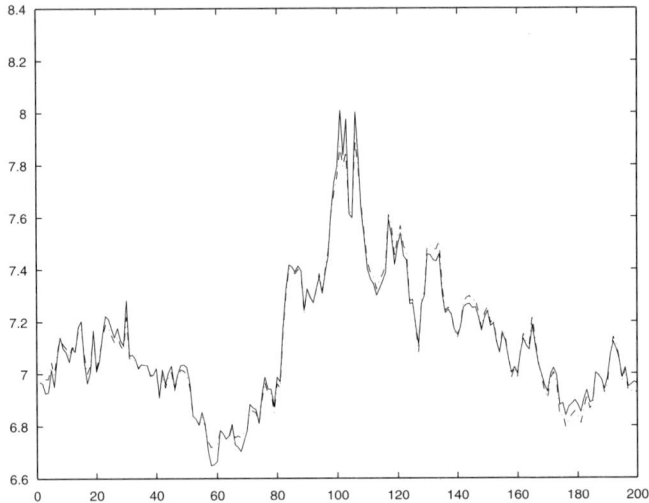

Fig. 19:   Error-based quadratic PLS.

outer mappings (Wold *et al.*, 1989). In this respect, the error-based approach discussed above for updating the input weights for the quadratic PLS approach has been recently extended to include the neural network and radial basis function PLS algorithms (Baffi, 1998).

Qin and McAvoy (1992) and Qin (1993) proposed a nonlinear PLS algorithm by combining the universal approximation capabilities of feedforward neural networks with the robustness of PLS regression, leading to a neural network PLS (NNPLS) algorithm. In their algorithm they use a set of centred sigmoid neural networks, one for each latent variable, to fit the nonlinear inner regression, while retaining the outer mapping of the linear PLS algorithm, i.e. without updating the weights of the outer input mapping. They state that the use of neural networks as inner regression models within the PLS framework makes the NNPLS regression approach generic for nonlinear modelling, even with respect to the quadratic PLS approach proposed by Wold *et al.* (1989), which uses or imposes a quadratic regression as an inner regression model. In fact, by using a sigmoid neural network no functional relationship needs to be assumed *a priori* when building the inner PLS models. Thus the nonlinear PLS model relies on a general tool with universal approximation capabilities, and training a sigmoid network for each component ensures that the nonlinear relationship between each pair of input–output scores can be approximated by a different network without affecting or being affected by the others. Furthermore, since one of the major advantages arising from the use of the PLS calculation algorithm is that each pair of input–output latent variables is not correlated with respect to the others, training a sigmoidal network for each component ensures that the nonlinear relationship between each pair of input–output scores can be approximated by a different network without affecting or being affected by the other networks. They also showed that all the individual single-input–single-output neural network models can be assembled with the weights and the loadings of the outer linear mapping, into a single global multiple-input–multiple-output neural network model. Wilson *et al.* (1997) proposed the use of radial basis function networks to regress the mapping between the input and the output score vectors in a similar manner to that of Qin and McAvoy (1992), i.e. without any updating of the input weights. In this respect it represents little new in terms of algorithm development, since the only difference between the two approaches is in the kind of nonlinear functional relationship used to fit the inner mapping. However, there are a number of advantages that arise from the use of RBF networks to fit the inner mapping in contrast to the use of sigmoidal feedforward networks. In fact, even though the approximation properties of radial basis function networks are typically comparable with those of sigmoidal networks, training algorithms for radial basis function networks are considerably simpler and faster than those used for sigmoidal networks. A different approach to neural network PLS modelling arises from the properties of autoassociative neural networks (Kramer, 1992). These are four-layer networks consisting of two nonlinear feature layers (layers 1 and 3), a linear bottle-neck layer (layer 2) and a linear output layer (layer 4). The input and the output data of autoassociative networks are the same. The two feature layers have the same number of units as the number of input–output variables, and provide a nonlinear mapping between the outer space (input variables) and the feature space (nonlinear transformation of the input variables). The bottle-neck layer is placed between the first and the second feature layers and consists of a smaller number of units than the feature layers, in order to force (project) the output from the first feature layer on to a lower-dimensional space. In this way the autoassociative networks provide a useful tool for projecting information from a high-dimensional dataset down on to a lower dimensional nonlinear space represented by the output of the bottle-neck layer, and hence can be used to develop nonlinear PLS algorithms. This can be achieved by replacing the input data matrix with the output data matrix on the output layer of the autoassociative network, and modifying the structure of the output layer of the network.

Nonlinear PLS algorithms based on the use of artificial neural networks might still be classified as projection-based regression techniques, even though the projection feature is no longer linear. This is due to the projective feature provided by the activation functions of the network nodes (typically sigmoid, centred sigmoid or Gaussian functions). In the case of the autoassociative networks the predictor and the response latent variables can be identified with the inputs to, and the outputs from, the activation functions of the bottle-neck layer respectively. In addition, in a similar way to PLS, the relationship between the predictor and the response variables can be modelled by means of a reduced number of activation functions. However, from a statistical point of view these algorithms cannot be classified as multivariate statistical regression techniques. Furthermore, they are only qualitatively comparable with the family of linear and nonlinear PLS algorithms based on the use of the PLS calculation algorithm (i.e. based on a sound statistical approach). In practice they lead to PLS models which have a global neural network structure and which are trained using neural network training algorithms, i.e. nonlinear optimisation routines.

Holcomb and Morari (1992) proposed a neural network implementation of the PLS algorithm based on the use of a feedforward neural network (PLS/neural), with both linear and centred sigmoid activation functions. They started from considering the linear PLS method as a two-layer neural network consisting of a linear feature layer and a linear output layer. The feature layer is a hidden layer which performs a linear combination of the input data in order to provide an orthonormal linear mapping from the input space to the feature space, and hence acts as a projection device within the network structure in a similar manner to the projection matrix in the linear PLS algorithm. The output layer then performs a linear combination of the output signal arising from the feature layer. The PLS model is subsequently built using an optimisation routine (e.g. backpropagation). Starting from this assumption, they defined a new PLS approach which is fully implemented as a multilayer feedforward network. In particular they defined the PLS/neural network structure as a three-layer network consisting of a linear feature layer and a two-layer feedforward network with mixed linear and centred sigmoidal activation functions for the hidden units, together with a linear activation function for the output layer. To train the three-layer network they proposed a hybrid approach based on the use of PCA to select the number of neurons of the feature layer (i.e. to select the number of latent variables to retain in the model). The weights of the feature layer are also initiated using PCA, using the directions of the principal components to initialise the directions of the latent variables. Finally the backpropagation algorithm is used to train the two-layer feedforward neural network and backpropagation again to train the overall three-layer network. The whole approach makes use of different learning parameters for the weights in different layers. They went on to suggest using PCA to determine the true dimension of the input dataset and hence the number of neurons to use in the feature layer. However using PCA to identify the true dimension of the system can lead to poor performance, because PCA identifies the number of PCs which are required to explain most of the variability of the input dataset, and not the latent variables which are required to model the correlation structure between the input and the output datasets. Autoassociative neural networks with a one-dimensional bottle-neck layer represent the main body of the nonlinear PLS algorithm proposed by Malthouse et al. (1997), which is fully implemented with a neural network. It reflects the properties of the linear PLS algorithms as a projection-based regression technique but with nonlinear features that provide both a nonlinear PLS model and a nonlinear PCA representation within the same network. The overall structure consists of a proper autoassociative network which performs a nonlinear compression of the input variables, and of the second half of another autoassociative network which performs the decompression of the output of the bottle-neck layer of the full autoassociative network on to the output variables. The second half-autoassociative network is linked to the first full-autoassociative network by means of another mixed linear and nonlinear hidden layer. This provides the mapping between the output from the input-variables feature layers to the input to the output-variables feature layer. The overall network structure is trained by means of nonlinear optimisation algorithms.

12   Dynamic PLS modelling

The importance of dynamic process modelling has been discussed earlier, particularly with respect to empirical models and neural network representation. Such models are built using monitored plant data collected either during normal operations or during a series of designed plant tests. The use of PLS, whether in linear or nonlinear form, is attractive in that it brings to dynamic modelling the structure and robustness of projection methodologies. Chemical processes have a large number of input variables, many of which undergo slow changes (low frequency) in a correlated manner. As a consequence collinearities might be evident not only between different variables but also between subsequent time measurements of the same variable. A number of different dynamic PLS algorithms have been proposed aimed at integrating dynamic features within a PLS framework to provide dynamic PLS models. Ricker (1988) proposed the use of PLS as a regressive tool for finite impulse response (FIR) modelling. Qin (1993) proposed the use of PLS within ARX and ARMAX structures. Here, both the algorithms use an augmented input matrix to predict the quality variables by means of PLS regression. The main difference between the above two approaches is that the FIR form uses only time lagged values of the input variables to the process, whilst the ARX/ARMAX approaches also use time lagged values of the output (quality) measurements. The ARX/ARMAX PLS approaches require fewer past input and output values than the FIR PLS approach, since the inclusion of time lagged output variables brings information about the system to the model.

It can be argued that to use ARX/ARMAX models it is necessary to have output (quality) measurements available on-line. However, instead of using the actual output values, the model can be fed with the predicted output values to achieve a multistep-ahead regression model. In practice the predicted output values need to be filtered, especially when the model is being used within a model-based control strategy. Any error or mismatch occurring in the prediction of the output values would tend to accumulate and affect the next model prediction. The use of a filter damps any fast variations in the predicted value and allows a reliable prediction of the actual output value to be made. The major drawback of the FIR and the ARX/ARMAX approaches is the selection of the time lags to be used. Nevertheless, since PLS ensures a dimensionality reduction and robustness to ill-conditioned data matrices, it is possible to include a large number of past data. Obtaining the correct number of past input and output measurements (i.e. the variable time delays required), can be achieved through the use of cross-validation to select the best model not only with respect to the number of latent variables but also to the number of time delays used for each variable. However, each variable should be tested independently from the others resulting in an extremely time-consuming procedure.

A different approach to dynamic PLS modelling is one proposed by Kaspar and Ray (1993a,b). This uses a linear dynamic filter (discrete transfer function) to pre-treat the input data prior to presenting them to the PLS model. In this way no delayed values are presented to the regression model. The structure of the dynamic filter must be carefully specified in advance, and this requires specific knowledge of the system which might not always be available. They overcome this limitation by incorporating the dynamic filter within the PLS model, that is, moving the dynamics of the trans-formation to the inner relation between the input and the output scores. In this way different dynamics can be defined for each latent variable. The major drawback of this approach is that the PLS calculation algorithm must be modified to optimise the order of the dynamic filter while regressing the PLS model parameters. In this respect the approach is probably more complex than the FIR or ARX/ARMAX approaches.

13   Nonlinear principal components analysis

Nonlinear principal components analysis (PCA) is an extension of the linear technique of PCA. Whilst PCA identifies the linear correlations between the variables, nonlinear PCA uncovers both the linear (second-order statistics) and nonlinear (higher-order statistics) correlations. This generalisation is achieved by projecting the process vari-ables down on to curves or surfaces instead of lines or planes. The approach is based upon solving a similar optimisation problem, i.e. minimising the mean-squared error $E\|X - \hat{X}\|^2$. A dataset X comprising n samples of m variables can be expressed in terms of k nonlinear principal components, with $k \ll m$:

$$X = F(T) + E,$$

where $T = [t_1, t_2, \dots , t_k]$ is the matrix of nonlinear principal component scores, F is the nonlinear function equivalent to the loadings in linear PCA and E is the matrix of residuals. By combining the principal curve algorithm (Hastie and Stuetzle, 1989) and the autoassociative neural network (Kramer, 1992), Dong and McAvoy (1994) proposed an algorithm for nonlinear PCA. Autoassociative networks are feed-forward networks whose inputs and outputs are identical, with network training aimed at approximating the resulting identity mapping between network input and network output. A typical network topology is shown in Figure 20, indicating the key feature of autoassociative networks, i.e. the data compression or bottle-neck inner layer. It is this network bottle-neck layer that provides the topology with the very powerful properties of feature extraction. Both the first and final hidden layers have dimensions greater than (or equal to) the input–output layer and significantly greater dimension than the middle feature layer. Following convergence, the network bottle-neck provides infor-mation which describes significant features, or signatures, of the data. The architec-ture of this network comprises five layers: an input layer, mapping layer, bottle-neck layer, demapping layer and an output layer. The outputs of the bottle-neck layer can then represent the nonlinear features contained within the data (erroneously called nonlinear principal components). The use of nonlinear features has been shown to

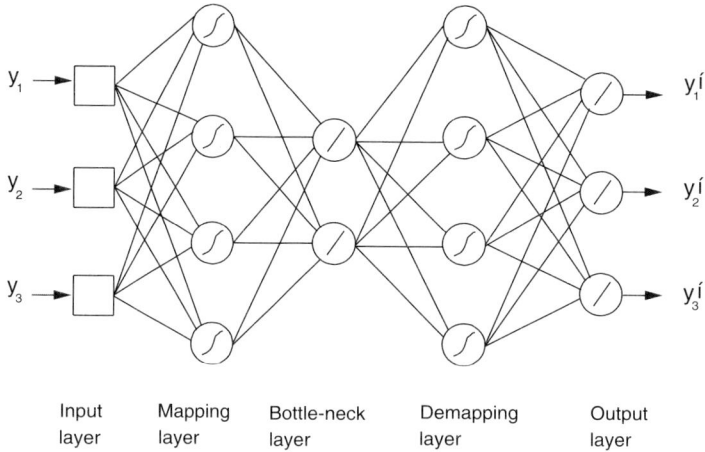

Fig. 20: Autoassociative neural network.

describe successfully the underlying structure of nonlinear data (Ignova *et al.*, 1994). Although the paper of Kramer presented some very interesting concepts a number of questions can be raised. First, since the network has five layers, it is very difficult to train. Secondly, the number of nodes in the mapping layer, the demapping layer and the bottle-neck layer are difficult to determine. For example, should engineering judgement be exercised to identify the number of features contained within the data? Finally, the theoretical meaning of the outputs of the bottle-neck layer is unclear since in practice they are not nonlinear principal components but nonlinear features. Extensions to the autoassociative neural network architecture to allow the generation of nonlinear principal components requires the use of the statistical procedure of principal curves (Hastie and Stuetzle, 1989). A linear principal component minimises the sum of the orthogonal deviations between a straight line and the data whilst the nonlinear approach summarises the data by a smooth curve (a principal curve). Principal curves are generalisations of principal components. The calculation of principal curves essentially comprises two steps, a projection step and a smoothing step. The calculation is generally started with the principal component as the initial curve. In the projection step, data points are projected down on to the curve. Then, in the smoothing step, the curve is smoothed using techniques such as the locally weighted regression smoother (Cleveland, 1979) or kernel smoothers (Gasser and Muller, 1979). The procedure iterates between the two steps until convergence results. In principal component analysis, the loading vectors are used to generate principal scores for new data. However, the principal curve procedure does not provide this facility, and hence nonlinear loadings cannot be derived. In industrial process applications, it is desirable to have a nonlinear principal model which can be used to generate nonlinear principal components for new data. Dong and McAvoy (1994) proposed using neural networks to learn a nonlinear principal component model by combining the principal curve algorithm and the autoassociative network method.

This makes the construction and training of the network easier. This approach has three key features: the principal curve algorithm can be applied sequentially to identify each principal factor; the principal factors can then be used to determine the bottle-neck layer of the two three-layer networks; finally, the two three-layer networks can be trained separately. Two networks are required. The first maps the input data (m dimensional) on to l-dimensional principal scores, evaluated from the principal curve. The second model defines the relationship between the principal scores and the m-dimensional corrected dataset. If a nonlinear function can be used to describe the curve, this function is equivalent to the principal loadings for linear PCA. When the data are projected down on to the curve, indices can be found which express the magnitude of the projected points. These are equivalent to the principal scores for linear PCA. In nonlinear PCA, just as with linear PCA, there is no response variable and hence it is more suited to feature extraction than prediction. However, there is a limitation to Dong and McAvoy's nonlinear principal component analysis (NLPCA) method because of its sequential computation. For this approach, the first nonlinear PC is computed from the observed data, the second nonlinear principal component is then calculated from the residuals of the previous computation, and so forth. For example, when two latent variables are required to reproduce the observed variables, a general nonlinear model for dimensionality reduction should be given by

$$t_k = \Phi_k(\lambda_1, \lambda_2) + e_k, \quad k = 1, \ldots, n.$$

However, the sequential procedure for computing principal components assumes that each of the two-variable functions, $\Phi_1, \Phi_2, \ldots, \Phi_n$, is the sum of two single-variable functions,

$$\phi_k(\lambda_1, \lambda_2) = \Psi_{k1}(\lambda_1) + \Psi_{k2}(\lambda_2).$$

Therefore, these functions only represent a very limited class of nonlinear models and thus it may restrict the types of nonlinear feature that can be captured. Another approach, called the input training network (IT-net) method has been proposed by Tan and Mavrovouniotis (1995). It appears to exhibit better nonlinear compression properties than the autoassociative network approach.

14   Input training network

Tan and Mavrovouniotis (1995) proposed a nonlinear PCA method based upon the concept of the input training neural network (IT-net), where each input pattern is not fixed but adjusted along with the internal network parameters to reproduce a corresponding output pattern based upon the steepest gradient descent network optimisation rule. They proposed that the scaled process observations X could be used as the output layer pattern, and, after the network has been trained, the nonlinear principal scores T can be identified from the input layer. The architecture of an IT-net

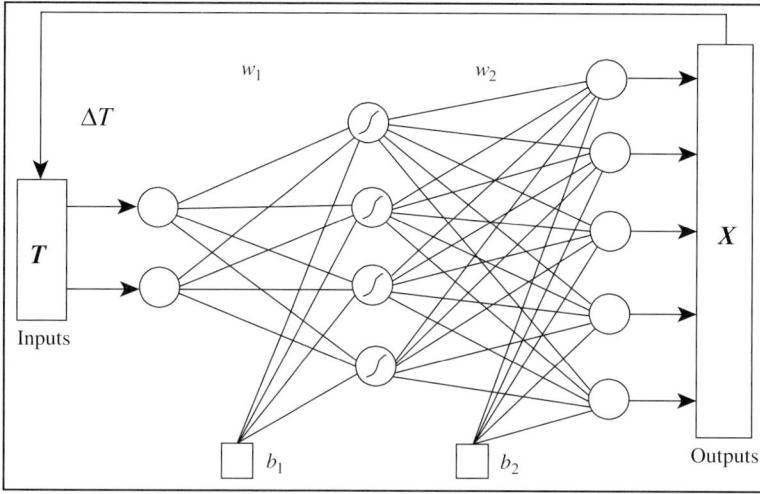

Fig. 21: The architecture of the IT-network.

is illustrated in Figure 21, where w_1 and b_1 and w_2 and b_2 are the weights and biases for the hidden and output layer, respectively.

Based upon the input training network, a nonlinear PCA algorithm has been developed. Three steps form the basis of the work. In the first step, linear PCA is applied to effect a linear transformation in which the observations are rotated to a new set of uncorrelated ordinates, allowing the main linear information to be extracted and compressed. The next step involves rescaling the linear principal component scores to unit variance to allow the recovery of the nonlinear structure in the new ordinate space of the transformed data. By compressing the linear structure, the nonlinear behaviour becomes more apparent. Finally, an improved network optimisation algorithm, the Levenberg–Marquardt (LM) algorithm, is used to capture the latent nonlinear structure in the transformed data. By this procedure, both the linear and nonlinear information can be captured in the final nonlinear principal component scores. In addition, if the problem addressed is linear, then the methodology collapses back to linear principal components analysis. The final nonlinear mapping function can be defined as

$$X = F(T)V^{1/2}P^T + E,$$

where $V(l \times l)$ is a diagonal matrix with the first l eigenvalues of the covariance matrix of the original scaled observations as the diagonal elements, P is the linear principal loading matrix comprising the eigenvectors of the covariance matrix of the process observations X with respect to retaining l linear principal components and E is the corresponding residual matrix.

15   Case study 4: examples using simulated data

15.1   Example 1: a circle

A similar example was presented by Kramer (1992) to demonstrate the various aspects of the autoassociative network method for the extraction of features from a nonlinear dataset. The training data are taken from a circle. A second dataset, the test set, is also generated from the same circle. An IT-net with a 1-5-2 structure is first trained on the circle data (Figure 22). The result obtained is then compared with that obtained from the autoassociative network. The training and test results obtained from the autoassociative network, the steepest descent input training (IT)-net approach and the Levenberg–Marquardt (LM) IT-net method are presented in Table 9. From these results, it is seen that the LM-IT net produces consistent results for the training and test datasets, whilst for the other two approaches the test data results were poorer. The autoassociative network and the back propagation (BP)-IT network gave correspondingly larger errors than the LM-IT network.

15.2   Example 2: a square

For the second example, training data were taken to be the coordinate values of 32 points on the perimeter of a square. The observed dimensionality is two and through nonlinear PCA it can be reduced to a one-dimensional problem. This dataset is more difficult than the circle example for neural networks because of the discontinuities at the corners of the square. The results from training a 1-6-2 IT-net and a 2-6-1-6-2 autoassociative network are plotted in Figure 23. The test data points were taken between the training data points. Table 10 compares the results for the three different methods. The superiority of the IT-net over the autoassociative network becomes even more noticeable in this example.

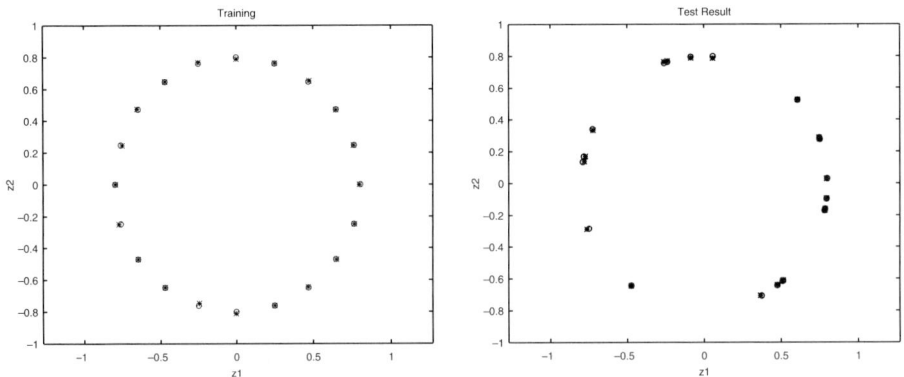

Fig. 22:   Training and testing results of the IT-net for Example 1.

Table 9 Root mean squared errors for Example 1.

	Autoassociative Net	BP-IT Net	LM-IT Net
Training Error	0.032	0.012	0.0047
Testing Error	0.070	0.015	0.0047

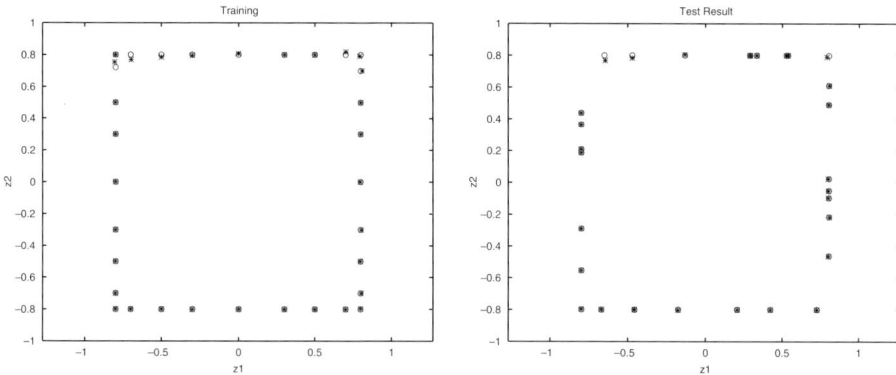

Fig. 23:   Training and testing results of IT-net for Example 2.

Table 10 Root mean squared errors for Example 2.

	Autoassociative Net	BP-IT Net	LM-IT Net
Training Error	0.047	0.023	0.007
Testing Error	0.174	0.026	0.005

16   Case study 5: application of nonlinear principal components analysis for fault detection

Nonlinear principal components analysis can be used in the application of multivariate process performance monitoring (multivariate statistical process control) to nonlinear processes. Movement of the principal component scores away from regions of nominal operation, together with increasing squared prediction error values, i.e. model mismatch, identifies changes in process operation that are other than 'common cause'. In many cases, especially with linear systems, the movements of the scores can be visibly different from those when the process is operating normally, with different faults causing the scores to move off in contrasting directions in the score space. However, in some situations, especially in nonlinear systems, the trajectories of the scores are not distinctive and it is proposed to use an accumulated scores plot to distinguish between different fault situations (Zhang *et al.*, 1996; Martin *et al.*, 1996b). The accumulated scores are defined as follows:

$$A = \int_0^t (x - \bar{x})dt,$$

where x is the nonlinear score, $\bar{x}$ is the mean of the nominal nonlinear score, and A is the accumulated nonlinear score. The accumulated scores for the nominal operating region cluster around zero, whilst during a process malfunction the accumulated scores move away from the region defining the nominal operating region. This approach is analogous to the cumulative sum (CUSUM) approach. Two processes are considered, to demonstrate the potential of nonlinear principal component analysis. The first is an industrial materials processing plant which exhibits a number of different regions of operation. The detection of process movement into these different regions of operation, as a result of changes in process physics and chemistry, is vital for the consistent production of material for subsequent processing. The second example relates to a continuous polymerisation reactor where four types of fault are considered, an impurity problem, a feed flow problem, a fouling problem and a combined impurity/fouling problem.

16.1   An industrial materials processing unit

An industrial processing unit provides material for further processing into machinable components (additional details are precluded for confidentiality reasons). The unit is subject to external raw material and energy supply changes as well as internal chemical changes. These changes affect process operation and consequently the resulting material. This has implications further down-stream on the subsequent processing and final product manufacture. The process operators are known to have observed between five and seven operating regions during production campaigns. It is difficult for the plant operators to control the movement between these different operating regions given the limited knowledge of process operation and the multitude of reasons for the changes. If the drift from one operating region to another could be identified and the cause–effect relationship established, then corrective action could be taken, leading to more consistent production and enhancing subsequent manufacturing operations.

Two sets of data, each of 2250 data samples, were available for analysis. Linear PCA was initially used to compress the dimensionality of the data. Figure 24 shows the linear scores plot of principal component 2 versus principal component 3. Here, the points denoted by 'o' are for the first dataset whilst the points denoted by '+' refer to the second set of data. In comparison, the results of the nonlinear PCA analysis are shown in Figures 25 and 26, which display the nonlinear scores plot of principal component 1 versus principal component 2 and the nonlinear scores plot of principal component 1 versus principal component 3, respectively. Six operating regions can now be identified, with the possibility of more during periods of operating region transition.

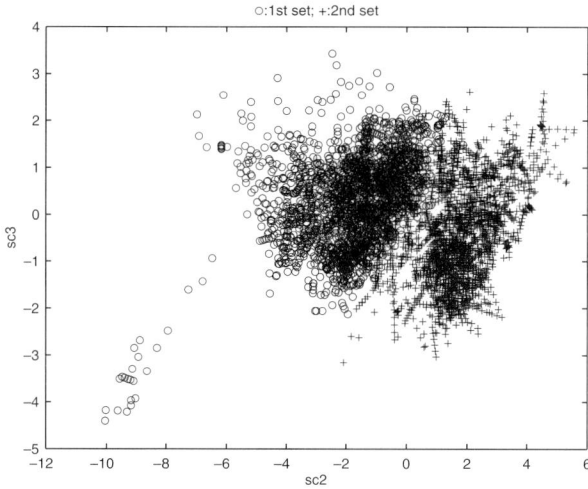

Fig. 24: Linear PCA scores plot. Principal component 2 versus principal component 3.

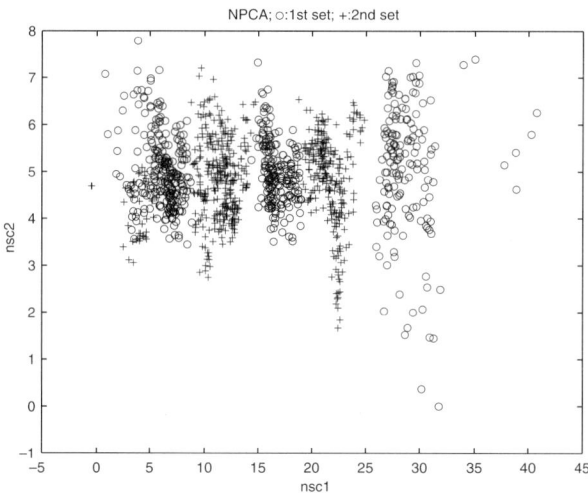

Fig. 25: Nonlinear PCA scores plot. Nonlinear principal component 1 versus nonlinear principal component 2.

Figure 27 shows the plot of the 'accumulated principal component score 2' for the two datasets. There is a clear indication of the evolution of the process operating conditions over the first 2250 samples before the process operating conditions were changed to bring the process back into nominal operation. It is encouraging that the nonlinear approach is able to provide the evidence of the plant behaviour that had been perceived by operating personnel.

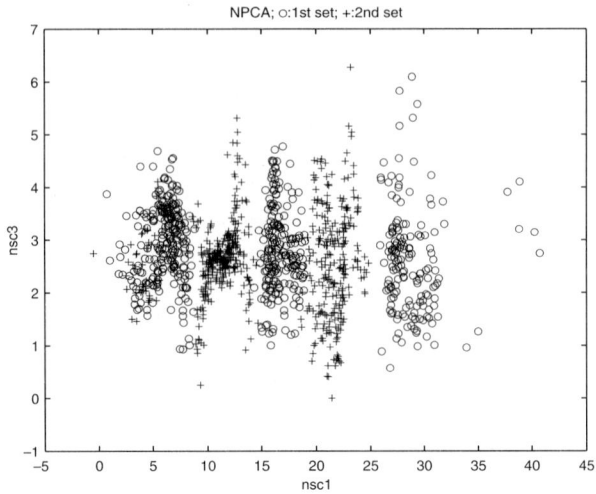

Fig. 26: Nonlinear PCA scores plot. Nonlinear principal component 1 versus nonlinear principal component 3.

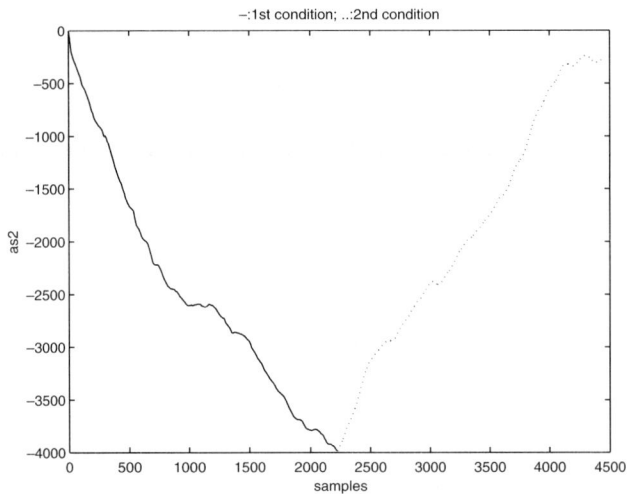

Fig. 27: Accumulated score 2 plot showing different operating regions.

16.2 A continuous manufacturing process

The polymerisation process considered here is a low-density polyethylene process. A comprehensive simulation based on detailed reaction kinetics has been developed by the Department of Chemical Engineering, University of Thessaloniki, Greece. There are 14 measured variables and four quality variables which are not measured. The simulation programme is capable of simulating the reactor under various operating conditions. The economic operation of a polymerisation process usually requires that unreacted species be recovered and recycled back into the process. Associated with the recycle of solvent and unreacted monomers is also the recycle of reactive impurities which are introduced into the system in the fresh feed or as a byproduct of chemical reactions. The levels of reactive impurities can be built up to the point where the reacting system is severely affected. Almost all types of polymerisation are sensitive to reactive impurities. In polymerisation processes, reactive impurities are usually traces of inhibitors or oxygen. The studies of Penlides *et al.* (1992) show that impurities in an emulsion system consume rapidly reactive free radicals, thus preventing particle generation and decelerating the growth of any polymer particles already present. Another problem affecting polymerisation is reactor fouling. Many polymers are viscous and can accumulate on the wall of a reactor vessel during polymerisation. Most of the polymerisation processes are operated under controlled temperature profiles, either constant or time varying. Reactor temperature is usually controlled by manipulating the flow rate or temperature of coolant through the reactor jacket. The accumulation of polymer particles on the reactor wall will reduce the heat transfer capability of the reactor and is known as reactor fouling. Reactor fouling will make the reactor temperature control system less effective. In this study, four fault situations are examined, namely reactive impurity, reactor fouling, solvent flow problems and a combined reactive impurity and reactor fouling problem. Process operating data for both normal and faulty operating conditions are available. Principal component analysis of the reactor data showed that two principal components can explain 49.2% of the overall data variability with three principal components explaining 65.4% of the total variability. In this case, there is a substantial amount of variance which cannot be explained by the first two, or indeed three, principal components. Nonlinear principal component analysis was then used to analyse the reactor data. Two nonlinear principal components explained 74.8% of the variability in the data whilst three nonlinear principal components can explain 89.8% of the data variance. This suggests that nonlinear principal component analysis is able to extract more information using fewer latent variables. The data were split into training, test and validation datasets with the results being presented for an 'unseen' validation dataset. A nonlinear principal component model with three principal components was developed. The first part of the autoassociative neural network model for the identification of the nonlinear scores has the structure (14-15-3) whilst, for the principal curves, (3-15-14). The structures for these network models was determined using cross-validation and higher-order correlation model validation tests (Morris *et al.*, 1994). In this study the use of nonlinear score plots to localise the four different fault situations resulted in the distribution observed in Figure 28. It can be seen that here the faults cannot be distinguished from within the nominal dataset scores.

In comparison, Figure 29 shows the plot of the accumulated scores. The different fault situations can now be clearly distinguished. In Figure 29 the upper left part of the plot shows the trajectories of the accumulated first and second scores. From this figure it can be seen that fouling is characterised by the score movement in the north-east direction while the combined impurity and fouling faults are characterised

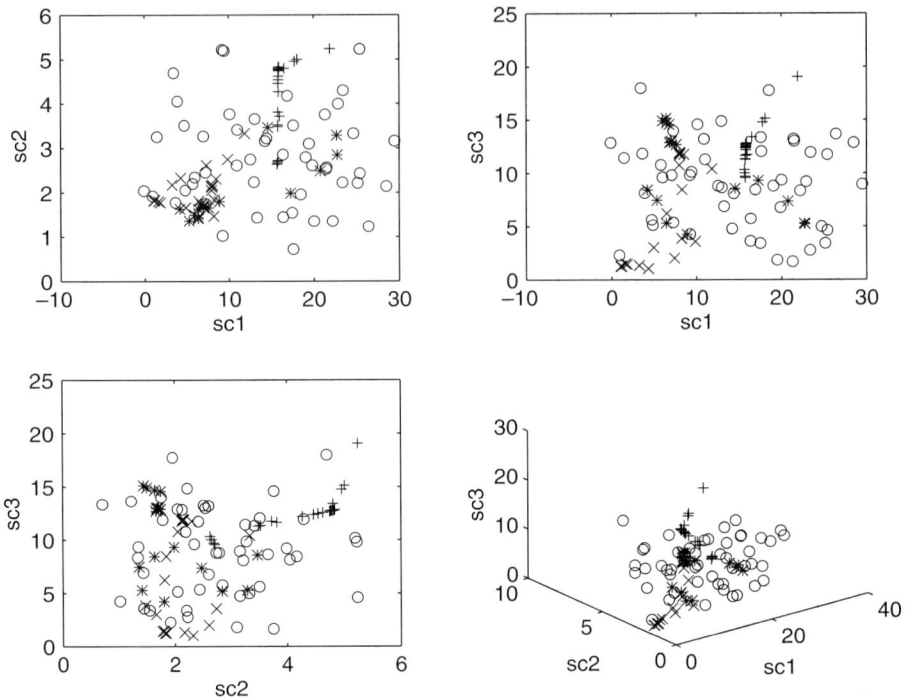

Fig. 28: Nonlinear score plots for different faults; ○, nominal; +, fouling; ×, impurity; *, solvent; ·, combined.

by the score movement in the north-west direction. Both impurity and flow problems result in the score moving in the south-west direction. However, they can be clearly distinguished from the lower left plot, which plots the integration of the second and third nonlinear scores. Here the impurity problem corresponds to the score movement in a south-west direction, while the flow problem corresponds to the score movement in a north-west direction. The results indicate that the accumulated scores can be effectively used to extract the new information resulting from the change in process operation, and as a result can contribute to the localisation of different process faults.

17   Conclusions

Artificial neural networks provide an exciting opportunity for the process engineer. It is possible to develop rapidly models of complex operations that would normally take many man months to model using conventional structured techniques. The significant potential of feedforward dynamic and recurrent networks as model-based predictors and model-based process controllers has been demonstrated by the authors, amongst others. However, it is essential to understand that neural network modelling is no replacement for good understanding of the process and the generated data.

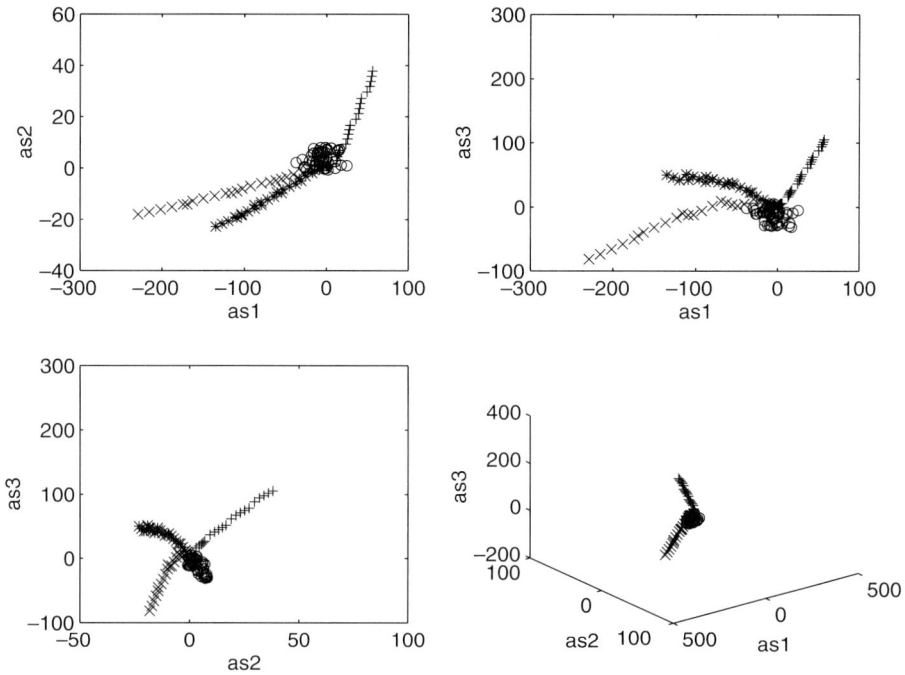

Fig. 29: Accumulated nonlinear scores plot; o, nominal; +, fouling; ×, impurity; *, solvent; ·, combined.

Careful study of the data using multivariate statistical methods provides a preliminary insight into the process behaviour and is an essential pre-requisite for good dynamic modelling.

There is no doubt that, for conventional classification problems, statistical approaches are both powerful and well founded. Neural network based classifiers can provide rapid solutions to complex nonlinear problems. Further research is required to understand the trade-off between statistical approaches and the use of artificial neural networks with their problems of optimisation, initialisation and multiminima performance. The combination of statistical methodologies and neural network techniques such as those discussed for nonlinear PLS and nonlinear PCA potentially provide a very powerful set of tools which are capable of modelling both the linear and nonlinear features within a dataset. Further research is being carried out at Newcastle in this area to identify the strengths and weaknesses of these techniques.

It is stressed though that, in contrast to some multivariate statistical methods, the study and application of artificial neural technology is still in its infancy and many questions still remain to be answered, including determination of the 'best' network topology, network initialisation, the convergence to local minima and solution sensitivity. What is best for one problem may well not be best for another. Currently, rather ad hoc procedures tend to be used. This is undesirable, and this arbitrary facet of an otherwise promising philosophy is an area of active research. There is also no established methodology for determining the robustness and stability of the networks when applied to typical industrial data problems. This is perhaps one of the most important issues that has to be addressed before their full potential might be realised. In the short-to-medium term they may indeed provide some interesting, useful and pragmatic solutions for solving process engineering problems. A panacea, however, they will never be.

18   Acknowledgements

The authors acknowledge the support of colleagues in the Centre for Process Analytics and Control Technology (CPACT), University of Newcastle, in particular Dr. Jie Zhang, Dr. Wan Luo, Dr. Adrian Conlin, Mr. Giuseppe Baffi and Mr. Feng Jia; also their colleagues in the EU BRITE/EURAM Projects 'INTELPOL—Intelligent Manufacturing of Polymers', Number BE 7009 and 'IMPROQ—Intelligent Equipment and Process Monitoring for Consistent Finished Product Quality', Number BE 8104. The EU and the UK EPSRC are also thanked for funding the neural network studies within CPACT.

19   References

Akaike, H. (1974). A new look at the statistical model identification. *IEEE Transactions on Automatic Control.* **19**, 716–723.

Baffi, G. (1998). Nonlinear Projection to Latent Structures. Ph.D. Thesis, University of Newcastle.

Baffi, G., Martin, E.B. and Morris, A.J. (1997). Nonlinear projection to latent structures revisited—Part I: the quadratic PLS algorithm. *Computers and Chemical Engineering* **23**, 395–411.

Barron, A.R. (1984). Predicted squared error: a criterion for automatic model selection. In *Self Organising Methods*, Ed. S.J. Farlow, pp. 87–103.

Bates, J.M. and Granger, C.W.J. (1969). The combination of forecasts. *Oper. Res. Q.* **20**, 451–468.

Bhat N. and McAvoy, T.J. (1990). Use of neural nets for dynamic modelling and control of process systems. *Computers Chem. Eng.* **14**, 573–583.

Bhat N., Minderman, P. and McAvoy, T.J. (1989). Use of neural nets for modelling of chemical process systems. *Preprints IFAC Symp. Dycord+89.* Maastricht, The Netherlands, Aug. 21–23, pp. 147–156.

Billings, S.A. and Voon, W.S.F. (1983). Structure detection and model validity tests in the identification of nonlinear systems. *Proc. IEE* D **130**, 193–199.

Billings, S.A. and Voon, W.S.F. (1986). Correlation based model validation tests for nonlinear models. *International Journal of Control* **44**, 235–244.

Billings, S.A. and Zhu, K.M. (1995). Model validation tests for multivariable nonlinear models including neural networks. *International Journal of Control* **62**, 749–766.

Breiman, L. (1992). Stacked regressions. Technical Report No. 367. Department of Statistics, University of California at Berkeley, USA.

Breiman, L. (1996a). Bagging predictors. *Machine Learning* **24**, 123–140.

Breiman, L. (1996b). Heuristics of instability and stabilisation in model selection. Technical Report No. 416. Department of Statistics, University of California at Berkeley, California, USA.

Brimmer, P.J. and Hall, J.W. (1993). Determination of nutrient levels in a bioprocess using near infra-red spectroscopy. *Canadian Journal of Applied Spectroscopy* **38**, 155–162.

Brown, G.K. (1992). On-line polymeric measurements in real time with NIR spectroscopy. *Proceedings of SPIE—The International Society for Optical Engineering* **1681**, 304–306.

Chen, S., Billings, S.A. and Luo, W. (1989). Orthogonal least squares methods and their application to nonlinear system identification. *Int. J. Control* **50**, 1873–1896.

Chen, S., Billings, S.A., Cowan, C.F.N. and Grant, P.M. (1990a). Nonlinear system identification using radial basis functions. *Int. J. Systems Sci.* **21**, 2513–2539.

Chen, S., Billings, S.A. and Grant, P.M. (1990b). Nonlinear system identification using neural networks. *Int. J. Control* **51**, 1191–1214.

Cleveland, W.S. (1979). Robust locally weighted regression and smoothing scatter plots. *J. Amer. Statist. Assoc.* **74**, 829–836.

Conlin, A., Martin, E.B. and Morris, A.J. (1998). Data augmentation: an alternative approach to the analysis of spectroscopic data. *Chemometrics and Intelligent Lab. Systems* **44**, 161–173.

Cybenko, G. (1989). Approximation by superposition of a sigmoidal function. *Math. Control Signal Systems* **2**, 303–314.

Di Massimo C., Lant P.A., Saunders A., Montague G.A., Tham M.T. and Morris A.J. (1992). Bioprocess applications of model based estimation techniques. *J. Chem. Tech. Biotechnol.* **53**, 265–277.

Dong, D. and McAvoy, T.J. (1994). Nonlinear principal component analysis based on principal curve and neural networks. In *Proc. ACC, Baltimore, USA*, pp. 1284–1288.

Efron, B. (1982). *The Jackknife, the Bootstrap and Other Resampling Plans*. Society for Industrial and Applied Mathematics, Philadelphia.

Efron, B. and Tibshirani, R. (1993). *An Introduction to Bootstrap*. Chapman and Hall, London.

Elman, J.L. (1990). Finding structures in time. *Cognitive Science* **14**, 179–211.

Frank, I.E. (1990). A nonlinear PLS model. *Chemometrics and Intelligent Lab. Systems* **8**, 109–119.

Frasconi, P., Gori, M. and Soda, G. (1992). Local feedback multilayered networks. *Neural Computation* **4**, 120–130.

Gasser, T. and Muller, H.G. (1979). Kernel estimation of regression functions. In *Smoothing Techniques for Curve Estimation*. Springer-Verlag.

Geladi, P. and Kowalski, B.R. (1986). Partial least squares regression: a tutorial. *Analytica Chimica Acta* **185**, 1–17.

Gemperline, P.J., Long, J.R. and Gregoriou, V.G. (1991). Nonlinear multivariate calibration using principal components regression and artificial neural network. *Anal. Chem.* **63**, 2313–2323.

Girosi, F. and Poggio, T. (1990). Networks and the best approximation property. *Biological Cybernetics* **63**, 169–179.

Goldberg, D.E. (1989). *Genetic Algorithms in Search, Optimisation and Machine Learning*. Addison-Wesley Publishing Company Reading, Mass.

Hansen, M.G. and Khettry, A. (1994). In-line monitoring of molten polymers: NIR spectroscopy, robust probes and rapid data analysis. In *Proceedings of Annual Technical Conference — ANTEC* **94**, pp. 2220–2224.

Hastie, T.J. and Stuetzle, W. (1989). Principal curves. *J. Amer. Statist. Assoc.* **84**, 502–516.

Henson, M.A. and Seborg, D.E. (1994). Adaptive nonlinear control of a pH neutralisation process *IEEE Transactions on Control Systems Technology* **2**, 169–182.

Holcomb, T.R. and Morari, M. (1992). PLS/Neural Networks. *Computers Chem. Engng.* **16**, 393–411.

Höskuldsson, A. (1988). PLS regression methods. *J. Chemometrics* **29**, 409–412.

Hunt, K.J. and Sbarbaro, D. (1991). Neural networks for internal model control. *IEE Proceedings* D **138**, 431–438.

Hunt, K.J., Sbarbaro, D., Zbikowski, R. and Gawthrop, P.J. (1992). Neural networks for control systems—a survey. *Automatica* **28**, 1083–1112.

IEEE (1988, 1989, 1990). Special Issues on Neural Networks, *Control Systems* magazines, nos. 8, 9, 10.

Ignova, M., Glassey, J., Montague, G., Morris, A.J. and Kiparissides, C. (1994). Neural Networks and Nonlinear SPC. In *3rd IEEE Conference on Control Applications, Glasgow*, pp. 1271–1276.

Jackson, J.E. (1991). *A Users Guide to Principal Components.* John Wiley and Sons, New York.

Jacobs, R.A., Jordan, M.I., Nowlan, S.J. and Hinton, G.E. (1991). Adaptive mixture of local experts. *Neural Computation* **3**, 79–87.

Johansen, T.A. and Foss, B.A. (1997). Operating regime based process modelling and identification. *Computers Chem. Engng.* **21**, 159–176.

Jordan, M.I. and Jacobs, R.A. (1994). Hierarchical mixture of experts and the EM algorithm. *Neural Computation* **6**, 181–214.

Kaspar, M.H. and Ray, W.H. (1993a). Partial least squares modelling as successive singular value decompositions. *Computers Chem. Engng.* **17**, 985–989.

Kaspar, M.H. and Ray, W.H. (1993b). Dynamic PLS modelling for process control. *Chemical Engineering Science* **48**, 3447–3461.

Kavšek-Biasizzo, K., Škrjanc, I. and Matko, D. (1997). Fuzzy predictive control of highly nonlinear pH process. *Computers Chem. Engng.* **21**, Suppl., S613–S618.

Kim, S.J., Lee, M., Park, S., Lee, S.Y. and Park, C.H. (1997). A neural linearising control scheme for nonlinear chemical process. *Computers Chem. Engng.* **21**, 187–200.

Koulouris, A., Bakshi, B.R. and Stephanopoulos, G. (1995). Empirical learning through neural networks: the wave–net solution. *Advances in Chemical Engineering* **22**, 437–444.

Kramer, M.A. (1992). Nonlinear principal component analysis using autoassociative neural networks. *AIChE J.* **37**, 233–243.

Ku, C.C. and Lee, K.Y. (1995). Diagonal recurrent neural networks for dynamic systems control. *IEEE Transactions on Neural Networks* **6**, 144–155.

Leontaritis, I.J. and Billings, S.A. (1987). Model selection and validation methods for nonlinear systems. *Int. Journal of Control* **45**, 311–341.

MacQueen, J. (1967). Some methods for classification and analysis of multivariate observations. In *Proceedings of the Fifth Berkeley Symposium on Mathematical Statistics and Probability*, pp. 281–297, University of California Press, Berkeley, CA.

Malthouse, E.C., Tamhane, A.C. and Mah R.S.H. (1997). Nonlinear partial least squares. *Computers Chem. Engng.* **21**, 875–890.

Marquardt, D. (1963). An algorithm for least squares estimation of nonlinear parameters. *J. Soc. Ind. Appl. Math.* **11**, 431.

Martens, H. and Foulk, S. (1990). NIR instruments and multivariate change thinking in process analysis. *Advances in Instrumentation (Proceedings)* **45**, 1517–1528.

Martin, E.B. and Morris, A.J. (1996a). Nonparametric confidence bounds for process monitoring charts. *J. of Proc. Cont.* **6**, 349–358.

Martin, E.B., Morris, A.J. and Zhang, J. (1996b). Process performance monitoring using multivariate SPC. *Proceedings of IEE Control Theory and Applications* D **143**, 132–144.

Mejdell, T. and Skogestad, S. (1991). Estimation of distillation composition from multiple temperature measurements using partial-least-squares regression. *Ind. Eng. Chem. Res.* **30**, 2543–2555.

Micchelli, C.A. (1986). Interpolation of scattered data: distance matrices and conditionally positive definite functions. *Constructive Approximation* **2**, 11–22.

Mockel, W.D. and Thomas, M.P. (1992). Determination of trans-esterification reaction endpoint using NIR spectroscopy. *Proceedings of SPIE—The International Society for Optical Engineering* **1681**, 220–230.

Montague, G.A., Tham, M.T., Willis, M.J. and Morris, A.J. (1992). Predictive control of distillation columns using dynamic neural networks. In *Proceedings of 3rd IFAC Symposium on Dynamics and Control of Chemical Reactors, Distillation Columns, and Batch Processes, Maryland, USA*, pp. 231–236.

Morris, A.J., Montague, G.A. and Willis, M.J. (1994). Artificial neural networks: studies in process modelling and control. *Transactions of the Institute of Chemical Engineers: Chemical Engineering Research and Design* A **72**, 3–19.

Motard, R.L. and Joseph, B. (1994). *Wavelet Applications in Chemical Engineering.* Kluwer Academic Publishers, Boston.

Narendra, K.S. and Parthasarathy, K. (1990). Identification and control of dynamical systems using neural networks. *IEEE Trans Neural Networks* **1**, 4–27.

Park, J. and Sandberg, I.W. (1991). Universal approximation using radial basis function networks. *Neural Computation* **3**, 246–257.

Pati, Y.C. and Krishnaprasad, P.S. (1993). Analysis and synthesis of feedforward neural networks using discrete affine wavelet transformations. *IEEE Trans. Neural Networks* **4**, 73–85.

Penlidis, A., Ponnuswamy, S.R., Kiparissides, C. and O'Driscoll, K.F. (1992). Polymer reaction engineering: modelling considerations for control studies. *Chemical Engineering Journal* **50**, 95–107.

Powell, M.J.D. (1985). Radial basis functions for multivariable interpolation: a review. In *Proceedings of the IMA Conference on Algorithms for the Approximation of Functions and Data, RMCS Shrivenham.*

Powell, M.J.D. (1987). Radial basis function approximations to polynomials. In *Proc. 12th Biennial Numerical Analysis Conference, Dundee*, pp. 223–241.

Puebla, C.G. (1992). Monitoring final quality of liquid dyestuffs through optical spectroscopy. *Proceedings of SPIE—The International Society for Optical Engineering* **1681**, 260–263.

Qin, S.J. (1993). Partial least squares regression for recursive system identification. In *Proceedings of the 32nd Conference on Decision and Control, San Antonio, Texas*, pp. 559–604.

Qin, S.J. and McAvoy, T.J. (1992). Nonlinear PLS modelling using neural networks. *Computers and Chemical Engineering* **16**, 379–391.

Raviv, Y. and Intrator, N. (1996). Bootstrapping with noise: An effective regularisation technique. *Connection Science* **8**, 356–372.

Ricker, N.L. (1988). The use of biased least-squares estimators for parameters in discrete-time pulse response models. *Ind. Eng. Chem. Res.* **27**, 343–350.

Rumelhart, D.E., Hinton, G.E. and Williams, R.J. (1986). Learning internal representations by error propagation. In *Parallel Distributed Processing*. Eds. D.E. Rumelhart and J.L. McClelland. MIT Press, Cambridge, MA.

Saunders, A. C. (1992). Process Monitoring and Nonlinear Observers. Ph.D. Thesis. Department of Chemical and Process Engineering, University of Newcastle upon Tyne, UK.

Scott, G.M. and Ray, W.H. (1993a). Creating efficient nonlinear network process models that allow model interpretation. *Journal of Process Control* **3**, 163–178.

Scott, G.M. and Ray, W.H. (1993b). Experiences with model-based controllers based upon neural network process models. *Journal of Process Control* **3**, 179–196.

Shao, R. Martin, E.B., Zhang, J. and Morris, A.J. (1997a). Confidence bounds for neural network representations. *Computers and Chemical Engineering Supp.* **21**, S1025–S1030.

Shao, R., Martin, E.B., Zhang, J. and Morris, A.J. (1997b). Confidence bounds for neural network representations. *5th International Conference on Artificial Neural Networks*, Cambridge, UK, 7–9 July.

Shimshoni, Y. and Intrator, N. (1998). Classifying seismic signals by integrating ensembles of neural networks. In *Proceedings of ICONIP, 18–21 September, Hong Kong*.

Sjoberg, J., Zhang, Q., Ljung, L., Benveniste, A., Delyon, B., Glorennec, P., Hjalmarsson, H. and Juditsky, A. (1995). Nonlinear black-box modelling in system identification: a unified overview. *Automatica* **31**, 1691–1724.

Sridhar, D.V., Seagrave, R.C. and Bartlett, E.B. (1996). Process modelling using stacked neural networks. *AIChE Journal* **42**, 2529–2539.

Su, H.T., McAvoy, T.J. and Werbos, P. (1992). Long term prediction of chemical processes using recurrent neural networks: a parallel training approach. *Ind. Eng. Chem. Res.* **31**, 1338–1352.

Tan, S. and Mavrovouniotis, M.L. (1995). Reducing data dimensionality through optimising neural network inputs. *AIChE Journal* **41**, 1471–1480.

Tham, M.T., Morris, A.J., Montague, G.A. and Lant, P.A. (1991). Soft sensors for process estimation and inferential control. *J Proc. Control* **1**, 3–14.

Thiebault, J. and Grandjean, B.P.A. (1991). Neural networks in process control — a survey. In *Proceedings of IFAC Conference on Advanced Control of Chemical Processes, Toulouse France,* pp. 251–260.

Tibshirani, R. (1996). A comparison of some error estimates for neural network models. *Neural Computation* **8**, 152–163.

Tsoi, A.C. and Back, A.D. (1994). Locally recurrent globally feedforward networks: a critical review of architectures. *IEEE Trans. Neural Networks* **5**, 229–239.

Turner, P., Montague, G.A. and Morris, A.J. (1996). Nonlinear and direction dependent dynamic process modelling using neural networks. *Proceedings IEE Control Theory and Applications* D **143**, 44–48.

Wang, Z., Di Massimo, C. Montague, G.A. and Morris, A.J. (1994). A procedure for determining the topology of feedforward neural networks. *Neural Networks* **7**, 291–300.

Wang, Z., Tham, M.T. and Morris, A.J. (1992). Multilayer neural networks: approximated canonical decomposition of nonlinearity. *Int. J. Control* **56**, 655–672.

Warwick, K., Irwin, G.W. and Hunt, K.J. (Eds.) (1992). *Neural Networks for Control and Systems*. Peter Peregrinus Ltd., London.

Webster, D.R. (1994). NIR calibration: going from a feasibility study to gaining information that can be used to improve a process. *NIR News* **5**, 15.

Werbos, P.J. (1990). Backpropagation through time: what it does and how to do it. *Proceedings of IEEE* **78**, 1550–1560.

Willis, M.J., Di Massimo, C., Montague, G.A., Tham, M.T. and Morris, A.J. (1991). Artificial neural networks in process engineering. *Proc. IEE D* **138**, 256–266.

Willis, M.J., Montague, G.A. and Morris, A.J. (1992a). Modelling of industrial processes using artificial neural networks. *Computing and Control Engineering Journal* **3**, 113–117.

Willis M.J., Montague G.A., Di Massimo C., Tham M.T. and Morris A.J. (1992b). Artificial neural networks in process estimation and control. *Automatica* **28**, 1181–1187.

Wilson, D.J.H., Irwin, G.W. and Lightbody, G. (1997). Nonlinear PLS modelling using radial basis functions. In *Proceedings of American Control Conference, Albuquerque, New Mexico*.

Wold, H. (1966). Nonlinear estimation by iterative least squares procedures. In *Research Papers in Statistics, Festschrift for Jerzy Neyman*, Ed. F. David, p. 411. Wiley, New York.

Wold, S. (1978). Cross-validatory estimation of the number of components in factor and principal component models. *Technometrics* **20**, 397–404.

Wold, S. (1992). Nonlinear partial least squares modelling II: spline inner function. *Chemometrics and Intelligent Lab. Systems* **14**, 71–84.

Wold, S., Kettaneh-Wold, N. and Skagerberg, B. (1989). Nonlinear PLS modelling. *Chemometrics and Intelligent Lab. Systems* **7**, 53–65.

Wolpert, D.H. (1992). Stacked generalisation. *Neural Networks* **5**, 241–259.

Zhang, Q. and Benveniste, A. (1992). Wavelet networks. *IEEE Transactions on Neural Networks* **3**, 889–898.

Zhang, J., Martin, E.B. and Morris, A.J. (1996). Fault detection and diagnosis using multivariate statistical techniques. *Trans. IChemE* A **74**, 89–96.

Zhang, J., Martin, E.B., Morris, A.J. and Kiparissides, C. (1997). Inferential estimation of Polymer quality using stacked neural networks. *Computers and Chemical Engineering* **21**, 1025–1030.

Zhang, J. and Morris, A.J. (1995a). Dynamic process modelling using locally recurrent neural networks. In *Proceedings of American Control Conference, Seattle, 21–23 June*, **4**, 2767–2771.

Zhang, J. and Morris, A.J. (1995b). Long range prediction models based on locally recurrent neural networks. In *Proceedings of IFAC Conference on Youth Automation, Beijing, 22–24 August*, **2**, 708–712.

Zhang, J. and Morris, A.J. (1995c). Sequential orthogonal training of mixed order polynomial neural networks. In *Proceedings of EUFIT'95, Aachen, Germany, 28–30 August*, **3**, 1581–1585.

Zhang, J. and Morris, A.J. (1996a). Process modelling and fault diagnosis using fuzzy neural networks. *Fuzzy Sets and Systems* **79**, 127–140.

Zhang, J. and Morris, A.J. (1996b). Neuro-fuzzy networks: Contributions to process modelling and fault diagnosis. Semi-plenary paper presented at *EUFIT'96, Aachen, Germany, 2–5 September*.

Zhang, J. and Morris, A.J. (1998). A sequential training strategy for single hidden layer neural networks. *Neural Networks* **11**, 65–80.

Zhang, J., Morris, A.J., Montague, G.A. and Tham, M.T. (1994). Dynamic system modelling using mixed node neural networks. In *Proceedings of IFAC Symposium ADCHEM'94, Kyoto, Japan, May 25–27*, pp. 114–119.

Index